WITHDRAWN
UST
Libraries

ORIGIN OF LAND PLANTS

ORIGIN OF LAND PLANTS

LINDA E. GRAHAM
University of Wisconsin
Madison, Wisconsin

John Wiley & Sons, Inc.
New York • Chichester • Brisbane • Toronto • Singapore

This text is printed on acid-free paper.

Copyright © 1993 by John Wiley & Sons, Inc.

All rights reserved. Published simultaneously in Canada.

Reproduction or translation of any part of this work beyond that permitted by Section 107 or 108 of the 1976 United States Copyright Act without the permission of the copyright owner is unlawful. Requests for permission or further information should be addressed to the Permissions Department, John Wiley & Sons, Inc., 605 Third Avenue, New York, NY 10158-0012.

This publication is designed to provide accurate and authoritative information in regard to the subject matter covered. It is sold with the understanding that the publisher is not engaged in rendering legal, accounting, or other professional services. If legal advice or other expert assistance is required, the services of a competent professional person should be sought. *From a Declaration of Principles jointly adopted by a Committee of the American Bar Association and a Committee of Publishers.*

Library of Congress Cataloging in Publication Data:

Graham, Linda E., 1946–
    Origin of land plants / Linda E. Graham.
       p.   cm.
    Includes bibliographical references and index.
    ISBN 0-471-61527-7 (acid-free)
    1. Paleobotany. 2. Plants—Evolution. I. Title.
QE905.G76   1993
561—dc20   93-18441

Printed in the United States of America

10 9 8 7 6 5 4 3 2 1

To Michael and Melissa

CONTENTS

Preface   xi

Acknowledgments   xv

1 An Introduction to Land Plant Origins   1

    What Are the Land Plants?, 3
    When Did Land Plants First Appear?, 4

2 Early Silurian and Late Ordovician Environments   15

    The Sun and the Atmosphere, 16
    Paleochemistry of Lakes and the Oceans, 20
    Geography and Climate in the Ordovician and Early
    Silurian Periods, 21
    Late Ordovician Terrestrial Biota and Soils, 24

**3 Approaches to an Understanding of Early Land
    Plant Evolution**   27

    A Historical Background, 28
    Modern Approaches to the Study of Land Plant Origins, 38
        Ultrastructure and Light Microscopy, 39
        Patterns of Enzyme Distribution and Occurrence of
        Resistant Biopolymers, 46
        Molecular Systematics, 47
        Approaches to Phylogenetic Tree Construction, 54

4 The Charophycean Algae — 56

Charales, 58
Coleochaetales, 61
Zygnematales, 67
Klebsormidiales, 70
Chlorokybales, 72
Features of a Hypothetical Ancestor of the Charophyceae, 73

5 Morphology, Ecology, and Physiology of Charophytes — 85

Morphological Characters in Relation to Environmental Variation, 86
Physiological Adaptations of Charophytes, 104

6 Gaps between Charophycean Algae and Land Plants — 115

Molecular Gaps, 117
Biochemical Gaps, 118
Morphological Gaps, 118
  Differences in Vegetative Structure, 119
  Reproductive Differences, 124

7 The Evolution of Plant Morphology: Cell Walls, Cytoskeleton, Cytokinesis, Intercellular Communication, and Histogenesis — 146

Cell Walls, 146
The Plant Cytoskeleton, 153
Phragmoplasts and Cytokinesis, 157
Plasmodesmata and Desmotubules, 164
Parenchyma, Meristems, and Morphogenesis, 165

8 Evolution of Plant Sexual Reproduction — 177

The Origin of Sexual Reproduction in Charophytes, 178
The Origin of Land Plant Gametangia, 185
Origin of the Land Plant Embryo, Sporophyte, and Life Cycle (Alternation of Generations), 190
  The Origin of Cell-to-Cell Interactions in Plant Embryogenesis, 191
  Developmental Interactions at the Placental Junction, 193

Control of the Onset/Delay of Meiosis, 197
The Origin of Walled Meiospores, 198

9 The Origin of Plant Signal Transduction Systems, Phytohormones, Photomorphogenesis, and Secondary Metabolism   209

Eukaryotic Signal Pathways, 210
Phytohormones, 212
Phytochrome and Photomorphogenesis, 214
Charophycean Phenolic Compounds: Possible Relevance to the Origin of Lignin, Flavonoids, Cutin, Cutan, and Sporopollenin of Embryophytes, 215
    Charophycean Phenolics, 215
    Lignin: Properties and Biosynthesis, 219
    Flavonoid Properties and Biosynthesis, 224
    Cutin and Cutan, 227
    Sporopollenin Composition and Synthesis, 231

10 Land Plant Origins—A Summary   233

What We Know About the Origin of the Plant Kingdom, 234
Major Gaps in Our Understanding of Plant Origins, 242
Seedless Plants and Algae as Model Systems for Understanding Higher Plant Processes, 246
    Noncharophycean Protists, 247
    Charophycean Green Algae, 247
    Seedless Plants, 249

Literature Cited   253

Index   275

PREFACE

More than 80 years ago, a similarly titled book, *The Origin of a Land Flora, A Theory Based upon the Facts of Alternation,* was published by F. O. Bower, Sc.D., F.R.S., Regius Professor of Botany at the University of Glasgow. In common with several plant scientists who had gone before and numerous theorists to follow, Professor Bower was fascinated by problems connected with the origin of terrestrial plants, and he concluded that the origin of the characteristic plant life cycle, alternation of generations, was inextricably linked to the early evolution of modern plants. Professor Bower documented his ideas in considerable detail, basing his conclusions on data that had emerged from the "golden age" of comparative plant morphology. Bower's determination that alternation of generations arose by delay in meiosis and that the sporophyte generation of plants had undergone progressive evolutionary elaboration was reiterated in a later book entitled *Primitive Land Plants,* published in 1935. Several significant textbooks and numerous papers have been organized around the ideas expounded in Bower's seminal work. Some subsequent authors have supported Bower's thinking, whereas others have offered alternative hypotheses.

During the past two decades significant advances have been made toward understanding the origin of plants, thanks to new paleontological information and the application of electron and fluorescence microscopy, modern biochemical techniques, and new methods of phylogenetic analysis to the problem. The emerging use of molecular systematics has recently resulted in amazing progress and promises to further reveal the history of plants stored in nucleic acid molecules. However, this new data is scattered throughout the literature of various fields and is not readily accessible. The purpose of this book is to gather into one place the information that has become available since Bower's time and to reexamine hypotheses related to plant origins in the light of recent evidence. I hope that the resulting summary will be useful to students and investigators entering the field, as well as to the writers of introductory biology and botany textbooks. The results may also prove valuable in focusing future research efforts toward elucidation of important events in early plant evolution.

This book, like Bower's, approaches the problem of the origin of plants primarily from the standpoint of research results emerging from the study of modern organisms, but as in the 1908 treatment, an effort has been made to incorporate paleobotanical evidence relevant to early plant evolution. Discussion of the fossil record of plant evolution is, however, limited to those recent research findings that are directly related to the earliest events in plant history, inasmuch as there are several excellent modern paleobotanical treatments of later vascular plant evolutionary radiations. Very early events in plant evolution are often treated quite briefly in most botanical texts (primarily because the earliest fossil record of plants is relatively sparse). It is my hope that the present integrative treatment will illuminate the substantial evidence provided by molecular, biochemical, physiological, developmental, structural, and ecological studies of extant organisms that supports modern hypotheses concerning plant origins that are based on paleobotanical work. An effort has been made to incorporate recent findings concerning paleogeography, paleochemistry of aquatic environments and the atmosphere, and paleoclimatology in an examination of issues related to early plant evolution within an appropriate environmental context.

This book was conceived as an exercise in synthesis and represents an effort to bring together significant results from fields as disparate as molecular biology and global biogeochemistry. No attempt was made to present an encyclopedic review of any single topic, and where multiple references existed, the most recent was generally chosen. A conscious effort was made to exhibit the findings of major workers in the field in order to facilitate bibliographic searches by readers desirous of more detail. The level and presentation were designed to be accessible to the undergraduate reader while also providing important details to active researchers in the field. It is inevitable that some important work will have been inadvertently overlooked or not included because of space constraints. It is also likely that my interpretations of the work published by others could be improved. I hope that readers will provide me with potential additions or corrections useful in production of future works on the subject of plant origins.

In Bower's time, many scientists were vitally interested in attempts to understand early plant history. Today, however, in an age of increasing scientific specialization and intense concern about immediate biological crises such as loss of biological diversity and other serious environmental issues, as well as intractable problems in biomedicine (not to mention poverty, war, and other societal problems), one might well question the relevance and relative value of research focusing on so remote an event as the origin of plants. What, if any, is the application to modern plant biology of basic research related to the evolutionary origin of plants? The central thesis of this book is that such knowledge is integral to the elucidation of poorly understood, yet fundamental, features of modern plants, upon which all of humanity and the biosphere of planet Earth depend. Evidence is provided for the relevance of early plant evolution to historical changes in global carbon cycling and climate; the ecology of modern freshwater and early terrestrial ecosystems; the origin of such fundamental plant physiological processes as carbon acquisition, photorespiration, morphogenetic responses to environmental signals, and the action of plant hormones; the basic cell biology underlying plant structure, function, and development—specifically, origin and evolution of the plant cytoskeleton; the origin of sexual reproduction in plants and their characteristic life cycle involving alternation of generations; the origin of plant secondary metabolism, including the production of resistant biopolymers; and the origin of genetic interrelationships between plant nuclear, mitochondrial, and chloroplast genomes. Actually, there are few facets of plant biology that cannot be illuminated by an understanding of early plant evolution and stimulated by investigations centered on the biology of simple organisms believed to be closely related to the direct ancestors of plants. In modern biology, model systems such as the fruit fly *Drosophila* and the "worm" *Caenorhabitis* are effectively used to develop an understanding of the fundamental properties of higher animals. In this book, the past and potential future use of simple plantlike organisms and seedless plants as model systems for the study of fundamental plant processes are explored in some detail. It is my hope that this book will stimulate the use of simple organisms related to plants in basic plant research and will, ultimately, have an impact on the solution of practical problems of concern to us all.

ACKNOWLEDGMENTS

The University of Wisconsin-Madison provided me with a semester's sabbatical leave that greatly facilitated completion of this project. My thanks to colleagues in the Department of Botany for supporting my sabbatical leave request and for supplying other assistance. Professor R. F. Evert and his staff (including Susan Eichhorn and William Russin) provided necessary access to excellent transmission electron microscopy and darkroom facilities. Professor Kenneth Keegstra contributed illustrations and helpful references. Special appreciation is due Botany Department artist Kandis Elliot, who provided most of the line illustrations, photographer Claudia Lipke, who did a superb job of photographing plant specimens and making copy negatives, and Barbara Schaack, who helped to process the manuscript. Present and past graduate students Dr. James Hoffman, Dr. James Kranzfelder, Dr. Lee Wilcox, Dr. Charles Delwiche, Wesley Ebert, Claude Taylor III, Gary Wedemayer, Patricia Arancibia, Scott Kroken, Martha Cook, and Madeline Fisher were all helpful in many ways: as research collaborators, diving buddies in cold Wisconsin waters, and teaching assistants in courses that included material related to this book. Martha Cook provided expert editorial assistance, but I hasten to claim responsibility for any errors of fact or English usage that may have escaped her notice. Those who read portions of the manuscript and provided valuable comments and additional references include Professors Dianne Edwards F.L.S. (University College, Cardiff, Wales) and Jane Gray (University of Oregon), Dr. John Obst (USDA Forest Products Laboratory, Madison, Wisconsin). Sonia Cook, Dianne Edwards, Paul Biebel, Roy Brown, and Betty Lemmon provided original photographs for use in this book, and Norma Johnson and Karen Renzaglia generously contributed research material. Several years ago, Mary M. Conway of John Wiley & Sons stimulated me to begin this book, and I am grateful for her interest and the assistance of her colleagues,

Philip Manor and Allison Ort Morvay. My deepest appreciation for scientific training is extended to H. Wayne Nichols (and Washington University, St. Louis), the late Harold C. Bold (and the University of Texas), and Gordon E. McBride (and the University of Michigan, Ann Arbor). My mother, the late Evelyn Edwards, supported this project in very significant ways. Most of all, I thank my husband and scientific collaborator, Dr. James M. Graham, who reviewed a portion of the manuscript, and who contributed valuable ideas, technical assistance, and several original art works.

1

AN INTRODUCTION TO LAND PLANT ORIGINS

At the edge of a lake, the contrast between the sparse plant growth on the wave-washed sandy shore and the abundant, green vegetation of adjacent terrestrial communities is readily apparent. The lake also appears in dramatic contrast to the land vegetation. The reflective water surface mirrors sky conditions, appearing steel-gray on a cloudy day or sparkling on a sunny one. But hidden beneath the opaque water surface, visible to the observer willing to become immersed, is another forest of vegetation. A canopy of tall pondweeds and an understory of quillworts and other rosette plants form patches of submerged aquatic macrophytes. These are interspersed with meadows of stonewort algae and an occasional forbidding morass of green-black cyanobacteria. Airy clouds of pondsilk algae swirl amid green towers and castles formed by sponges harboring vast populations of algal cells within their tissues. Aquatic macrophytes themselves anchor a complex tangle of microscopic vegetation, consisting primarily of cyanobacteria, diatoms, and a variety of green algae that live much like epiphytic bromeliads, orchids and ferns on tropical forest trees. The myriad algae of near-shore communities are variously and admirably adapted to life in shallow waters. A few have ventured forth across the shoreline barrier to dry land, but these thrive only when moist and, with few exceptions, are not apparent to the casual observer. But half a billion years or so ago, certain forms of green algae made a more permanent transition from the water to dry land, ultimately giving rise not only to land vegetation, but also to aquatic "higher plants" descended from terrestrial ancestors, which later returned to life in the water. Their momentous journey changed the biosphere of planet Earth forever, setting in motion a chain of events leading to the colonization of land by animals, culminating, for better or worse, in the origin of the terrestrial creatures known as *Homo sapiens*.

2 AN INTRODUCTION TO LAND PLANT ORIGINS

The origin of land vegetation has been an exciting and controversial topic for well over a century, and much has been written about the "greening of the planet Earth." In this book I review the body of recent discoveries relating to land plant origins and use this evidence to evaluate various hypotheses that have been proposed to explain the process of early plant adaptation to terrestrial life. A brief discussion of the history of ideas related to land plant origins is provided in Chapter 3. This is followed by description of the various modern approaches to the study of plant origins: paleobotanical studies; comparative molecular, biochemical, and cellular investigations of modern algae and "lower" plants, together with cladistic and phenetic analyses of phylogenetic relationships; and experimental work. To set the stage for these topics, in Chapter 2 we examine the most recent evidence concerning the physical and biotic environment that existed on earth at the time of origin of land plants. An environmental description, however, depends on our ability to focus on a particular geological time period for the early origins of plants. Pinpointing the time of origin of the earliest land plants has been difficult, and the evidence related to this controversial subject is presented below. Our first priority will be to define what is meant by the term "land plant."

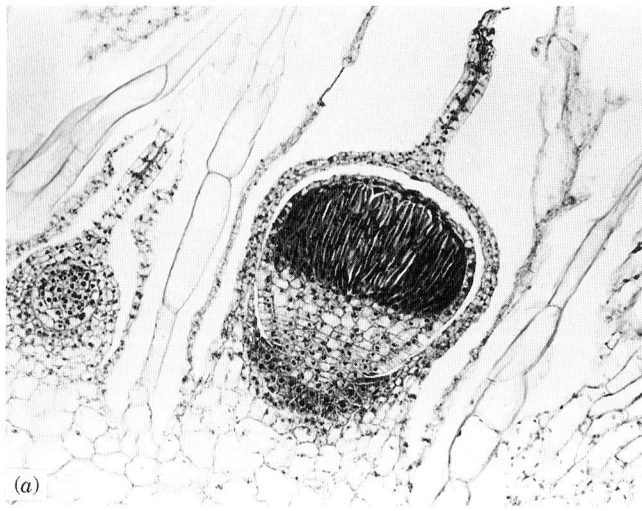

Fig. 1-1. Land plant embryos. *a.* Embryos are characteristic of even the simplest land plants—mosses, liverworts, and hornworts. These developing sporophytes of the liverwort *Marchantia* illustrate typical growth of embryos within the remains of the female gametangium (archegonium) and close association of haploid and diploid tissues at the placental region. Darkly stained placental cells are specialized to facilitate transport of photosynthate from gametophyte to sporophyte, thereby supporting development of the embryo and subsequent spore production. *b.* Close association between gametophyte and sporophyte generations also occurs during early embryogenesis in seed plants. This eight-celled embryo of *Arabidopsis thaliana* represents a young sporophyte emerging from the female gametophyte (embryo sac). Photo courtesy of Kay Robinson-Beers (University of California-Davis).

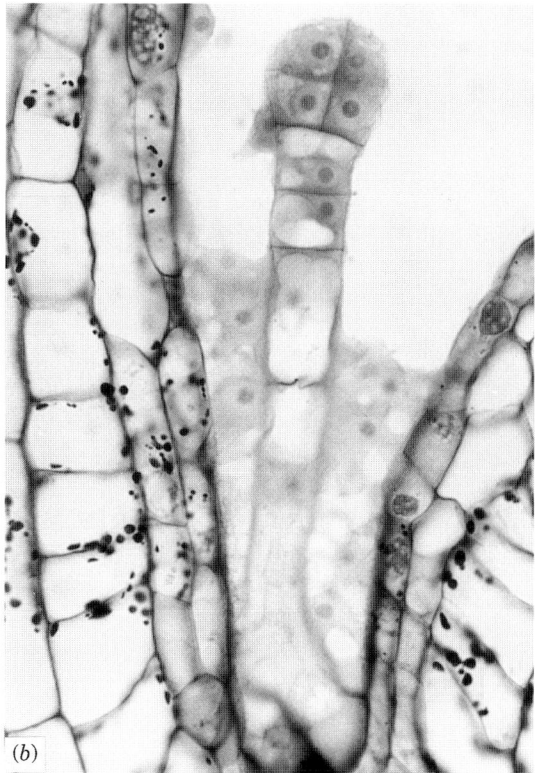

Fig. 1-1. *(Continued)*

WHAT ARE THE LAND PLANTS?

Faced with only modern terrestrial photosynthesizers (and heterotrophic organisms clearly derived from them, such as the liverwort *Cryptothallus* or the orchid *Corallorhiza*), it would be easy to define the land plants as equivalent to the embryophytes, plants having an embryo (a multicellular, nutritionally dependent sporophyte generation). This is because all modern plants having numerous adaptations to life on land (the bryophytes and vascular plants) also have embryos (Fig. 1-1). In contrast, all modern algae, whether living in water or in moist terrestrial habitats, lack embryos. The problem arises in considering the array of early land inhabitants that may have existed in the past and the sequence of acquisition of various adaptations that characterize modern plants. For example, there is evidence that certain aquatic algae, known to be related to the ancestry of plants, have taken the first critical steps toward evolutionary development of a nutritionally dependent embryo, in response to selective pressures of shallow-water environments (Graham, 1985). If the earliest photosynthetic land colonizers already possessed embryos (acquired while still in the water), but did not yet have adaptations that could be unequivocally related to terrestrial selection pressures, these organisms would be embryophytes, but not yet land plants. If, however, the ancestors of plants included terrestrial algae that had acquired unambiguous adaptations to land prior to origin of the embryo (as suggested by Stebbins and Hill, 1980), then these early land colonizers could be described as land plants, but not embryophytes. Although there is considerable evidence that modern embryophytes are descended from a common ancestor (Graham et al., 1991), in view of uncertainties about the nature of the first land plants, it is probably best to uncouple the terms "land plant" and "embryophyte" when considering early plant evolution. Some paleobotanists recommend that land plants should not be considered as a clade (a group of organisms descended from a common ancestor), but as a grade level of structural and biochemical organization. Unfortunately, as we shall see, it is not easy to define a group of biochemical and structural characters that can be as readily applied to fossil remains as to modern plants.

Recognizing the ambiguity of the term "land plant," and the tendency of many biologists to equate this term with the vascular plants, Gray and Boucot (1977) defined land plants as photosynthetic organisms that customarily live on land and whose relations are to other plants living on land. Applied to modern plants, their definition includes the embryophytes, i.e., vascular plants and bryophytes, but not fungi or most terrestrial algae. However, there is paleobotanical evidence for past existence of organisms that show considerable evidence of terrestrial adaptation, and thus are commonly considered to be land plants, but do not seem to be related to the mainstream of land plant evolution (i.e., the very strange Nematophytales [Strother, 1988]). In addition, there are a few morphologically simple terrestrial algae that are almost certainly related to the ancestry of land plants. Gray and Boucot recognized that the boundary between algae and land plants may have been indistinct during the transition to land, and, if interpreted broadly, their definition can be useful until more evidence is available concerning the sequence of acquisition of adaptations to land by the first plants.

WHEN DID LAND PLANTS FIRST APPEAR?

In the future it may be possible to use the ticking of molecular clocks (Zuckerkandl and Pauling, 1965) in the nucleic acids of modern organisms to obtain information on the time of origin of one or more possible embryophyte clades. A prerequisite would be information on relative rates of nucleotide substitution in green algae and "lower" land plants having similar generation times, and many sequences would be needed in order to reduce the chance of stochastic error in estimation of divergence time (Li and Grauer, 1991). In the absence of such information, the best present evidence comes from paleobotanical studies. This is not to say, however, that molecular systematics is ultimately likely to replace paleontology. Past morphological transition series, for example, cannot be readily predicted from the nucleic acid sequences

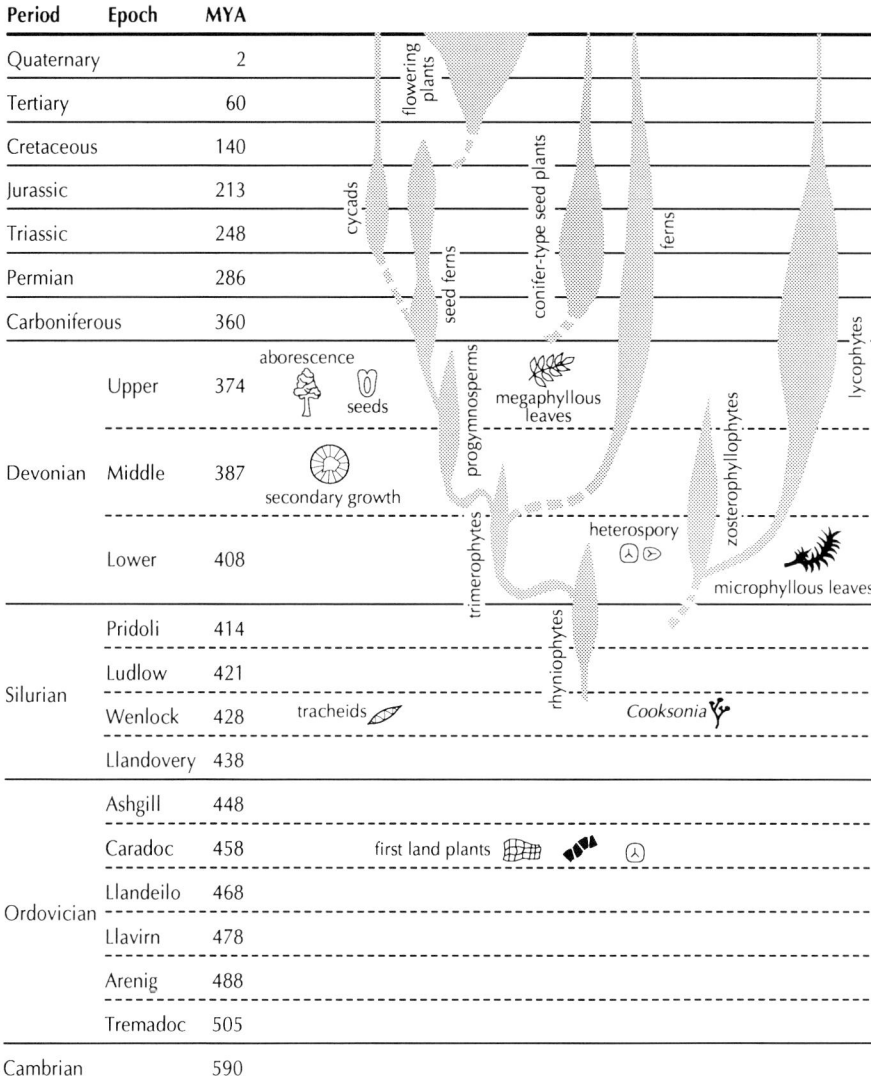

Fig. 1-2. This geological time chart, adapted from Gensel and Andrews (1987), illustrates the appearance of the benchmark vascular plant *Cooksonia* in the Silurian period and remarkable diversification of vascular plants in Devonian epochs. Microfossil evidence from the Ordovician period suggests that the first land plants appeared considerably earlier than *Cooksonia*. The earliest land plants may have been similar in structure to modern bryophytes, hence less resistant to microbial degradation and fossilization processes than vascular plants. This may explain the fragmentary nature of the fossil evidence related to early colonization of land by plants. Diagram reprinted by permission of *American Scientist,* journal of Sigma Xi, The Scientific Research Society.

of modern organisms, and modern sequences cannot be used to predict the existence of extinct organisms.

Paleobotanists often use the mid-Silurian appearance of the fossil plant *Cooksonia* Lang as a benchmark for the first occurrence of vascular plants (Fig. 1-2). Early plants that came after *Cooksonia* left a relatively diverse fossil record, with the result that evolutionary radiations of the vascular plants are comparatively well understood (Gensel and Andrews, 1984; Gensel and Andrews, 1987; Stewart and Rothwell, 1993; Taylor and Taylor, 1993). There is an increasing body of data relating to "pretracheophyte polysporangiophytes," fossil taxa that had branched sporophytes like modern vascular plants, but for which vascular tissue similar to that of modern plants is lacking (Kenrick and Crane, 1991). However, the fossil record of *Cooksonia*'s antecedents is sparse and fragmentary.

Cooksonia, named for the eminent palynologist Isabel Cookson, and studied extensively by Dianne Edwards and associates, was only a few centimeters tall, with dichotomously branched leafless stems bearing sporangia at the tips (Fig. 1-3). Hence, these branched axes represent the sporophyte generation; the nature of *Cooksonia*'s gametophyte generation is unknown. Peripheral regions of *Cooksonia* stems were characterized by sterome, a decay-resistant tissue composed of longitudinally oriented thick-walled cells. Sterome is believed to have functioned in structural support in early land plants having poorly developed vascular systems (Edwards et al., 1986). *Cooksonia* possessed a vascular system in which tracheids were relatively small and few in number (Edwards et al., 1992), and the epidermis bore typical stomata (Fig. 1-3). The presence of vascular tissue and stomata indicate that *Cooksonia* was highly adapted to maintain homeohydry, an equilibrium level of tissue hydration. The homeohydric condition is a major attribute of the sporophyte generation of vascular land plants (Raven, 1984). Sporangia of *Cooksonia* contained small spores with resistant walls, an unequivocal adaptation for wind dispersal in the dry terrestrial habitat. *Cooksonia*'s spore walls were also marked by what is called a trilete scar (Fig. 1-4), which is important as evidence that the spores were generated by the process of meiosis.

Study of some modern plants has shown that trilete marks form after meiotic division of spore mother cells, in the regions where the tetrad of daughter cells come into contact, while walls develop around the individual spores (Fig. 1-5). The meiospores of modern land plants typically contain sporopollenin, considered to be the most resistant biopolymer in nature. Sporopollenin is identified by characteristic solubility properties and absorption spectra (Southworth, 1974; 1990). Although most fossil spores have not been tested chemically for the presence of sporopollenin, the fact that they are highly resistant to microbial and chemical degradation is strong evidence that their walls, like those of modern land plant spores, contain sporopollenin. Trilete spores having resistant sporopollenin-containing walls are produced by various modern bryophytes, in addition to many vascular plants. The occurrence of resistant-walled spores occurring in tetrads and/or having trilete marks in the fossil record (Fig. 1-6) provides the best available evidence for the time of origin of land plants before the appearance of *Cooksonia*.

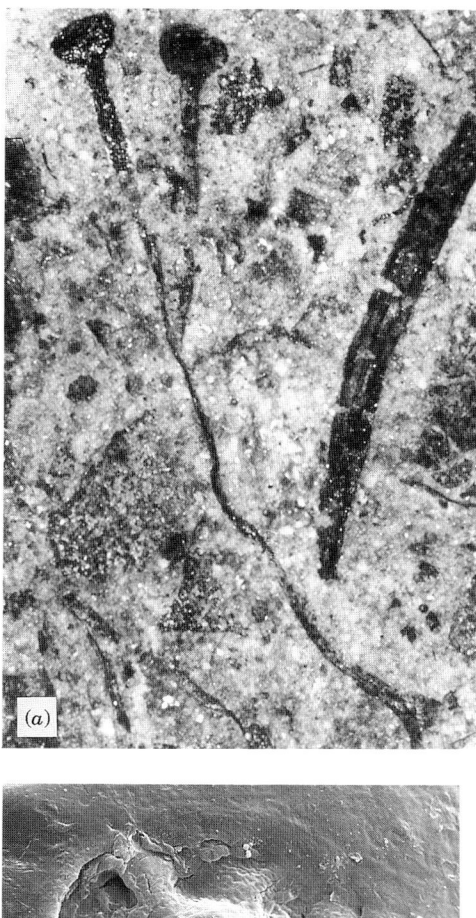

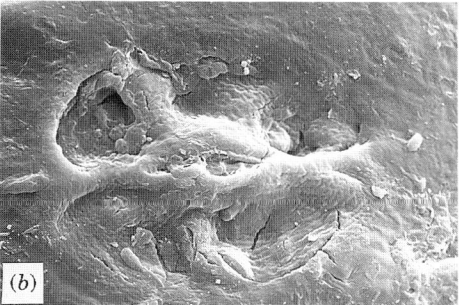

Fig. 1-3. *Cooksonia pertoni* is found in deposits of the Freshwater East Dyfed Pridoli (latest Silurian), about 405 to 400 million years of age. In describing *Cooksonia* the eminent paleobotanist W. H. Lang honored Isabel Cookson. The specific epithet refers to Perton Lane, the original location of this fossil. *a. Cooksonia* consisted of dichotomously branched axes with terminal sporangia. *b.* The presence of typical stomata on *Cooksonia* axes, and the presence of internal conducting cells resembling tracheids of modern vascular plants, indicate that *Cooksonia* was fully homeohydric, that is, capable of maintaining an equilibrium level of tissue hydration. These adaptations to the desiccating terrestrial environment contrast with poikilohydry of modern bryophytes, typified by gain and loss of water in response to degree of environmental hydration. The earliest land plants most likely resembled bryophytes in their water relations. Photos courtesy of Dianne Edwards (University College, Cardiff, Wales).

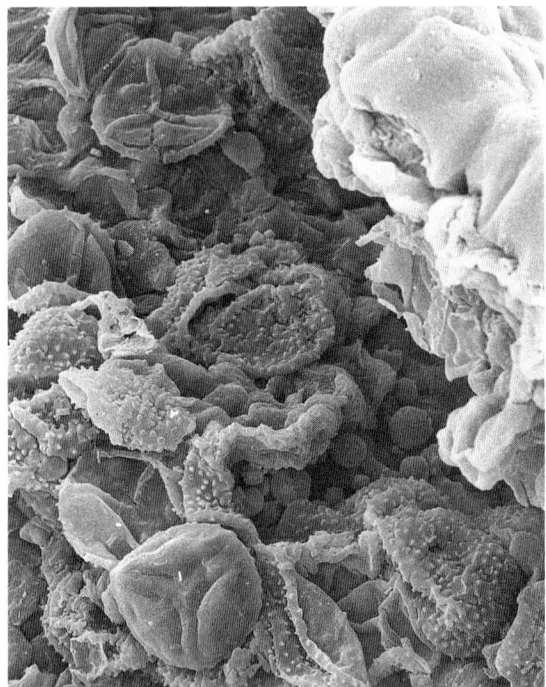

Fig. 1-4. Spores of *Cooksonia pertoni* from Welsh Borderland, Lower Devonian, deposits reveal occurrence of resistant walls and trilete marks. Photo courtesy of Dianne Edwards (University College, Cardiff, Wales).

The megafossil record of land plants, the occurrence of fossils large enough to see without the use of a microscope, is almost nonexistent in rocks older than those in which *Cooksonia* may be found. This gives a false impression that colonization of the land by plants occurred rather suddenly and that organisms resembling *Cooksonia* represent a starting point for land plant diversification. There are two lines of fossil evidence that suggest the existence of a well-established land flora millions of years before the time of *Cooksonia*. This evidence consists of plant microfossils and the earliest known fossils of land animals, which would probably have been dependent on a terrestrial food source.

Plant microfossils are obtained by crushing rocks into small pieces, then treating with strong acids, including hydrofluoric acid, which will dissolve silica (Gray and Shear, 1992). Any biologically derived material that survives this treatment, not to mention degradative processes occurring during fossilization, must necessarily be composed of highly resistant materials. Microfossils associated with plants fall into three general categories: spores that are resistant to degradation by means of their (presumed) sporopollenin walls, sheets bearing the impressions of cell walls or resistant cell layers that are presumed to represent fossil cuticles, and tubes with helical ornamentations that resemble the water-conducting cells

WHEN DID LAND PLANTS FIRST APPEAR? 9

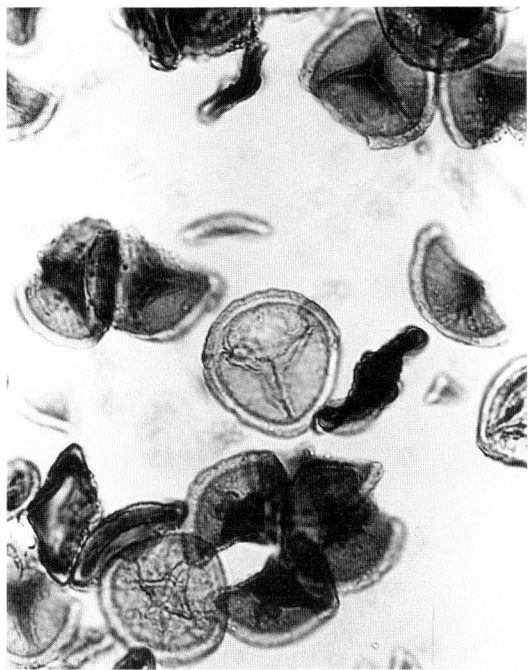

Fig. 1-5. Trilete marks on walls of modern plant spores result from compression against the other three cells of spore tetrads during spore wall development. Tetrads result from meiotic division of spore mother cells (sporocytes); hence, trilete marks on spores are widely regarded as evidence of meiotic origin.

(tracheids) of modern land plants. Many modern plant cuticles and higher plant fossil cuticles contain the biopolymer cutan, which is more resistant to chemical hydrolysis and microbial degradation than cutin (Tegelaar et al., 1991). The cell walls of modern tracheids are strengthened by incorporation of a resistant, complex polyphenolic material, lignin. Together, these three highly resistant polymers, sporopollenin, cutan, and lignin, are primarily responsible for the existence of plant remains in the fossil record. Moreover, occurrence of these three "plastics" in spores, cuticles, and water-conducting cells is generally regarded as evidence of adaptation to life on land. Evidence often cited for this conclusion is that resistant-walled spores, waxy cuticles, and lignified vascular systems are not present in algae, most of which are aquatic, and that resistant biopolymers tend to be less abundantly synthesized by embryophytic plants that have returned to an aquatic habitat.

The most productive student of the earliest microfossils that can be related to plant remains is Jane Gray at the University of Oregon. She and her associates have pointed out that the microfossil record is complementary to the record of plant megafossil remains and that it provides different information (Gray, 1984, 1985; Gray and Boucot, 1978). Her studies have revealed fossil spore evidence for

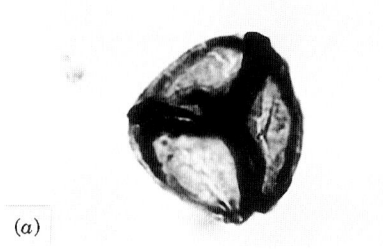

(a)

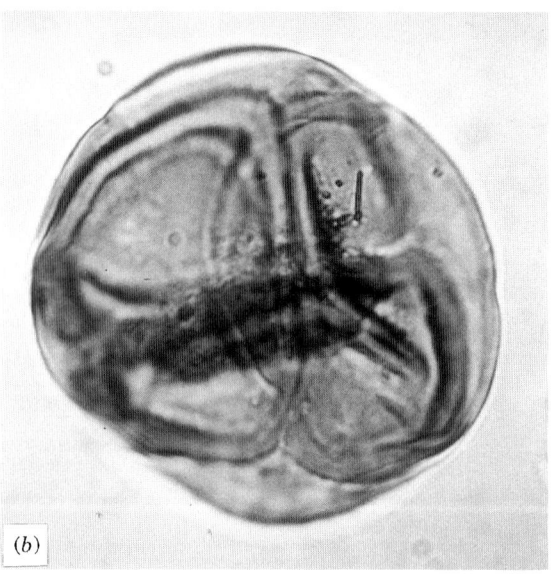

(b)

Fig. 1-6. Microfossil evidence related to early plants. *a.* This tetrahedral aggregate of presumed spores was derived from shale beds in the Early Silurian (Llandoverian) Tuscarora Formation in central Pennsylvania, which has primarily land-derived fossil remains. The preparation (made by Norma Johnson) involved, in part, treatment of material with 50 percent HF for several days. This suggests that presumed spore walls are composed of highly resistant sporopollenin. Tetrads of spores such as these are believed to originate from very early land plants. *b.* An Ordovician spore tetrad from the Elkhorn Formation (Ashgill), Kentucky, is enclosed in a resistant envelope and is about 42 micrometers in diameter. *c.* Sheet of cuticle-like material showing possible cell patterning from the later Ordovician (Caradoc) Melez Chograne Formation, southern Rhadames Basin and Murzuk Basin, western Libya. The nature of such cuticle-like sheets is poorly understood. Photos *b* and *c* by K. Colbath, generously provided by Jane Gray, cover photo accompanying Gray, Massa, and Boucot, 1982. Caradocian land plant microfossils from Libya. *Geology* 10:197-201.

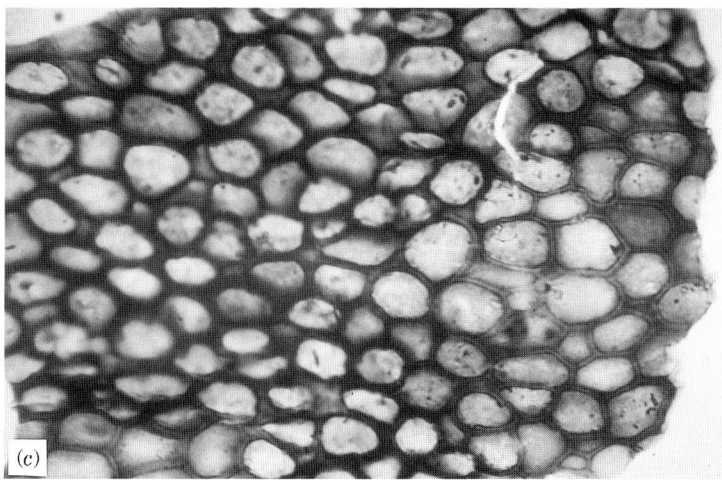

Fig. 1-6. *(Continued)*

the occurrence of two major phases of land plant adaptive radiation prior to the Devonian. In deposits of late early Silurian (Llandovery) age there appear single trilete spores similar to those of modern vascular, nonseed plants such as ferns and other pteridophytes. In contrast, spores extracted from Ordovician or earliest Silurian deposits are generally smaller and may occur in groups of two (dyads) or, more commonly, four (tetrads), sometimes enclosed by a resistant envelope. Some of the earliest known spores were found in Arabian rocks dated to the mid-Ordovician period (Llanvirn) (Gray et al., 1982). Similar spore groups occur in Silurian deposits formed in what are presently North America, Europe, South Africa, South America, and Australia, suggesting that early land plants were widespread as founder populations at this time (Gray et al., 1982; Gray, 1985a).

Although Gray has provided evidence that the tetrads are adherent meiospores, in cases where a triradiate mark is not clearly visible, it has been argued that the tetrads might represent groups of algal cells similar to modern *Chlorella* (Fig. 1-7). This argument is based on the observation that several types of algae that can be isolated from terrestrial habitats (including some species of *Chlorella*) deposit sporopollenin in their cell walls (Atkinson et. al., 1972) (Fig. 1-8). When *Chlorella* reproduces, each of four daughter cells derived by mitotic divisions may develop a cell wall while still enclosed within the parental wall. Hence, if resistant, the sporopollenin-containing parental walls enclosing four autospores of *Chlorella* might superficially resemble fossil tetrads. However, because the algal cell tetrads are not meiotic products, they do not exhibit triradiate marks, and *Chlorella* tetrads are much smaller than Gray's early spores.

Also bolstering Gray's claim that the fossil tetrads represent land plant spores is the co-occurrence in Silurian samples of cuticle-like sheets and tubes that somewhat resemble tracheids (Pratt et al., 1978). Niklas and Pratt (1980) reported chemical evidence for association of "ligninlike" compounds with banded tubes

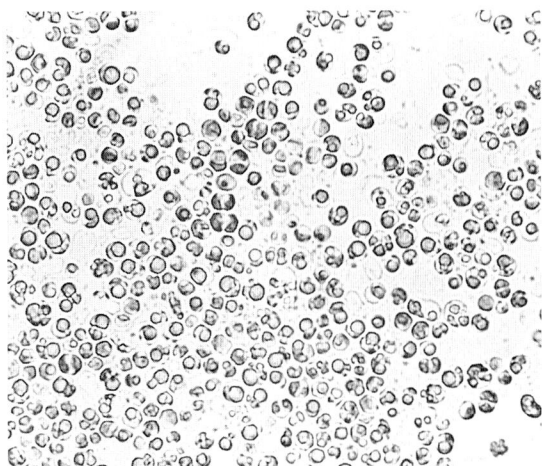

Fig. 1-7. Autospores of the chlorophyte *Chlorella* develop in groups of two or four enclosed within the cell wall of parental cells. Although it is possible that fossil analogues or homologues to enclosed autospores of *Chlorella* might exist, these, like the autospores of modern *Chlorella,* would not likely bear trilete marks, inasmuch as they are not products of meiosis. *Chlorella* autospores are also much smaller than the early spore tetrads or fossil spores which bear trilete marks. These observations strongly suggest that Ordovician spore tetrads and trilete spores are land plant remains rather than fossil algae.

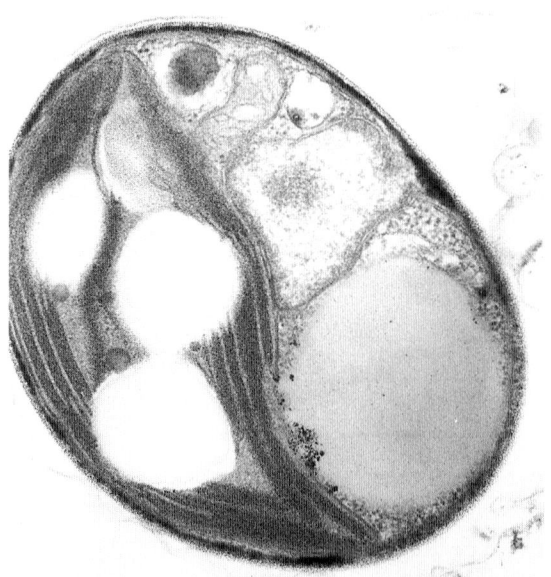

Fig. 1-8. Vegetative (nonreproductive) cell walls of *Chlorella* and some other chlorophytes that may occur in both aquatic and terrestrial habitats are resistant to acetolysis because they contain the resistant polymer sporopollenin. Although it is possible that land plant sporopollenin arose independently in charophycean ancestors (an example of parallel or convergent evolution), the occurrence of sporopollenin in chlorophytes and some other protists, as well as in plants, suggests that sporopollenin is of ancient origin.

from early Silurian stream-deposited sediments. The nature of the organisms that produced the tubes and sheets is highly conjectural; the tubes and sheets may have originated from organisms distinct from the spore producers. However, it has been suggested that the spore remains are derived from early land plants at the same level of organization as modern liverworts (Gray and Shear, 1992) (Figs. 1-9 and 1-10). Modern bryophytes are generally regarded as possessing cuticles, and various extant liverworts produce specialized water-conducting cells (Hebant, 1977). Most likely, early plants at the bryophyte grade would not often leave intact megafossils. But fragments of resistant parts, such as spores, cuticle, or conducting cells impregnated with polyphenolic compounds similar to lignin, might survive as microfossils. Under conditions in which even nonresistant plant remains are preserved by petrifaction, megafossil remains might survive. However, no Ordovician or early Silurian petrifaction fossils of intact early plants have yet been discovered. In the absence of information about the form of the organisms that generated the most ancient spores, sheets, and tubes, resources should continue to be directed toward further study of Ordovician and Silurian microfossils.

The earliest fossil record of land animals was recently extended back to the upper Silurian period, about 414 million years ago, by the finding of the remains of unequivocally terrestrial animals, two centipedes and an arachnid. The occurrence of predatory arthropods at this time suggests that complex terrestrial ecosystems were already in place and, combined with evidence provided by plant microfossils, implies that a significant terrestrial vegetation had become established much earlier,

Fig. 1-9. The existing microfossil evidence suggests that the first land plants may have outwardly resembled modern prostrate, thalloid liverworts. *Ricciocarpus* has a simple, dichotomously branching structure. Very little of such a land plant, other than trilete spores and perhaps a thin cuticular layer, would be expected to survive microbial degradation and to have persisted to the present time as fossil remains, except under the most fortuitous of preservational conditions.

14 AN INTRODUCTION TO LAND PLANT ORIGINS

Fig. 1-10. Modern liverworts such as *Pallavicinia* are known to produce central regions of specialized conducting cells. Although these are not lignified, it is possible that Ordovician or Silurian analogues developed prototypes of tracheids characteristic of modern and fossil vascular plants.

between the mid-Ordovician and the middle Silurian periods (Jeram et al., 1990; Gray and Shear, 1992). This early land community need not necessarily have included vascular plants; early land animals may have been supported by vegetation at the bryophyte level, a diversity of terrestrial algae, and perhaps even fungi and lichens. However, combined with the microfossil findings, recent reports of early land animals constitute a significant body of paleobiological evidence that plants were on land much earlier (Gray and Shear, 1992) than the benchmark provided by the appearance of *Cooksonia* (Edwards and Fanning, 1985; Edwards et al., 1992) during the mid-Silurian period (Taylor, 1982).

The paucity of fossil evidence related to pretracheophyte diversification is analogous to the poor fossil record of the earliest eucaryotes. Molecular phylogenies suggest that eucaryotes may be much older than indicated by the fossil record; this discrepancy has been at least partially attributed to low preservation potential (Knoll, 1992). When molecular evidence related to the time of divergence of embryophytes is available, it will be interesting to determine how closely it correlates with microfossil data. In the meantime, it is possible to consider hypotheses related to early plant adaptation, bearing in mind Knoll and Niklas's (1987) injunction that these must be considered within the context of environment. In the next chapter, recent concepts regarding physical and biotic conditions on earth at the time of land plant origins are discussed. The paleobiological evidence discussed in this first chapter suggests that focus should be placed on the time period from the mid-late Ordovician (perhaps 450 to 470 million years ago) through the early Silurian period.

2
EARLY SILURIAN AND LATE ORDOVICIAN ENVIRONMENTS

Recent advances in the fields of stellar development, paleooceanography, paleoclimatology, paleontology, and paleogeography allow the development of working hypotheses concerning the radiation environment, atmospheric composition, ocean chemistry, continental and polar positions, temperature regimes, characteristics of early soils, and aquatic and terrestrial biota present on earth during the geologic time periods associated with the earliest origins of land plants. As is true of today's biosphere, the physical and biological features characteristic of the late Ordovician–early Silurian biosphere were highly interdependent. For example, the relative proportions of oxygen and carbon dioxide in the atmosphere were dependent on organismal photosynthesis and calcification, which in turn were related to solar radiation and climatic conditions. Climate appears to have been highly connected to polar location, which has varied throughout the history of the earth as a result of continental movements. Biological activity has a feedback effect upon atmospheric chemistry, the radiation environment on earth, and the composition of soils. Emerging new methods for understanding atmospheric chemistry rely on the study of early soils. Thus, in developing a picture of physical and biotic conditions on earth at the time when land plants first originated, it must be kept in mind that the models on which the composite view is based may be altered by new information related to any of the factors mentioned above, or perhaps by factors and relationships that are as yet unrecognized. In the discussion that follows, the nature of solar output and atmospheric chemistry is addressed first, followed by the evidence related to ocean chemistry, the location of continents and relevance to climatic and sea-level conditions, aquatic and terrestrial biota, and the characteristics of early soils. All of these environmental factors are relevant to the development and evaluation of concepts concerning the origins of very early plant adaptations to terrestrial life.

THE SUN AND THE ATMOSPHERE

Today we recognize the great importance of upper atmospheric interactions between solar radiation and atmospheric gases, particularly the photochemical production of ozone (O_3) from oxygen (O_2), a reaction catalyzed by ultraviolet (UV) radiation from the sun. Because the ozone layer absorbs further UV radiation that would otherwise prove extremely harmful to life on earth, there is much current concern about the biological effects of breaches in the ozone layer resulting from anthropogenic chemical emissions. For example, Smith et al. (1992) reported that a 50 percent thinning of the stratospheric ozone layer over the Antarctic (the "ozone hole") is associated with increased UVB radiation (wavelength of 280 to 320 nm) at the surface of the southern ocean, and an estimated 6 to 12 percent reduction in primary productivity. Relationships between solar radiation and the early earth atmosphere were similarly important in the origin of terrestrial plant life.

Astronomers have measured the UV output of young T-Tauri stars, similar to our sun at a few million years of age, and found that young stars emit 10,000 times more UV than the sun does at present. It is thus assumed that the early earth received a very high UV radiation dose (Canuto et al. 1982). It is also widely agreed that the young sun was considerably less luminous than it is today and that progressive increase in solar luminosity has occurred through time. Planetary scientists believe that luminosity was sufficiently low for the first half of earth history that (assuming modern albedo and atmospheric composition) the earth's mean surface temperature should have been below freezing. However, the most ancient rocks provide evidence that liquid water was present as early as 3.8 billion years ago. This suggests that either the earth's albedo (reflective luminosity) was different from today's or, more probably, that atmospheric carbon dioxide levels were then much higher than they are today, perhaps as much as 100 to 1,000 times the present level. This higher carbon dioxide concentration in the atmosphere presumably had a substantial warming effect, compensating for the lower solar luminosity hypothesized for the Archean period (Kasting, 1987). It is estimated that solar luminosity had increased to about 93 percent of present levels by 900 to 600 million years ago, which predates our guess for the time of origin of the earliest land plants at 450 to 470 million years or so ago. Increasing levels of photosynthetically active radiation and high atmospheric carbon dioxide concentrations no doubt favored the growth of marine photosynthesizers (cyanobacteria and algae), some of which could remove dissolved carbon from the system by coupling photosynthesis to the formation of calcium carbonate, a long-lived carbon reservoir (Pentecost, 1985). The present sedimentary storage of carbon is estimated at 10^{22} g accumulated since the origin of life (Schlesinger, 1991). Further historical reductions in the partial atmospheric pressure of CO_2 (pCO_2) are associated with the development of a cover of higher land plants (Holland et al., 1986), some of which have been sequestered as coal for hundreds of millions of years. However, despite these reductions, atmospheric carbon dioxide levels probably remained significantly (three to ten times) higher than today's levels through the end of the Paleozoic era, some 250 million years ago (Kasting, 1987). Substantial evidence for these higher carbon dioxide levels is provided by recent measurements of stable isotope composition of ancient soils.

Cerling (1991) developed a model that has been used to estimate atmospheric carbon dioxide levels from late Paleozoic to modern times, based on the carbon isotopic composition of soil carbonate. This model was used by Mora et al. (1991) to infer that Silurian and Devonian atmospheric CO_2 levels were significantly elevated as compared with today's, based on stable isotope compositions of carbonates from paleosols of the lower Bloomsburg Formation (Upper Silurian) and the alluvial Catskill Formation (Upper Devonian) of central Pennsylvania. These results were in agreement with Ordovician atmospheric carbon dioxide levels estimated from rates of weathering of silicate rocks (Berner, 1991). Yapp and Poths (1992) have provided chemical evidence that 440 million years ago, the atmospheric partial pressure of CO_2 may have been 16 times higher than today. This conclusion was based on measurements of concentration and carbon isotope content of $Fe(CO_3)OH$ in the common mineral goethite (alpha-FeOOH) from an Upper Ordovician Neda Formation (Wisconsin, USA) ironstone. At this near-sealevel tropical site, the goethite was formed at a moderate temperature (24 degrees C), in a terrestrial soil. Yapp and Poths' (1992) evidence also indicates that some of the CO_2 in this soil arose from *in situ* oxidation of organic matter, suggesting the presence of terrestrial life predating the appearance of vascular plants.

However, biological activity, reflected in the rise of vascular plants in the Silurian/Devonian period and the rise of angiosperms in the Cretaceous period, has played a large role in weathering rates, and estimations of earlier (i.e., Ordovician) atmospheric carbon composition are limited by lack of information regarding the nature of the organisms that preceded vascular plants on the continents and how they affected chemical weathering rates (Berner, 1991). In summary, however, the aggregate of evidence indicates that atmospheric carbon dioxide levels were significantly higher during the colonization of land than at present. Later in this book a case will be made for correlation of higher atmospheric CO_2 composition with the origin and early success of the first land plants. To this end, the acquisition and fixation of CO_2 by plants and algae will be explored in greater detail in Chapter 5.

The depth of the ozone layer in today's upper atmosphere is related to modern levels of atmospheric oxygen. Biologists are interested in the history of atmospheric oxygen because it is assumed that early terrestrial life could not have survived without the presence of an ozone layer of some depth. It is generally believed that oxygen was absent from the earth's atmosphere before the emergence of oxygenic photosynthesis (Kasting, 1987), as early as 2,700 million years ago (Buick, 1992). Some have suggested that heavy UV bombardment from the early sun may have resulted in O_2 and O_3 production in the prebiological atmosphere by photolysis of water and CO_2 (Canuto et al., 1982). In contrast, it has been argued that photolytically produced O_2 would have been rapidly consumed in the oxidation of reduced gases and crustal minerals (Schlesinger, 1991).

It is apparent that the atmospheric histories of O_2 and CO_2 cannot be studied in isolation; however, estimation of early pO_2 levels is difficult because there is no related environmental parameter, such as climate in the case of CO_2, from which correlations can be made. Furthermore, a good history of past pCO_2 is needed to estimate past pO_2 levels (Kasting, 1987). There is geological evidence for the early occurrence of substantial atmospheric oxygen production in the form of banded iron formations generated in the deep ocean, and the simultaneous development of terrestrial red beds. These red iron deposits resulted from the reaction of ferrous iron with O_2, producing poorly soluble iron oxide, better known as rust (Fig. 2-1). Because the mixing time of the deep ocean is on the order of 1,000 years, the depths could have remained anoxic longer than surface waters or the atmosphere. As long as the deep ocean waters remained reducing, oxygen in downwelled surface water would have been consumed by ferrous iron. The end of the development of deep-banded iron formations marks the establishment of oxidizing conditions in the deep ocean about 1.7 billion years ago (Kasting, 1987). Upon saturation of the deep ocean oxygen sink (ferrous iron and other reduced species), oxygen could begin to accumulate in the atmosphere. It is estimated that of all the O_2 ever evolved from photosynthetic organisms, only 4 percent remains in the atmosphere today; the rest is bound in oxidized sediments. The great length of time required to saturate these sediments explains the long domination of simple procaryotic organisms. Knoll (1992) noted that the earliest fossil evidence attributable to eucaryotes (1.7 to 1.9 billion years in age) is approximately coincidental with the rise in atmospheric oxygen to 15 percent of modern levels, which is indicated by the weathering of iron formations to have occurred about 1.9 million years ago.

It is likely that an effective ozone screen would have been in place at an oxygen level of one-tenth present levels (or about 2.1 percent). The resulting increase in aquatic photosynthesis would generate even more oxygen and ozone, in a positive feedback loop. The problem arises in determining when this critical level of oxygen and ozone was achieved, or in other words, how fast atmospheric oxygen accumulated (Kasting, 1987). Some authors have argued that the buildup of ozone (dependent on atmospheric oxygen levels) controlled the timing of colonization of the land by plants (Lowry et al., 1980; Chapman, 1985). However, others have argued that if oxygen levels rose sharply—to near present levels by Devonian-Carboniferous time, as has been suggested by Kasting (1987)—ozone levels may have been high enough to screen out damaging short-wavelength (220 to 290nm) UV radiation. In view of the continued uncertainty regarding oxygen, ozone, and UV radiation levels at the time of origin of the first plants, at present it is not possible to choose between these two hypotheses on the basis of available paleochemical evidence. However, new methods have recently been developed for estimating ancient atmospheric oxygen levels by study of paleosols (ancient soils). These methods, based on the differential behavior of iron in soils formed on granitic and basaltic rocks (Holland et al., 1986), may shed light on rates of oxygen accumulation in the early atmosphere. Further, new models of atmospheric oxygen concentration changes have been generated on the basis of rates of burial

Fig. 2-1. Iron oxide-rich geological strata known as "red beds" may be found in various locations, including the Grand Canyon in the southwestern United States. Red beds are evidence of oxygen production (generally presumed to have originated primarily from aquatic photosynthesis) to levels saturating iron-containing sediments, thus allowing atmospheric oxygen partial pressure to increase.

and weathering of organic carbon and pyrite sulfur, which suggest that near modern levels of oxygen occurred in the Ordovician period, with substantial variation above and below this level occurring more recently (Berner and Canfield, 1989). Such new approaches will allow increasingly more accurate estimates of ozone depth to be made.

PALEOCHEMISTRY OF LAKES AND THE OCEANS

Although there is little supporting empirical evidence, Beerbower has (1985) suggested that low-nutrient (oligotrophic) lakes were more common in pre-Devonian times than at present. Occurrence of oligotrophic lakes is relevant to the development of hypotheses concerning the physiology of possible freshwater transmigrants, which is considered in greater detail in Chapter 5. There is more evidence available to indicate the chemical status of marine waters. Until recently, investigations into the chemical composition of seawater in the past have been based on analysis of deposits formed by evaporation (marine evaporites). Fluid inclusions in halite (NaCl) may represent samples of ancient seawater from which a determination of changes in seawater chemistry over time may be made. Unfortunately, at the present time the fate of trapped brine is not well understood. In addition, it is possible for the concentration of major salts to have varied by a factor of 2 to 3 in either direction without affecting mineral sequences in marine evaporites. Despite the uncertainties described above, the evaporite evidence suggests that the composition of seawater has been constant for the last 900 million years (Holland et al., 1986). Recently, Johnson and Goldstein (1993) discovered apparently unaltered Cambrian sea water samples preserved in marine low-magnesium calcite cements. Analysis of these samples suggested that the dominant carbonate material precipitated from sea water was different 500 million years ago (perhaps related to changes in atmospheric CO_2 levels), but that salinities were similar to modern sea water. Since we are assuming that land plants appeared no earlier than the Ordovician period, this new evidence suggests that an ocean chemistry similar to that of the present had been established for millions of years before the ancestors of modern plants made the transition from aquatic to terrestrial habitats. This is relevant to the question of whether plants colonized land from fresh waters or from the oceans. Although it is true that a few modern marine algae are able to adapt to freshwater conditions within periods of time as short as a few generations (Sheath and Cole, 1980), most extant seaweeds cannot survive in fresh water. In addition, migration of the seaweeds *Cladophora glomerata* and *Bangia atropurpurea* from marine to fresh waters appears to be accompanied by loss of capacity for sexual reproduction (Graham, 1982a). If the sea had been substantially less salty 500 million years ago than at present, the importance of marine versus freshwater origin of algal land plant ancestors would be much reduced. However, the emergent picture of Ordovician ocean chemistry suggests that marine and freshwater algae of that time were subjected to quite different environments, and that the evidence related to marine versus freshwater plant origins must be evaluated (see Chapter 4).

GEOGRAPHY AND CLIMATE IN THE ORDOVICIAN AND EARLY SILURIAN PERIODS

Paleogeographers have used paleomagnetic data to construct maps showing the relative positions of continental fragments present 458 to 421 million years ago (the Ordovician through the early Silurian periods) (Fig. 2-2). Although the paleomagnetic data are reasonably good for latitudinal positions of land masses, longitudinal

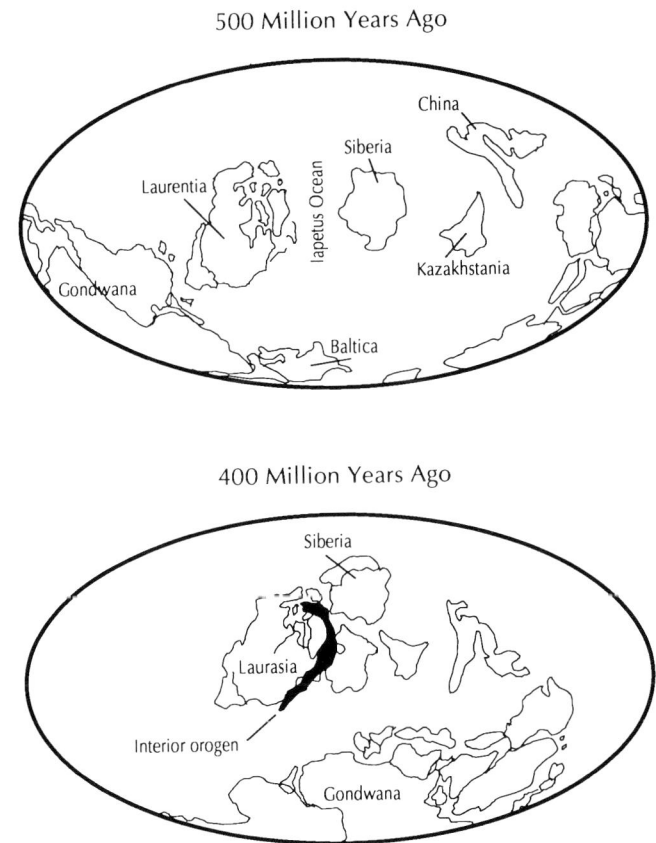

Fig. 2-2. At the time of plant colonization of terrestrial habitats, approximately 450 to 470 million years ago, changes in the relative positions of continental plates were leading toward formation of Laurasia and, ultimately, Pangaea. From ("Mountain Belts and the Supercontinent Cycle" by J. Brendan Murphy and R. Damian Nance). Copyright © (1992) by *Scientific American, Inc.* All rights reserved.

positions are less certain. The global reconstruction for this time shows that there was a Northern Hemisphere ocean covering about half the earth's surface. A band of land and shallow seas partly circled the earth around the equator. One of these seas was the proto-Atlantic ocean, also called the Caledonian or Iapetus Ocean, located between what is now North America and Western Europe. The major landmass, Gondwana, was associated with the South Pole. In the late Silurian or early Devonian period, the Laurentian and Baltican landmasses became sutured, resulting in elimination of the Iapetus Ocean (Holland, 1985; Livermore et al., 1985; Scotese et al., 1985; Murphy and Nance, 1992) (Fig. 2-2). These tectonic movements formed the Old Red Sandstone continent (Laurussia) which is famous as the location of pretracheophyte and early tracheophyte fossils such as *Cooksonia, Rhynia,* and *Psilophyton* (Edwards and Fanning, 1985). Another center of early vascular plant diversification is demonstrated by the so-called *Baragwanathia* flora of what is now Africa and Australia, which was then Gondwana. The pre-Devonian megafossil record suggests that the Baragwanathian flora was perhaps of earlier origin than that of Laurussia (Garatt et al., 1984). However, all these events, interesting as they are, occurred significantly later than the earliest appearance of land plants. Therefore, in considering paleoclimates, we primarily focus on data relevant to Ordovician and early Silurian times.

Development of climate models is more difficult for such ancient periods than for more recent times. However, there is considerable geological evidence for widespread glaciation of a previously ice-free earth in the late Ordovician period (440 million years ago). Glacial fluctuations at this time were on the same scale as those of the Pleistocene epoch. Rapid growth of an extensive ice sheet centered over what is now Saharan Africa was probably triggered by unusually rapid tectonic plate movements. Sea level dropped by 70 meters, draining shallow seas and resulting in a mass extinction event in which 50 percent of some groups and 12 percent of families were lost. Two searches for an associated iridium layer were negative, suggesting that the late Ordovician extinctions did not result from an extraterrestrial impact event such as that thought to have occurred at the Cretaceous–Tertiary boundary (Brenchley, 1989). This conclusion is supported by Wang et al. (1992), who used neutron activation analysis to probe the Ordovician/Silurian boundary at two locations in the Yangtze Basin. These workers could actually observe the extinction event as dramatic differences in the relative abundances of graptolites, brachiopods, trilobites, nautiloids, ostracods, and bivalves at the base of the zone defined by the graptolite *Glyptographus persculptus*. They found a coincident iridium maximum, which, at 0.23 ppb was almost as large as that marking the late Eocene impact event. However, they also noted evidence for correlation between Ir abundance and sedimentation rate variation sufficient to explain the Ir levels without invoking the occurrence of a cataclysmic extraterrestrial impact.

Because extinction events may be related to subsequent evolutionary rise and diversification of major groups (such as insects, birds, mammals, and flowering plants in the case of the Cretaceous–Tertiary event), it is quite possible that the late Ordovician glaciation extinctions were related to the appearance of land plants. Some specific adaptations that plant ancestors might have acquired in response to climatic conditions in the Ordovician period are discussed in Chapter 8.

Evidence that atmospheric CO_2 levels were considerably higher 400 to 500 million years ago than at present seems, at first consideration, to be incompatible with the evidence for late-Ordovician glaciation, because modern data indicate that there is a direct relationship between increasing atmospheric CO_2 levels and global warming—the greenhouse effect. However, Crowley et al. (1987) have explained this apparent paradox as a function of changes in polar location through time. These workers used a two-dimensional energy balance model to simulate seasonal temperature on a supercontinent-sized landmass and found that seasonal cycles are greatly influenced by land-sea distribution patterns. Paleomagnetic data indicate that the South Pole was offshore in the Cambrian, located in North Africa during the Ordovician, moved into midcontinental Africa in the Devonian, and located in what is now Antarctica by the late Paleozoic period. The onset of widespread glaciation in the late Ordovician period is postulated to be related to the location of the South Pole at the edge of a large continental landmass (Gondwana). The mechanism of the continental-edge effect is that the very large heat capacity of the mixed-layer ocean (as compared with the lower heat capacity of land) reduces the magnitude of summer warming in nearby land areas. Glacial growth is also favored by the increased moisture characteristic of coastal areas, and the probability of glacial development is greater during Milankovitch cool-summer orbits (relatively short-lived events resulting from periodic oscillations in the earth's orbit). The model predicts that glaciation becomes possible when the pole moves to within 800 km of the continent edge, resulting in summer temperatures near freezing. Ice-albedo feedback (increased reflection of heat into space) may then force climate into a glaciated state. When the pole is more than 1,000 km from the edge of the continent, summer temperatures may be too high for glaciation to occur. The simulation predicts that areas of low latitude could have maximum summer temperatures as high as 34°C, even higher than are characteristic of present climate models. This suggests the occurrence of arid conditions in midcontinent interiors and possibly strong monsoon conditions in coastal areas (Crowley et al., 1987). The hypothetical explanation for a glaciated world in the late Ordovician period is supported by evidence of a subsequent glacial event in the late Paleozoic (Permian epoch), following a period of warm, moist conditions in the Devonian and Carboniferous periods. During plant evolution these climate changes have been associated with lush growth of mesic vegetation, such as forests of arborescent lycopods and other large pteridophytes in the Carboniferous period, followed by the dominance of gymnosperms, which are characterized by adaptations to cooler and more xeric conditions of the Permian period. The model reported by Crowley et al. (1987) suggests that location of the South Pole in a coastal Antarctic location in the late Paleozoic epoch is related to the cooler climate at that time and suggests that movement of the pole offshore in the late Permian period set the stage for warmer conditions in the Cretaceous era. Although a decreasing CO_2 level is thought to have contributed to the onset of Permian glaciation, these authors hypothesize that land-sea distributions may be more important than CO_2 in dictating the time of glacial onset. Their model may explain the co-occurrence of high atmospheric CO_2 levels and high-latitude glaciation in the late Ordovician period, the likely time for the emergence of the first land-adapted plants.

LATE ORDOVICIAN TERRESTRIAL BIOTA AND SOILS

There is considerable fossil evidence for the occurrence of procaryotic, microbial life forms as far back as 3 billion years ago, in the Precambrian era. By late Precambrian times, organisms very similar to modern cyanobacteria (blue-green algae) were abundant, forming large numbers of calcareous stromatolites in shallow, protected seas. Similar growths can be observed today along the shores of Western Australia (Bold and Wynne, 1985). The activities of the cyanobacteria were of tremendous significance in the origin of other forms of life, in that this group, from which the ancestors of chloroplasts arose (Giovannoni et al., 1988) generated molecular oxygen from water for more than a billion years before the appearance of the first eucaryotic cells. Eucaryotic protists, the vast majority of which require aerobic, aquatic habitats, have been around for perhaps 1.7 or even 1.9 billion years (Knoll, 1992), and must have achieved considerable diversity by the time the first land plants appeared. Several authors have suggested that terrestrial surfaces had become colonized by green blankets of cyanobacteria, eucaryotic algae, and perhaps even lichens, long before the appearance of the first embryophytes (Jeffrey, 1962; Stebbins and Hill, 1980; Beerbower, 1985; Wright, 1985; Selden and Edwards, 1989; Gray and Shear, 1992). Three lines of evidence support this scenario: the great tolerance of extreme environments exhibited by many modern terrestrial algae, fossil remains, and some evidence derived from study of early soils.

A considerable diversity of cyanobacteria (Fig. 2-3) and eucaryotic algae may be isolated from just about any terrestrial habitat on earth. They occur in the air, on persistent snow, in rocks of the dry Antarctic, in soil, and on a wide variety of other living and inert terrestrial substrates (Bold and Wynne, 1985; Graham et al., 1981). Soil crusts of cyanobacteria are common and ecologically important as sources of fixed nitrogen. Cyanobacteria have recently been discovered to produce sheath pigments, such as scytonemin, that provide protection against UV radiation (Garcia-Pichel and Castenholz, 1991). It has previously been mentioned that some terrestrial eukaryotic algae have incorporated the highly resistant polymer sporopollenin into their cells walls, presumably as protection against desiccation and/or parasitism. Because sporopollenin also occurs in dinoflagellates and other algae whose habitats are exclusively aquatic, the resistant polymer probably represents a very ancient eucaryotic innovation. The terrestrial epiphyte *Phycopeltis* (Good and Chapman, 1978), and the soil algae *Chlorella* and *Scenedesmus* (Fig. 2-4), which also inhabit fresh waters, are green eucaryotes that no doubt find the polymer useful on land as well.

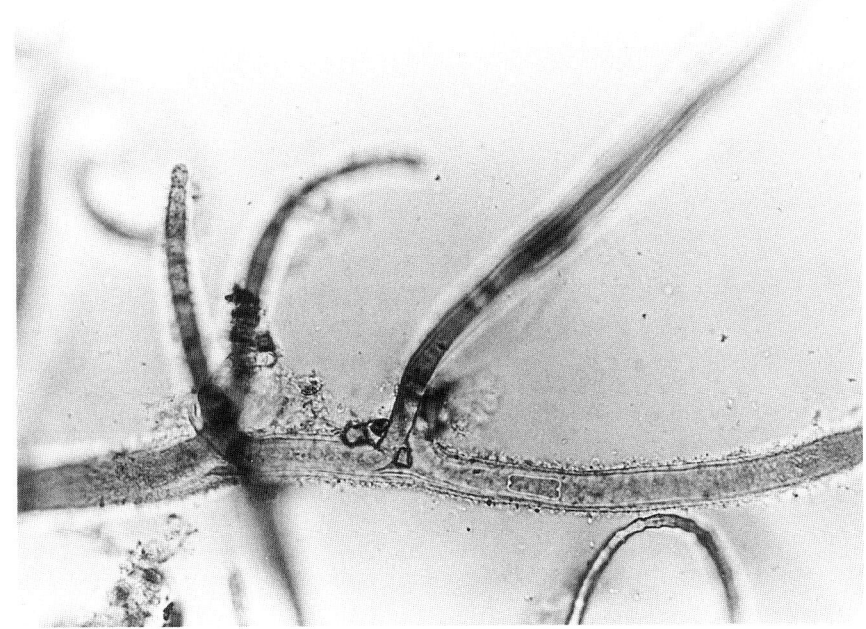

Fig. 2-3. Many cyanobacteria (blue-green algae) such as *Scytonema* are capable of living in soil and other terrestrial habitats. Adaptations such as UV-absorbing sheath pigments increase their fitness in the relatively exposed land environment.

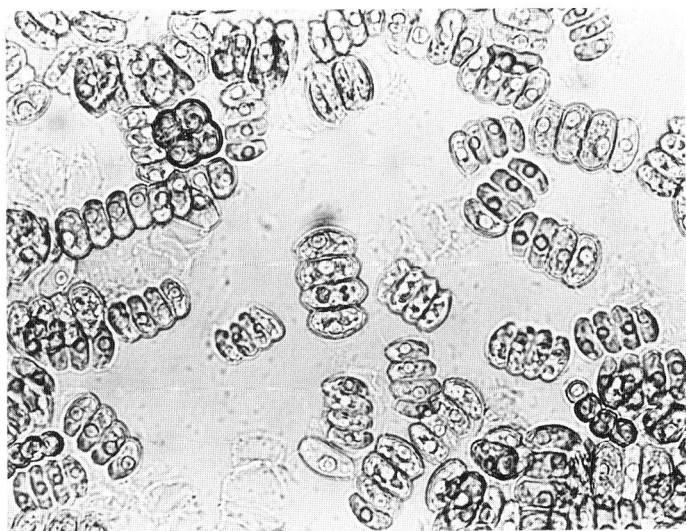

Fig. 2-4. The chlorophyte *Scenedesmus*, which occupies freshwater lakes and soils, produces sporopollenin in cell walls. Sporopollenin presumably functions in desiccation resistance, but may also retard microbial attack. The existence of resistant-walled eucaryotic algae, such as *Chlorella* and *Scenedesmus*, and terrestrial cyanobacteria indicates that the land surface may have been quite green prior to the origin of embryo-producing plants.

Fig. 2-5. It is possible that association of fungi with blue-green and green algae to form lichens such as *Cladonia* (illustrated here) may also have characterized early terrestrial habitats, but fossil evidence is lacking.

The fossil cyanobacterium *Eosynechococcus moorei* is thought to resemble closely the modern soil inhabitant *Gloeothece coeruleus,* suggesting that biologically active soil crusts existed as far back as the Precambrian period (Wright, 1985). Retallack (1985) compared a late Ordovician paleosol with a late Silurian soil and two younger soils, all formed in presumed subhumid, seasonally dry, subtropical climates on sedimentary and metamorphic rock. As expected, the younger soils exhibited better structure. However, the late Ordovician paleosol was unexpectedly better developed than envisioned for a landscape lacking in vascular plants. The most surprising feature was the occurrence of numerous deep burrows, evidence for the presence of animals, 3 to 16 mm wide, of some kind. No evidence was found for the presence of plant megafossils or microfossils (such as spores), suggesting that the burrowers relied on terrestrial algae, or perhaps lichens (Fig. 2-5), as a food source. Because other sites for early paleosols are known, perhaps their study will further illuminate the nature of the earliest terrestrial communities (Retallack, 1985). Gray and Shear (1992) provide additional details regarding the fossil evidence for early terrestrial communities.

We have now defined what is meant by the term "land plants," described the evidence related to time and place of their first appearance, and begun an exploration of possible environmental correlates aimed toward understanding why land plants appeared when they did. To this list of what, when, where, and why, we must now add the question of how land plants appeared. The evidence contributing to an answer is the subject of the rest of this book.

3

APPROACHES TO AN UNDERSTANDING OF EARLY LAND PLANT EVOLUTION

As we look toward the turn of the century (and a new millennium), it is apparent that newly developed techniques for study of nucleic acids provide tremendous potential for advance in many areas of biology, including the study of plant origins. This is an exciting time for plant biologists. The period between the mid-1800s and early 1900s must have been similarly stimulating; the excitement of discoveries related to plant genetics, life cycles, reproduction, physiology, evolution, and fossil history permeates F. O. Bower's classic 1908 book entitled *The Origin of a Land Flora* and a later (1935) text. By 1851, Hofmeister had recognized the ubiquity of alternation of the sporophyte and gametophyte generations in plant sexual reproduction. Strasberger (1884) discovered the correlation between alternation of generations and changes in chromosome number (Fig. 3-1). It was recognized that explanations for the origin of land plants must necessarily also address the question of the origin of alternation of generations (Celakovsky, 1874). The importance of the origin of plant reproduction in the development of ideas related to plant origins is reflected in the subtitle of Bower's 1908 book, which reads *A Theory Based upon the Facts of Alternation*. Now, of course, many would argue that "hypothesis" would be a better description for Bower's body of ideas, which were largely untested and for which there was not widespread acceptance. Nevertheless, Bower's 1908 treatise was, and still is, recognized as a significant contribution because of the care and detail with which his ideas were documented. Bower primarily used data resulting from studies conducted during the "golden age" of comparative plant morphology, combined with emerging information from the infant field of paleobotany. Nearly a century later, a much wider variety of approaches are available, including electron microscopy, fluorescence microscopy and immunolocalization techniques, improved methods for biochemical analysis and examination of fossils, nucleic acid sequencing, and

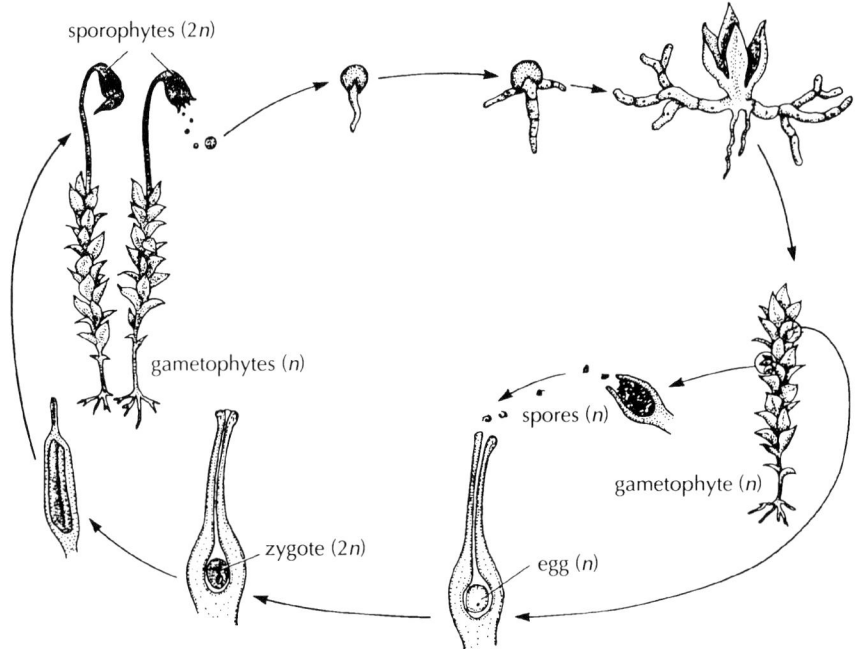

Fig. 3-1. The life cycle of embryophytes consists of alternation of generations in which the chromosome number of cells of the gametophyte generation is typically half that of cells belonging to the sporophyte. Embryophyte life cycles are also characterized by nutritional and developmental dependency of early embryos (young sporophytes) on the gametophyte generation. In modern bryophytes the sporophyte remains dependent throughout its lifetime, but sporophytes of modern vascular plants ultimately become independent of the gametophyte. Over evolutionary time, vascular plant gametophytes have become progressively smaller and less conspicuous than sporophytes. For example, flowering trees may be hundred-foot-tall sporophytes that alternate with few-celled gametophytes hidden within ovules and walls of pollen grains. In flowering plants, as well as in the simplest bryophyte, early embryogenesis involves close nutritional and developmental association of gametophyte and sporophyte generations. Illustration from Graham (1985), *American Scientist* 73:179.

computerized data analysis techniques. In the rest of this chapter, a brief description of the history of ideas related to the origin of land plants and their characteristic life cycle is followed by a discussion of modern approaches to the study of land plant origins.

A HISTORICAL BACKGROUND

Hofmeister's elucidation of the fundamental features of alternation of generations in plants, published eight years prior to Darwin's *Origin of Species,* did not address evolutionary questions about the origin of plant reproduction. However,

Celakovsky (1874), realizing that the plant life cycle must have been derived from previously existing reproductive modes, developed two alternative hypotheses for the origin of alternation in plants. These are referred to as the "homologous" or, sometimes, "transformation" hypothesis on the one hand, and the "antithetic" or "interpolation" hypothesis on the other. These alternatives are often called the homologous and antithetic "theories," but as indicated previously, the term "theory" should be restricted to well-documented, widely accepted ideas. Because there is controversy even today as to which of Celakovsky's concepts is more correct, they are more appropriately termed "hypotheses."

Succinct descriptions of differences between the two concepts can be found in most plant morphology textbooks, for example, in Bold et al. (1987). Today we recognize that hypotheses for the evolution of alternation of generations in plants must include an explanation of the origin of the embryo, which in all modern land plants is nutritionally and developmentally dependent on the gametophyte generation. Basically, the homologous hypothesis proposes that the ancestors of land

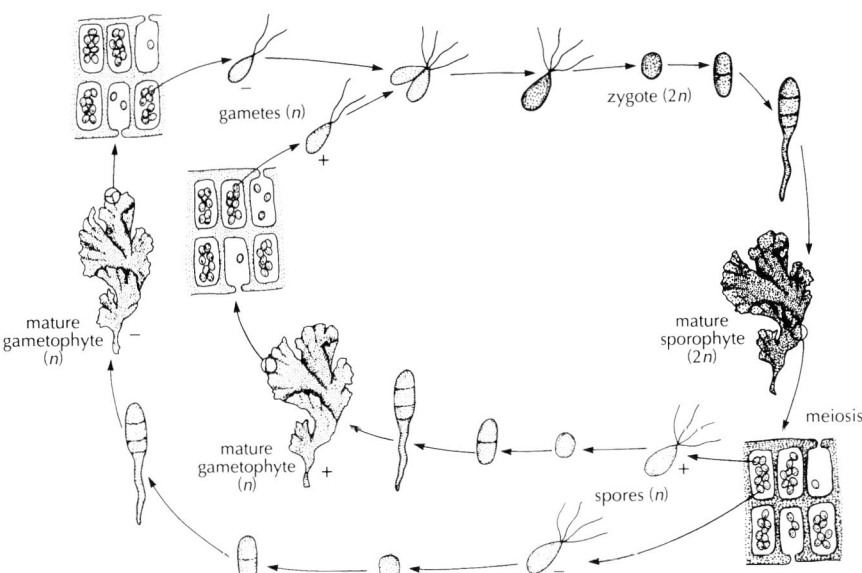

Fig. 3-2. Green seaweeds such as *Ulva*, the sea lettuce, have life cycles involving alternation of sporophyte and gametophyte generations coupled by syngamy (gamete fusion) and meiosis, as do embryophytes. Zygotes and early sporophyte stages of green seaweeds, however, are not nutritionally and developmentally coupled to parental gametophytes, as is characteristic of embryo-producing plants. This fact, and other evidence, demonstrates that alternation of generations has evolved independently in green seaweeds and embryophytes. However, the evolutionary mechanism that most likely gave rise to alternation of generations, delay in meiosis, is probably common to both lineages. Diagram from Graham (1985), *American Scientist* 73:181.

plants were algae that had a life cycle involving alternation of a multicellular spore-producing generation (sporophyte) with a multicellular gamete-producing generation (gametophyte). An example of an alga having this kind of life cycle is the sea lettuce, *Ulva* (Fig. 3-2). The homologous hypothesis assumes that a formerly independent sporophyte generation developed a parasitic relationship with the gametophyte generation (Fig. 3-3). In contrast, the antithetic hypothesis suggests that the

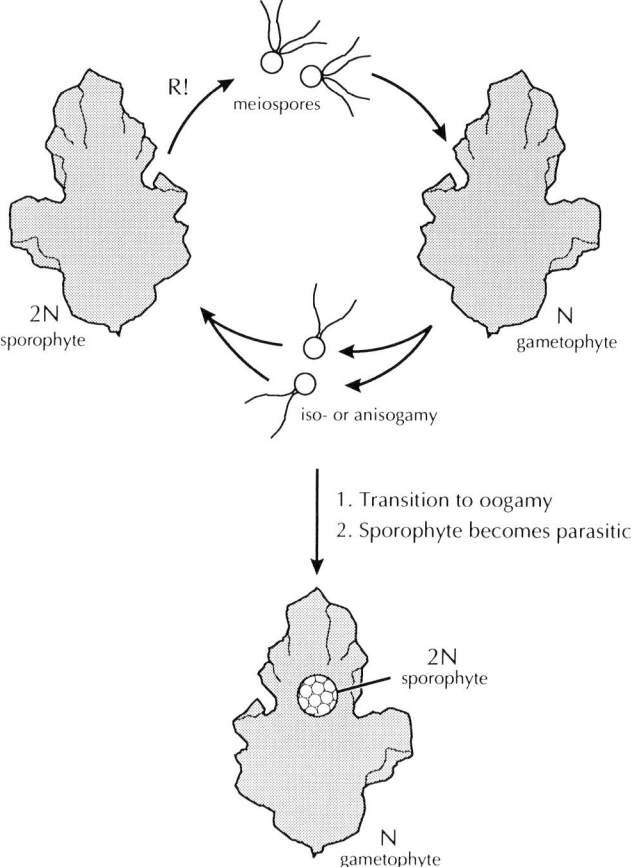

Fig. 3-3. The homologous hypothesis is based on the premise that ancestors of embryophytes had alternation of isomorphic (structurally similar), free-living generations similar to *Ulva*'s life cycle, and that the sporophyte became secondarily associated with the gametophyte generation with regard to nutritional and developmental interactions. Proponents of this hypothesis usually suggest that this association occurred when sporophytes acquired "parasitic" capabilities. However, evolutionary change from autotrophic, free-living habit to parasitic habit would almost certainly necessitate some reduction in sporophyte size, hence reduction in reproductive potential (fitness). Selective advantage conferred by such hypothetical transition is unclear, and parallel model systems useful in testing the homologous hypothesis have not been identified.

algal ancestors of plants lacked alternation of generations, and that the multicellular spore-producing generation of land plants (the sporophyte) was derived by delaying zygotic meiosis. The freshwater green alga *Coleochaete,* whose life cycle is shown in Fig. 3-4, is often cited as an example of an ancestral type. Origin of the nutritional and developmental relationship between generations observed in embryophytes is explained by the occurrence of mechanisms for export of photosynthate from source to sink, between cells of a single organism (Fig. 3-5), and does not require that the diploid phase acquire parasitic capabilities during the transition to land (Graham, 1985). Debate as to which of these hypotheses is most acceptable continues to the present day, influencing the direction and focus of modern plant research.

Bower (1908) found the antithetic hypothesis compelling, and made its defense the focus of his book. He was strongly influenced by the work of phycologists (students of algae). Among these was Pringsheim, who in 1860 had published some detailed studies of reproduction and development in *Coleochaete,* pointing out interesting features of its postfertilization development. Bower specifically cited *Coleochaete* as a model of the kind of alga from which land plants might have arisen by slight modification of the reproductive process that Pringsheim and others had described. Interestingly, however, Pringsheim (1878) was later regarded as a supporter of the homologous hypothesis (Fritsch, 1916). The famous phycologist F. E. Fritsch (1916) also wrote authoritatively on the

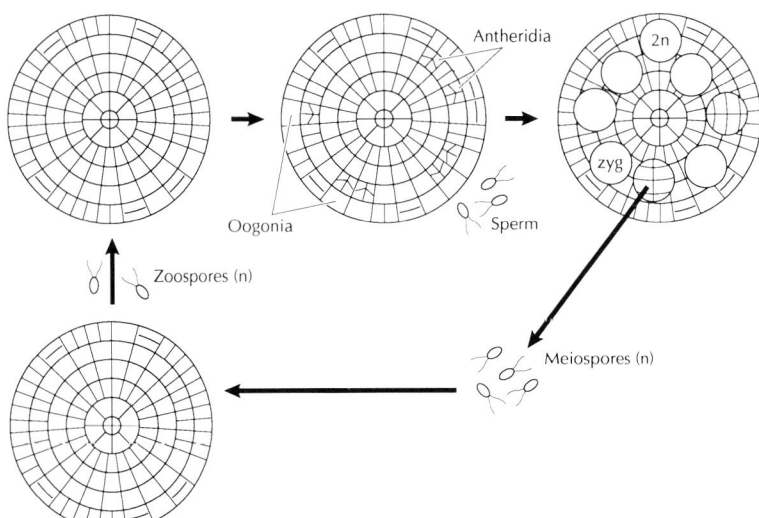

Fig. 3-4. The freshwater green alga *Coleochaete* is characterized by a life cycle in which the diploid phase (zygote) is unicellular. Meiosis occurs at zygote germination in *Coleochaete* and all other charophytes that have sexual reproduction. Zygotic meiosis contrasts with sporic meiosis (production of spores by meiosis), which evolved independently in seaweeds such as *Ulva* and land plants. *Chara* and relatives have a life cycle similar to that of *Coleochaete,* except that zygotes are retained upon the haploid parental thallus through sporogenesis in *Coleochaete,* but not in Charales.

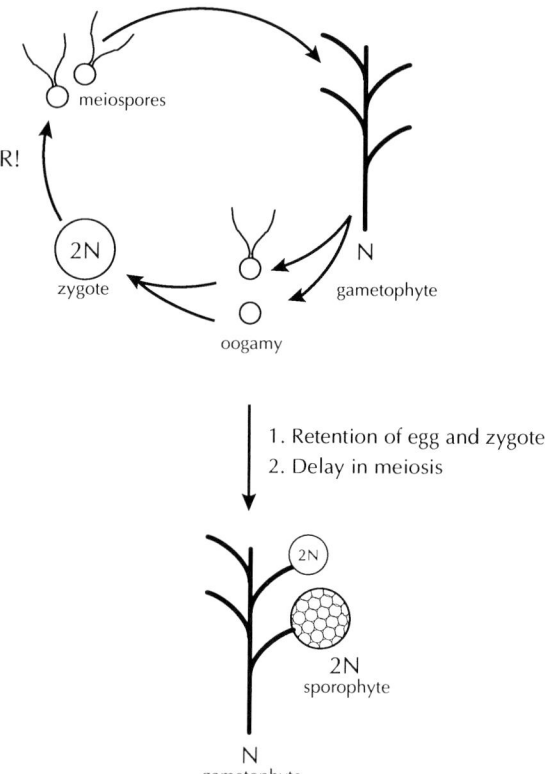

Fig. 3-5. The antithetic hypothesis proposes that delay in meiosis occurred in the life cycle of an alga similar to *Coleochaete*, generating a small, multicellular diploid generation—a sporophyte. Retention of zygotes upon parental thalli, coupled with nutritional and developmental interactions between zygotes and gametophytes of charophycean algae illustrate putative preadaptations leading to evolution of the embryo. The antithetic hypothesis does not postulate changes more dramatic than delay in meiosis, a process which must have occurred many times during protist evolution. *Coleochaete* and relatives thus provide useful model systems for testing validity of the antithetic hypothesis.

origin of land plants from algae, professing to be more interested in thinking about the characters of ancestral types than choosing sides in the alternation of generations controversy. However, he proposed that the bryophytes arose from algae like *Coleochaete* by elaboration of the zygote (as is postulated by the antithetic theory), but that pteridophytes (ferns and other nonseed vascular plants) originated from branched green algae having spatial separation of spore and gamete production, referred to as "pseudo-homologous alternation" (Fritsch, 1916). Later Zimmermann (1959), originator of the famous "telome theory" of the origin of plant organs, illustrated some of Fritsch's ideas concerning the evolution of land plants from algae. Zimmermann's diagrams have been widely reproduced in botany textbooks and papers on plant origins (Fig. 3-6).

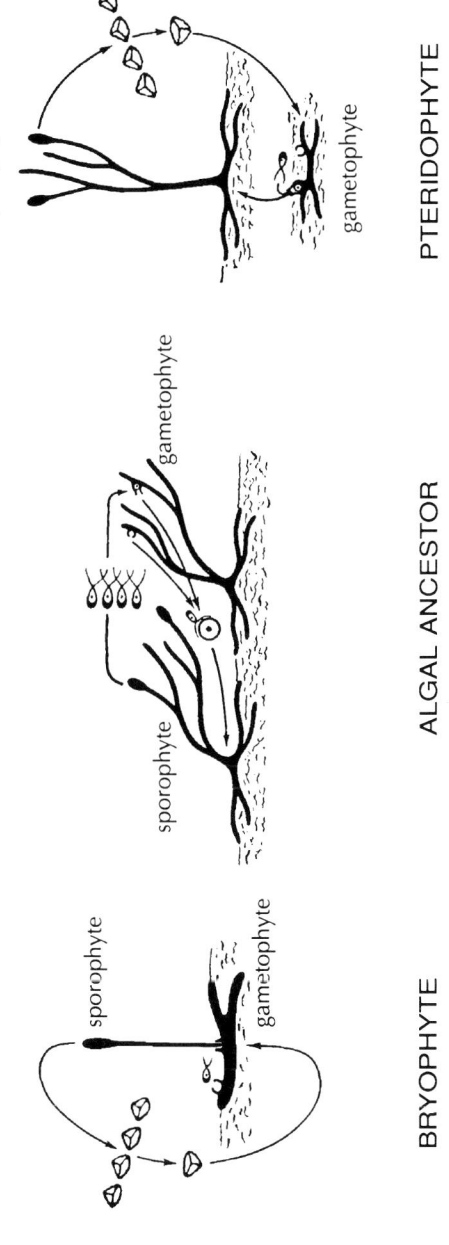

Fig. 3-6. A commonly reproduced diagram (adapted from Zimmermann, 1965), illustrating hypothetical origin of the life cycles of bryophytes on one hand, and pteridophytes on the other, from ancestral algae having alternation of independent sporophyte and gametophyte generations and free-living zygotes. Very little evidence supports this model; in contrast, the balance of modern investigations supports a hypothesis of land plant origins from algae having zygotic meiosis (illustrated by modern *Chara* or *Coleochaete*).

A. H. Church, a contemporary of Fritsch, published a short (but famous) book entitled *Thalassiophyta and the Subaerial Transmigration* (1919). Church was greatly impressed with the high degree of morphological complexity of the brown algae. He concluded that if the brown algae could acquire such impressive morphologies in the sea, green seaweeds may also have had such potential. Church suggested that a hypothetical group of morphologically complex, but now extinct, green seaweeds, the Thalassiophyta, colonized the land and gave rise to land plants. More recently, Church's ideas have fallen into disfavor, primarily because no one has discovered evidence that the Thalassiophyta actually existed. Later, Jeffrey (1968) and Meeuse (1966) discounted a marine origin for plants, arguing instead for descent from freshwater green algae. Jeffrey's (1962, 1969) papers were notable for stimulating later writings on the importance of transition to homeohydry by acquisition of desiccation-avoidance adaptations that must have occurred during the evolution of land plants from aquatic algae, and for developing interesting new ideas concerning other plant adaptations to land.

In 1972, Jeremy Pickett-Heaps and Harvey Marchant ushered in the modern era of the study of plant origins. They used transmission electron microscopy to establish that cell division and reproductive cell ultrastructure of certain freshwater green algae, particularly *Coleochaete* and the stonewort *Chara*, were remarkably similar to land plant cytokinesis and sperm structure, yet distinctively different from those of other algae. This evidence was used to construct a phylogenetic tree showing a relationship between these green algae and the ancestry of modern land plants (Fig. 3-7). Ultrastructural study of the related alga *Chaetosphaeridium* led Moestrup (1974) to the same conclusions. *Coleochaete*, *Chara*, and *Chaetosphaeridium* are primarily freshwater forms that exhibit sexual reproduction, but lack alternation of generations. In 1980, Stebbins and Hill generated the interesting idea that land plants originated from unicellular soil algae that acquired adaptations to the terrestrial environment in situ, rather than from multicellular algal invaders from the aquatic habitat. They accepted the ultrastructural evidence for relationship of multicellular aquatic algae such as *Coleochaete* and *Chara* to land plants, but suggested that these and related forms had originated from terrestrial ancestors, then

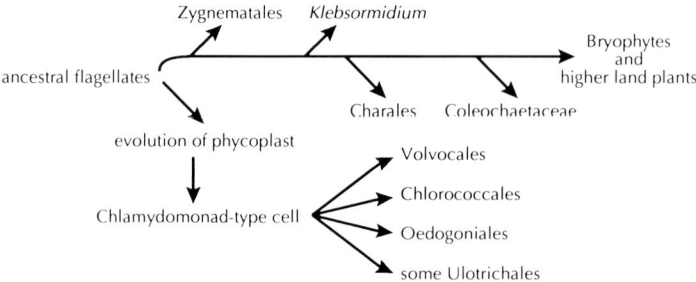

Fig. 3-7. Ultrastructural studies of green algae led to the proposal that some green algae (now known as charophytes) were more closely allied to the ancestry of land plants than others. Diagram from Pickett-Heaps (1976), © 1976, American Institute of Biological Sciences, with permission of the author.

reentered the water analogously to modern aquatic embryophytes. Stebbins and Hill also came down squarely on the side of the homologous hypothesis for the origin of the land plant life cycle. Another strong supporter of the concept that land plants arose from green algae having alternation of generations was R. M. Schuster, the eminent student of liverwort and hornwort morphology and systematics. Schuster (1984a, b) produced a detailed argument against concepts of land plant origins proposed by the famous plant morphologist D. H. Campbell (1912) and the well-known phycologist G. M. Smith (1955).

Subsequently, the merits of the antithetic hypothesis were argued on the basis of ultrastructural data derived from study of reproduction in *Coleochaete* (Graham, 1985). This argument focused on an explanation for the origin of nutritional and developmental relationships between generations that characterize land plant embryogenesis. Mishler and Churchill (1984) applied cladistic techniques to the body of ultrastructural and biochemical data related to green algal and land plant relationships. As a result of this analysis, they concluded that the land plant life cycle arose by elaboration of the zygote in an organism similar to *Coleochaete*. In 1985 the same workers published another cladistic analysis, this time reaching the controversial conclusion that a species of *Coleochaete* was the sister group to the entire extant plant kingdom (the embryophytes). Subsequent cladistic analyses incorporating additional data have upheld the finding that *Coleochaete* and the stoneworts (Charales) are closely related to land plants (Bremer et al., 1987; Graham et al., 1991) (Fig. 3-8). The relevance of *Coleochaete*'s reproductive characteristics to the

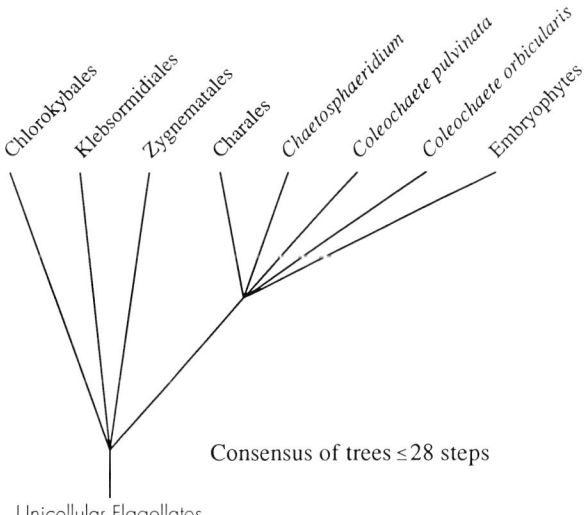

Fig. 3-8. Cladistic analyses of a data set composed of morphological and biochemical characters supports alliance of charophycean algae with the ancestry of embryophytes and suggests a concept of rapid radiation of embryophytes and higher charophytes from a common ancestor that was an oogamous, branched filament. Diagram from Graham, Delwiche, and Mishler (1991), *Advances in Bryology* 4:213-244, with permission of Gebruder Borntraeger.

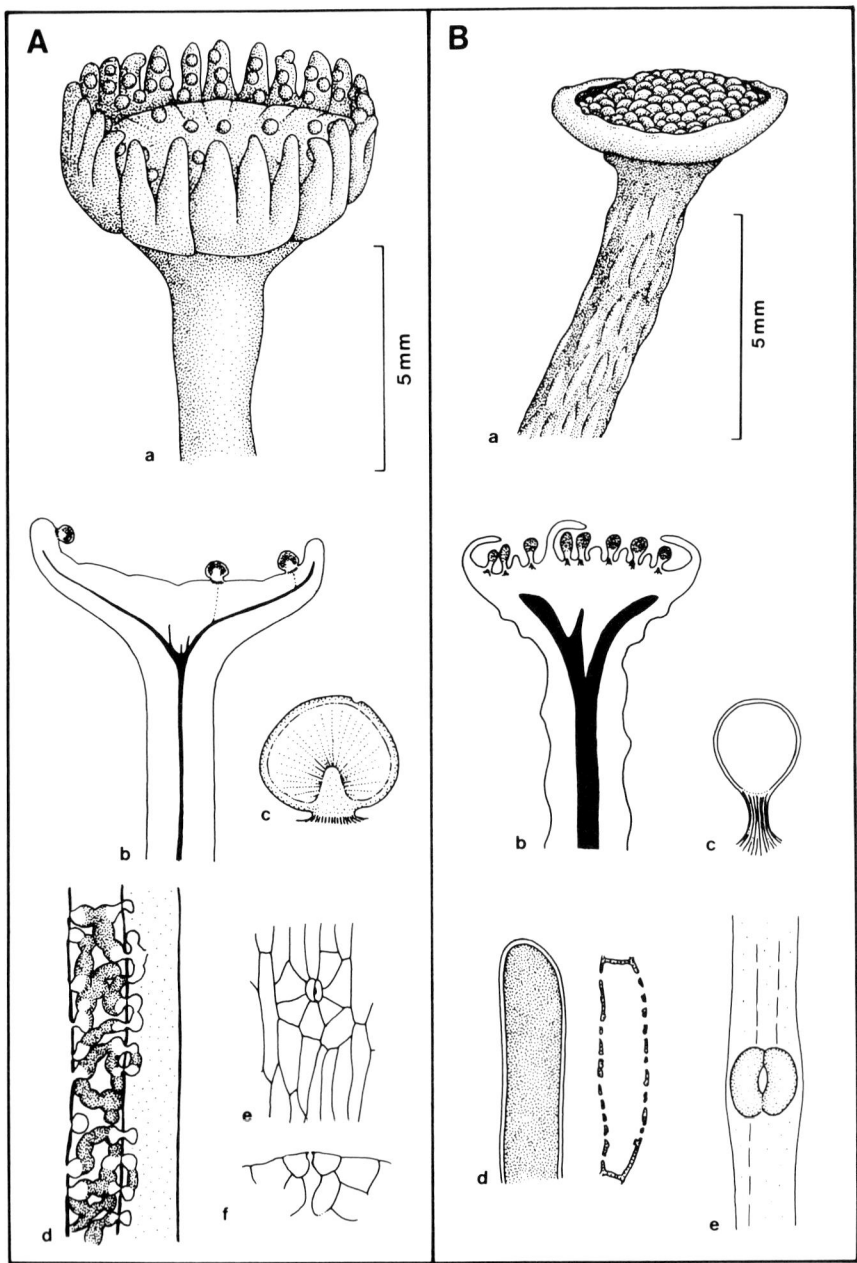

Fig. 3-9. *Lyonophyton* (A), *Kidstonophyton* (B), and *Langiophyton* (C) are extremely well preserved gametophytes from the Lower Devonian of Scotland (the famous Rhynie Chert deposits which have yielded a diverse array of early plant fossils). In each case, an unbranched stalk bore terminal, radially symmetrical structures upon which were attached

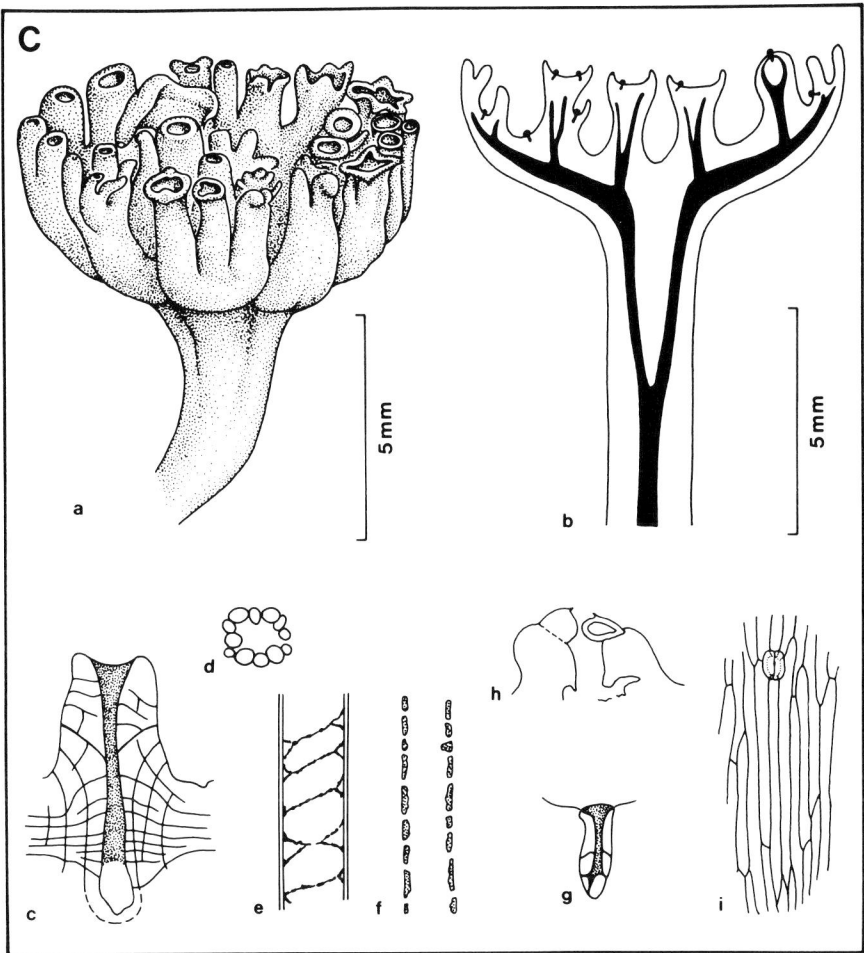

Fig. 3-9. *(Continued)* short-stalked antheridia (A-c, B-c), or archegonia (C-c). The archegonia of *Langiophyton* are notable in that the number of neck cells is larger than is typical of archegonia of extant plants (C-d). The central stalk tissue (shown as dark material in A-b, B-b, and C-b) was composed of elongated cells interpreted as conducting cells (A-d; B-d; and C-e, f). *Langiophyton* possessed two distinct types of conducting cells (C-e,f), and glands (C-g). This evidence, together with presence of stomata (A-e,f; B-e; and C-h,i), suggests a high degree of adaptation to terrestrial life. There are structural similarities between each of these gametophytes and specific sporophyte taxa. For example, the distinctive glands exhibited by *Langiophyton* are similar to those of the sporophyte *Horneophyton*. This evidence strongly suggests that at least some early Devonian plants had alternating generations that were similar in size and structure. Diagrams from W. Remy, P. Gensel, and H. Hass (in press). The gametophyte generation of some early Devonian land plants, *International Journal of Plant Sciences* (in press), kindly provided by the authors.

origin of the embryophytes and their life cycle was later bolstered by discovery of distinctive, chemically similar materials at the junction between haploid and diploid generations in *Coleochaete* and bryophytes (Delwiche et al., 1989). Further compelling evidence for close relationship of *Coleochaete,* Charales, and related green algae (Mattox and Stewart, 1984) and land plants has been provided by molecular analyses (Baldauf et al., 1990; Manhart and Palmer, 1990). This evidence is described in more detail later in this chapter.

In 1980a, b, Remy and Remy reported the finding from the Rhynie chert of Devonian gametophytes that were unusually well preserved as petrifactions in silica. The newly discovered fossils, named *Lyonophyton* (and *Langiophyton* and *Kidstonophyton,* discovered later), exhibited cellular details including convincing antheridia containing developing spermatozoids and presumptive archegonia (Fig. 3-9), as well as regions likely to represent conducting tissues. Although these fossils are probably millions of years too recent to represent remains of the earliest embryophytes, Remy and Hass (1986; 1988) considered them to be evidence in support of the homologous hypothesis of plant life cycle origins. These authors regard *Lyonophyton* (and similar fossils), as the gametophyte generations in nearly isomorphic alternation of generations with morphologically similar spore-producing plants. It is evident that consensus has not yet been reached among plant scientists regarding the origin of plants and their life cycle.

MODERN APPROACHES TO THE STUDY OF LAND PLANT ORIGINS

Authorities on plant origins have often taken the position that the fossil record holds the only evidence of the true course of plant evolution. Gray and Shear (1992) concluded that all hypotheses about the ecology and evolutionary history of terrestrial life must be tested against the fossil record. Yet, as we have seen, with the exception of Jane Gray's microfossils, the record of the earliest events in land plant evolution is virtually nonexistent (Thomas and Spicer, 1987). Moreover, unless remains preserved at least as well as the petrified Devonian gametophytes are found, the fossil record will not be able to provide the level of cellular detail necessary to understand the early stages of embryo and sporophyte evolution. Does that mean that the search for knowledge of early plant origins is doomed? Not at all. In spite of inadequate funding levels for paleobotany, exciting new fossil evidence appears with regular frequency. In addition, modern cellular, biochemical, and molecular studies have revealed that the evolutionary relationships of protists (including algae) and "lower" plants can be revealed by three major approaches: (1) comparative study of conservative ultrastructural characters such as those related to cell division and reproduction, (2) comparison of enzyme distribution patterns and study of enzymatic pathways of secondary metabolism (Stafford, 1991), and (3) analysis of differences among homologous DNA sequences. Once relationships are established, the evolutionary steps leading to origin of the embryo and other land plant features can be deduced. Mishler and Churchill (1984) have argued that fossil remains are not truly interpretable until hypotheses of phylogeny have first been determined from analysis of the characters of modern organisms. Granted, care must be taken to distinguish the properties of organisms that have emerged as the result of evolutionary diversification from those reflecting environmental or developmental influences. It is also necessary to use multiple lines of evidence—morphology as well as various molecules—to deduce patterns of descent in the distant past (Crawford, 1990; Chapman and Buchheim, 1991; Li and Grauer, 1991). But it is possible to deduce the major events in plant evolution. The contributions of past and emerging new technologies in light and electron microscopy, biochemistry, and molecular systematics to knowledge of plant origins are discussed below.

Ultrastructure and Light Microscopy

In the late 1960s and 1970s, a number of workers, including Jeremy Pickett-Heaps and his colleagues, examined the ultrastructure of a wide variety of green algae (Pickett-Heaps, 1976). They found that most green algae did not undergo mitosis and cytokinesis in the same way as higher plants. This was surprising inasmuch as green algae had long been thought to be ancestral to land plants by virtue of similarities in photosynthetic pigments and other cellular features. One major difference observed was that the nuclear envelope of most green algae remains intact or develops only localized discontinuities (fenestrations) (Fig. 3-10), whereas the nuclear envelope of dividing land plant cells breaks down completely (Fig. 3-11). These character states are referred to as "closed" or "fenestrated" versus "open" mitosis. At cytokinesis, the spindle of many green algae (including the well-known *Chlamydomonas*) collapses before cytokinesis occurs. In order to keep daughter nuclei well separated while cytokinesis occurs, these algae produce a system of microtubules oriented parallel to the plane of division. This microtubular system is called a phycoplast. Some other green algae, related to the seaweed *Ulva,* do not form phycoplasts; rather, spindle microtubules persist while a precocious furrow divides the cytoplasm. In contrast, embryophyte cytokinesis is typically associated with a system of perpendicularly oriented microtubules, which together with actin microfibrils and associated vesicles is called a phragmoplast (Fig. 3-12). Two green algal genera, *Coleochaete* and *Chara,* produce an array of microtubules and vesicles at cytokinesis very similar to land plant phragmoplasts (Pickett-Heaps, 1976; Marchant and Pickett-Heaps, 1973; Brown et al., 1993). The pondsilk algae *Spirogyra* and *Mougeotia* were discovered to possess what appeared to be an evolutionary precursor to the phragmoplast (Fowke and Pickett-Heaps, 1969; Galway and Hardham, 1991). *Coleochaete, Chara, Spirogyra* and a few other genera (*Stichococcus, Raphidonema, Klebsormidium,* and *Chlorokybus*) exhibit an open and persistent spindle apparatus, as do land plants. These green algae thus form a natural group with respect to mitotic and cytokinetic features.

In parallel with this ultrastructural survey of cytokinesis, Pickett-Heaps and others also conducted ultrastructural studies of green algal reproductive cells: zoospores, meiospores, and gametes. Some of the algal reproductive cells bore

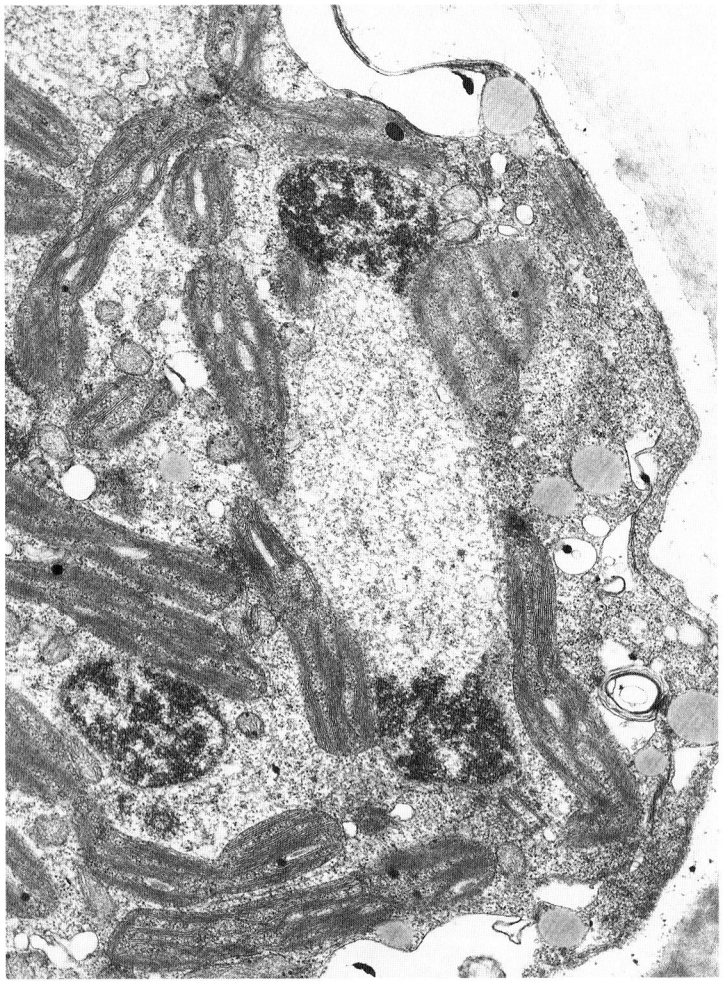

Fig. 3-10. Closed mitosis is characterized by persistence of an intact (or nearly intact) nuclear envelope throughout the mitotic process, as illustrated here by a telophase nucleus in a sporangium of the green alga *Trentepohlia aurea*. Photo from Graham and McBride (1978), *Journal of Phycology* 14:133.

similarities to the sperm of higher plants and, not surprisingly, these came from the same algal genera whose mitosis and cytokinesis were also plantlike. Specifically, *Chara, Nitella, Coleochaete, Klebsormidium,* and *Chlorokybus* had either flagellate spores or gametes, or both, that contained a multilayered structure (MLS) (for example, Sluiman, 1983) (Figs. 3-13 and 3-14), or its homologue, a "degenerate MLS." MLSs are also known to occur in the flagellate spermatozoids of land plants, ranging from bryophytes to some gymnosperms. In contrast, MLSs were not present in most other green algae. Some workers also noticed that

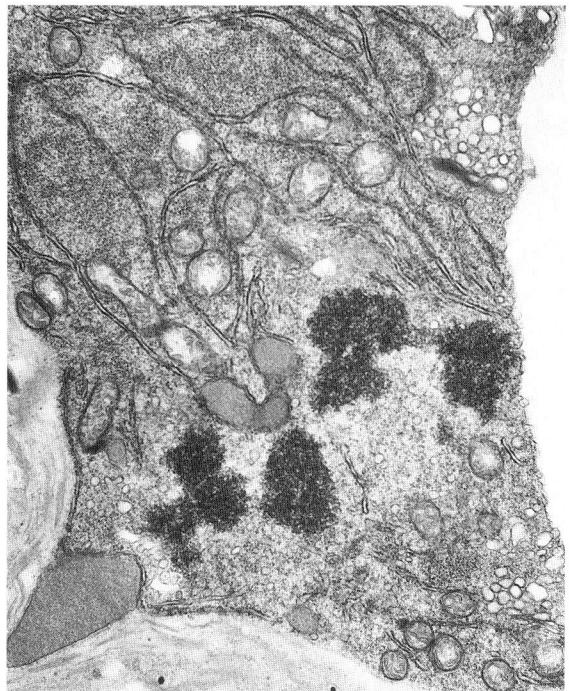

Fig. 3-11. In contrast to closed or fenestrated mitosis, open mitosis, illustrated here by metaphase in *Coleochaete orbicularis,* is characterized by disintegration of the nuclear envelope early in mitosis, coupled with reformation at telophase. Although nuclear envelope breakdown occurs during mitosis in animals as well as in charophytes, some other protists, and land plants, little is known about the controlling mechanism or the evolutionary history of open mitosis.

the same green algae that resembled land plants in cell division and reproductive cell structure also had peroxisomes (Fig. 3-15) that were structurally and biochemically similar to those of land plants (Stewart et al., 1972), whereas microbodies of other green algae were different. Subsequently, persistent and open mitotic spindles, MLSs, and peroxisomes came to be regarded as ultrastructural markers for relationship to land plants. As a group, these characteristic structures are still considered reliable indicators of common ancestry.

In 1984, Mattox and Stewart used ultrastructural evidence to establish the green algal class Charophyceae, to include the Charales (*Chara, Nitella,* and several other genera), Coleochaetales (*Coleochaete* and *Chaetosphaeridium*), Zygnematales (a very large taxon including *Spirogyra* and *Mougeotia* and other Zygnemataceae, and the desmids), Klebsormidiales (*Klebsormidium, Stichococcus,* and *Raphidonema*), and Chlorokybales (*Chlorokybus*). The concept of a distinct lineage of green algae, the Charophyceae, that is more closely related to the

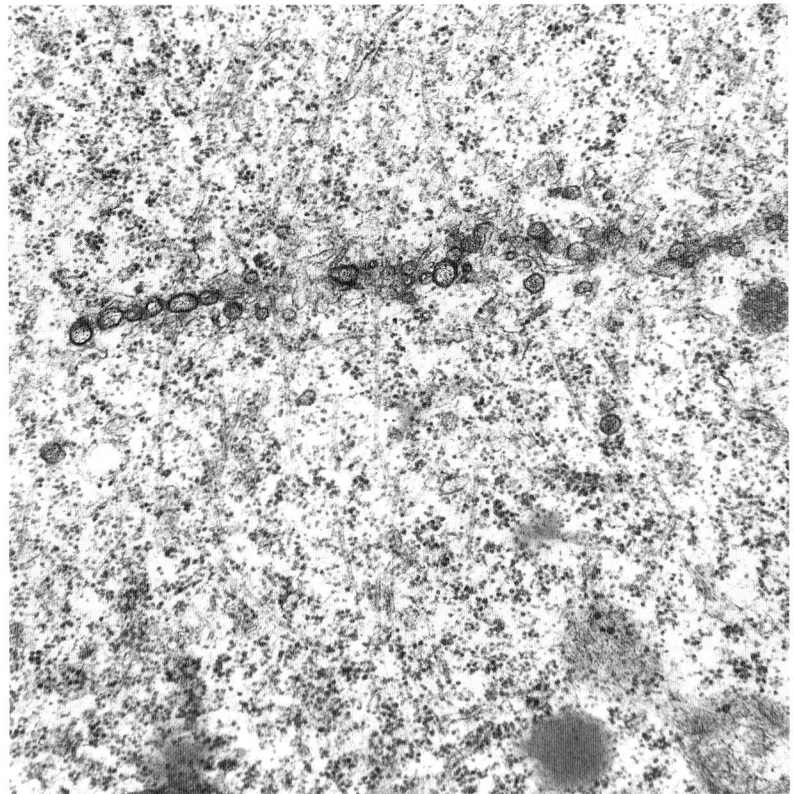

Fig. 3-12. Phragmoplasts of land plants, illustrated here by a dividing corn (*Zea mays*) root cell, like those of charophycean algae, consist of a system of parallel microtubules and dictyosome-produced vesicles. Microtubules are thought to guide vesicles to the cytokinetic plane, where they fuse to form the cell plate. In land plants, endoplasmic reticulum trapped within the developing cell plate forms plasmodesmatal desmotubules, but the occurrence of desmotubules in charophycean plasmodesmata is controversial.

ancestry of land plants than are other green algae continues to be robust (Chapman and Buchheim, 1991). Members of the entire class Charophyceae are referred to as "charophycean algae" or "charophytes," whereas members of the order Charales are "charalean algae." Often the term "charophyte" has been used in the literature to refer exclusively to charalean algae. However, ultrastructural and other evidence clearly indicates that Mattox and Stewart's (1984) concept of inclusion of Charales within a larger, presumably monophyletic group, the Charophyceae, broadens the definition of the term "charophyte." Mattox and Stewart (1984) also recommended formation of the class Ulvophyceae to include green algae having closed or fenestrated mitosis and lacking phycoplasts (ex. *Ulva*), and the class Chlorophyceae to include phycoplast-forming green algae

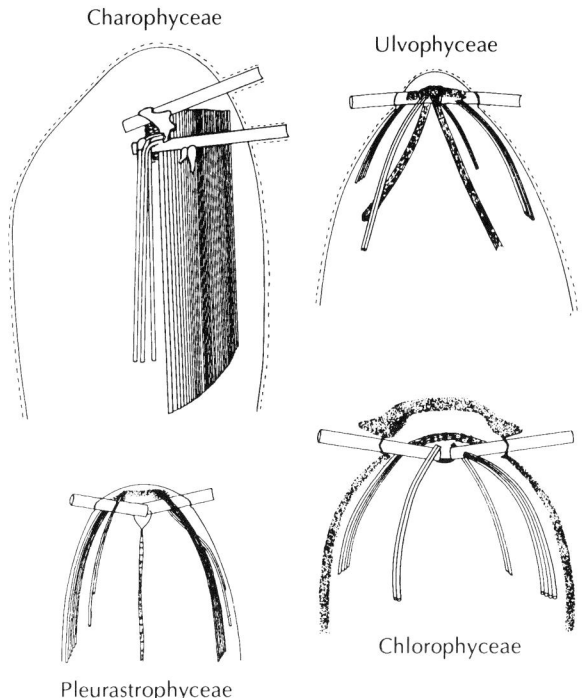

Fig. 3-13. The flagellar apparatuses of green algae exhibit variation that correlates with mitotic and cytokinetic differences. The flagellar apparatus of green algae grouped with the Charophyceae is most similar to that of flagellate land plant sperm. Reproduced from K. R. Mattox and K. D. Stewart. 1984. Classification and Evolution in Green Algae, In: Irvine and John (eds.), Systematics of the green algae, Academic Press, with permission.

(ex. *Chlamydomonas, Chlorella*) (Fig. 3-16). Recent molecular evidence suggests some doubt about the monophyly of some of Mattox and Stewart's green algal classes (Chapman and Buchheim, 1991). However, it is clear that green algae placed in the classes Ulvophyceae or Chlorophyceae are not as closely related to the ancestry of land plants as are the charophycean algae.

Until recently, ultrastructural studies of green algae have relied on chemical fixation techniques that have not always resulted in acceptable preservation of labile elements such as the cytoskeletal apparatus. Better results have been obtained with cryofixation and freeze substitution (Lokhorst et al., 1988). No doubt future ultrastructural studies of green algae will make increasing use of new freezing technologies. There is also great potential for application of immunolocalization procedures at the transmission electron microscopic level in investigations of green algal phylogeny, but to date this technology has been underutilized. However, rapid progress is being made in the application of immunofluorescence and confocal microscopy to green algal and bryophyte cell

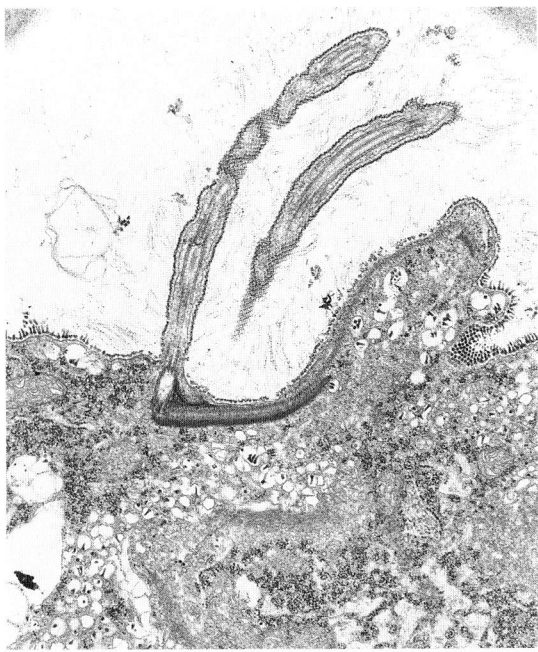

Fig. 3-14. The multilayered structure (MLS) of *Coleochaete scutata* zoospores, like that of other charophytes, is attached to flagellar bases by a complex of striated fibers. MLSs are conjectured to function as organizing centers for cytoskeletal microtubules. MLSs consist of an upper layer of adherent microtubules that extend down into the cell forming a "spline," one or more strata of vertically oriented lamellae, and an underlying layer of more or less homogeneous, electron-dense material. The layered portion of MLSs is called the lamellar strip (LS), and the term "blepharoplast" refers to the entire complex of MLS, flagellar basal bodies, and associated structures. MLSs of embryophytes are similar to those of charophytes in many respects, but differ in the angle of attachment of the microtubular layer to lamellae in the LS. Embryophyte blepharoplasts also differ in development and mature structure from those of charophyte algae. Photo from Graham and McBride (1979), *American Journal of Botany* 66:890.

biology and phylogeny. Examples include investigations into cytoskeletal proteins involved in cytoplasmic streaming in charalean algae (Williamson, 1992), involvement of actin in chloroplast rotation (Wagner and Grolig, 1992) and microtubule dynamics (Galway and Hardham, 1991) in zygnematalean algae, comparative mitosis in bryophytes and *Coleochaete* (Brown and Lemmon, 1990c), and other cytoskeletal work on bryophytes (Doonan and Duckett, 1988). It has been proposed that further immunofluorescence studies of cell division in green algae may shed considerable light on the origin of poorly understood plant development processes (Graham and Kaneko, 1991), and these prospects will be considered further in Chapter 7.

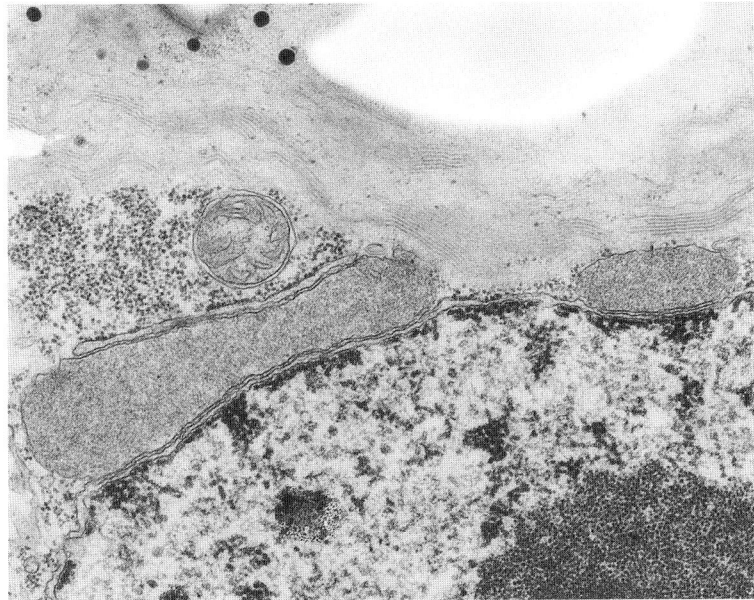

Fig. 3-15. Peroxisomes similar in structure and biochemical character to those of land plants are also present in charophytes. As illustrated by a cell of *Coleochaete orbicularis*, the peroxisome typically lies between the nucleus and plastid. Photo from Graham and Kaneko (1991), *Critical Reviews in Plant Science* 10:329.

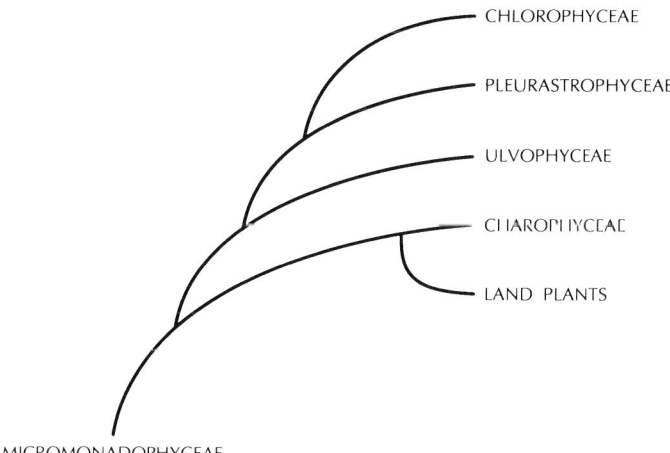

Fig. 3-16. Mattox and Stewart (1984) defined five classes of green algae based on ultrastructural and biochemical features. Their hypotheses regarding diversification of green algal lineages is currently being tested with the techniques of molecular systematics. Illustration adapted from Mattox and Stewart (1984), Classification and evolution in green algae. In: Irvine and John (eds.), *Systematics of the Green Algae*, Academic Press, pp. 29–72, with permission of Academic Press.

Patterns of Enzyme Distribution and Occurrence of Resistant Biopolymers

In a ground-breaking phylogenetic survey of the occurrence of two photorespiratory enzymes, Frederick et al. (1973) reported that charophycean algae and land plants possess a glycolate oxidase, whereas other green algae examined are characterized by glycolate dehydrogenase. Both enzymes function in scavenging carbon that would otherwise be lost by excretion of glycolate from photosynthetic cells, but they differ in cellular location and in other respects. Glycolate oxidase, a flavoprotein, is located in peroxisomes, transfers electrons to molecular oxygen to form hydrogen peroxide, oxidizes L- but not D-lactate in addition to glycolate, and is insensitive to 2 mM cyanide. In contrast, glycolate dehydrogenase is not located in peroxisomes, is not coupled to molecular oxygen, oxidizes D- but not L-lactate, and is inhibited by 2 mM cyanide. It has been suggested that the presence of glycolate oxidase in ancestral charophycean algae played an important role in the success of the earliest land plants. Consequently, the relationship between glycolate oxidase occurrence in charophytes and carbon dioxide and oxygen abundance in aquatic and terrestrial environments are explored later in greater detail (see Chapter 5).

Copper/zinc superoxide dismutase (Cu/Zn SOD) is another enzyme characteristic of land plants and charophycean algae, but absent from all other green algae tested (deJesus et al., 1989). This enzyme eliminates highly reactive, damaging oxygen radicals from plant cells, with the formation of hydrogen peroxide. Other superoxide dismutases having iron or manganese in reaction centers may be found in plants and the cells of other organisms. However, differences in the amino acid sequence and protein structure of the copper/zinc SOD as compared to other SODs suggests that these enzymes are not homologous. In angiosperms there are two forms of Cu/Zn SOD, one located in the plastid stroma and lumen; the other cytosolic and in glyoxysomes. The cellular location of Cu/Zn SOD in charophycean algae is unknown, and could be phylogenetically significant in view of the suggestion that Cu/Zn SOD may be associated with the biochemical pathway leading to lignin biosynthesis (Chapman, 1985; deJesus et al., 1989).

The distribution of class I aldolases (Jacobshagen and Schnarrenberger, 1990) and the urea-degrading enzyme urease (Syrett and Al-Houty, 1984) are also of significance in the origin of plants from green algae, but insufficient numbers of taxa have been examined to allow broad phylogenetic conclusions to be made at present. No doubt there are other phylogenetically informative enzymes whose study could shed light on the patterns of evolutionary divergence leading to land plants. For example, surveys should be conducted in charophycean algae for the occurrence of enzymes in biosynthetic pathways leading to formation of resistant biopolymers (sporopollenin, lignin, cutin, and cutan) and flavonoids.

As mentioned earlier, production of resistant biopolymers such as sporopollenin, lignin, and cutan is regarded as a hallmark of adaptation to the terrestrial environment. Sporopollenin occurs in the zygote walls of *Spirogyra* (deVries et al., 1983), *Chara* (Blackmore and Barnes, 1987), and *Coleochaete* (Delwiche et al., 1989;

Graham, 1990). Because sporopollenin is produced by various other green algae and dinoflagellates, it is undoubtedly ancient in origin. Unfortunately, little is known about the composition and biogenesis of sporopollenin (Southworth, 1990), and comparative enzymology is completely unexplored. An important technological advance, however, is the ability to obtain high-quality infrared spectra useful in identification of sporopollenin from samples as small as single cells, by use of Fourier Transform infrared (FTIR) microscopy (Delwiche et al., 1989).

Resistant polyphenolic compounds have been demonstrated to occur in *Coleochaete* and bryophytes by correlative TEM, fluorescence, and FTIR microscopy (Delwiche et al., 1989), and lignans are known to occur in hornworts (Takeda et al., 1990), but larger phenolic polymers that are unambiguously assignable to lignin have not been demonstrated to occur in plants below the level of the tracheophyte grade (Lewis and Yamamoto, 1990). It has been suggested that at least one enzyme associated with the phenylpropanoid pathway, phenylalanine ammonia lyase (PAL), must occur in algae because these organisms produce ubiquinone (Chapman, 1985). However, no surveys of enzymes directly associated with lignin biosynthesis have yet been conducted in charophycean algae. For some years it was believed that there was evidence for occurrence of flavonoids in charalean algae (Markham and Porter, 1969). This is significant because flavonoid biosynthesis is related to early steps in the phenylpropanoid pathway (Chapman, 1985); however, this finding has recently been questioned (Kubitzki, 1987; Markham, 1988). Land plants, including bryophytes, are generally regarded as possessing a cuticle (Holloway, 1982), whereas cuticle has been regarded as absent from green algae (Mishler and Churchill, 1985). However, Graham and Delwiche (1992) reported the occurrence of a resistant surface layer on thalli of certain *Coleochaete* species. The structure of this layer is considered in Chapter 9, but its chemistry has not yet been studied. No surveys of enzymes associated with biosynthetic pathways leading to cutin or cutan (the biopolyester components of cuticle) have yet been conducted in charophycean algae.

Molecular Systematics

Molecular studies of evolution are based on the premise that the genetic material of living organisms contains records of evolutionary history that can be used to deduce patterns of descent. Zuckerkandl and Pauling (1965) pointed out that evolutionary information is lost during transcription of messenger RNA from DNA, during RNA processing, and again in translation of messages to protein sequences as a result of degeneracy of the genetic code. Thus, DNA sequences represent higher information content than RNA or protein sequences. For many groups of organisms, including higher plants, there is a substantial body of systematic information relating to comparative study of proteins derived from isozyme analyses, amino acid sequencing, and serology, and data from indirect methods of assessing DNA divergence such as restriction fragment length polymorphism (RFLP) analysis or DNA-DNA hybridization (Crawford, 1990). However, molecular studies related to the evolutionary transition from green algae to land plants are so recent that investigators have almost exclusively used DNA sequencing in analyses of phylogenetic relationships. DNA sequences that are known to have low nucleotide substitution rates can be used to examine very ancient divergences (Li and Grauer, 1991), such as those occurring in the diversification of the charophyceae and the origin of land plants perhaps 450 to 470 million years ago. Progress in the field has been rapid and dramatic. Before examining the major breakthroughs, however, it is appropriate to consider some caveats and possible pitfalls in the use of DNA sequences to trace evolutionary pathways.

In the phylogenetic analysis of DNA sequence data, it is important that homologous regions are compared. Errors can result from comparison of paralogous sequences resulting from gene duplication and subsequent divergence. For this reason, single-copy regions are preferred. Although DNA sequence data are widely regarded to be less subject to the effects of parallel and convergent evolution (homoplasy) than are morphological data, cases of parallel nucleotide substitution in response to similar selective pressures have been documented (Li and Grauer, 1991). Problems commonly occur in achieving correct alignment of putatively homologous sequences; deletions or insertions of nucleotides can make alignment difficult. Although algorithms for sequence alignment are available, proper alignment may depend on the assumptions used for the frequency of gaps in the sequence relative to the frequency of point mutations (Li and Grauer, 1991). Even if alignment is relatively straightforward, other assumptions used in analyzing nucleotide substitution data can affect the topology (branching patterns) of phylogenetic trees. For example, some nucleotides are more mutable than others; substitutions between purines (A and G) or pyrimidines (C and T), called transitions, occur more often than transversions, substitution of a purine for a pyrimidine or vice versa. The transitions C to T and G to A are especially common (Li and Grauer, 1991). In addition, rates of nucleotide substitution may vary among taxa whose sequences are being compared. Whether algorithms compensating for these effects have been used must be considered in evaluation of phylogenetic analyses based on DNA sequence data. Last, but not least, phylogenies based on molecular data are often assumed to represent phylogenies of the organisms from which sequences were derived. However, when a phylogeny is constructed from one gene sequence per species, the resulting tree really represents the history of the gene, not necessarily the organisms (Crawford, 1990; Li and Grauer, 1991). Such trees are referred to as "gene trees" (see Fig. 3-17 for an example) to distinguish them from trees purporting to represent the phylogeny of organisms, called "species trees," which typically utilize a variety of types of characters (Fig. 3-18 is an example). Construction of gene trees is particularly vulnerable to error when the interval between species divergence is short; because nucleotide substitutions occur randomly, it is possible for earlier diverging species to accumulate fewer nucleotide substitutions (and thus erroneously appear more derived) than later diverging species. An extensive genetic data set on inbred mice allowed Atchley and Fitch (1991) to test the concordance of various gene trees and species trees. These workers found considerable variation in the

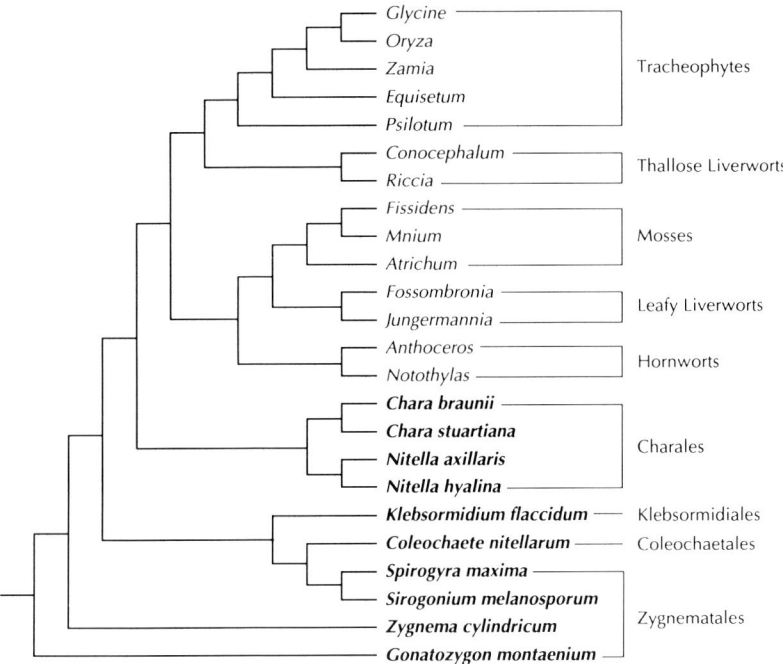

Fig. 3-17. This example of a "gene tree," a cladogram generated from nuclear encoded ribosomal RNA sequences by Chapman and Buchheim (1991) (published in *Critical Reviews in Plant Science* 10:361, used with permission), indicates close relationship between charophycean algae and embryophytes. The data also suggest that embryophytes are monophyletic, that is, derived from a common ancestor. However, separation of the genera *Spirogyra* and *Sirogonium* from presumed relatives *Zygnema* and *Gonatozygon* is not supported by other characters. A more complete understanding of the branching order of charophytes will require additional data.

ability of gene trees to provide an accurate picture of known relationships, with some, but not all, gene trees accurately reflecting relationships. The use of larger amounts of data, longer sequences, and as many genes as possible in phylogeny reconstruction helps to reduce the effects of stochastic error (Li and Grauer, 1991). At present, published species trees focused on phylogeny of charophycean algae and divergence of land plants are based on relatively few morphological (including ultrastructural) and biochemical characters, and published gene trees reflect comparisons of only one or a few sequences. Combined use of sequences of several genes, preferably from various genomes, as well as combined use of structural, biochemical, and molecular characters, is highly desirable, but so far has not been accomplished.

There are three genomes of potential use in tracing the evolution of land plants from charophycean ancestors: mitochondrial, nuclear, and chloroplast. In higher plants mitochondrial nucleotide substitution rates are typically lower than those

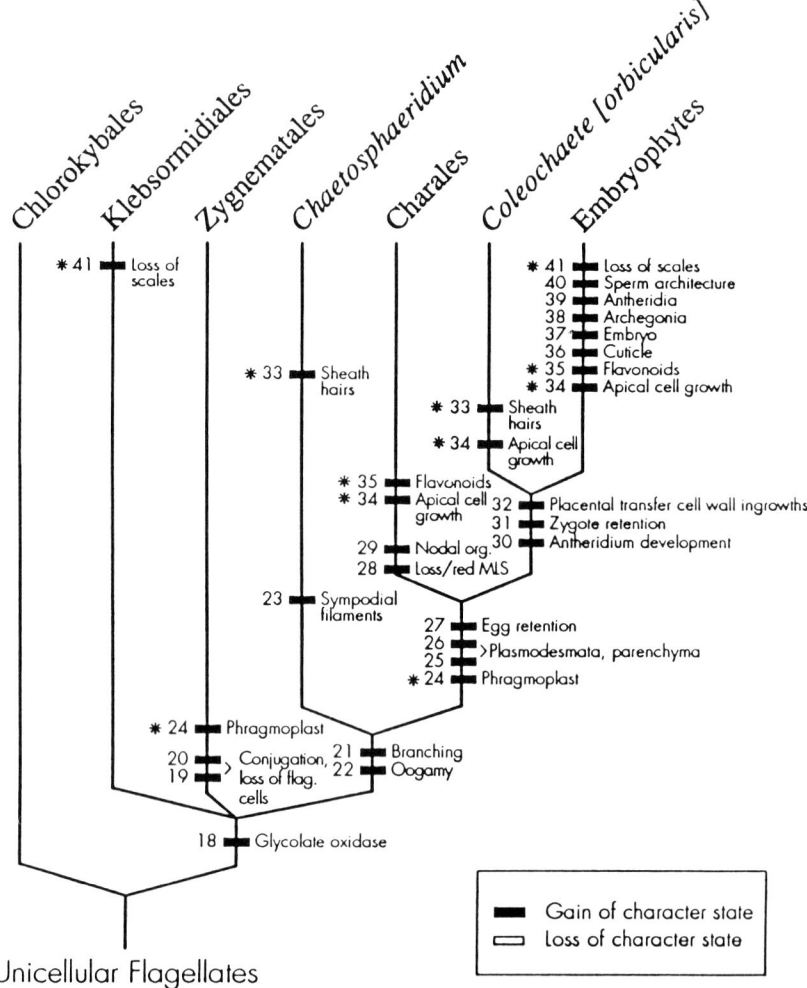

Fig. 3-18. This example of a "species tree," first published by Mishler and Churchill (1985) and redrawn to different conventions by Graham et al. (1991), represents parsimony analysis (a cladogram) based on structural and biochemical characters. Hypothesis of monophyletic origin of embryophytes from an ancestral charophyte is supported. However, branching order should be viewed conservatively because the number of structural and biochemical characters available for analysis is quite small. Reproduced from Graham, Delwiche, and Mishler (1991), Phylogenetic connections between the 'green algae' and the 'bryophytes,' *Advances in Bryology* 4:233, with permission of Gebruder Borntraeger.

in the chloroplast genome, which in turn are lower than nucleotide substitution rates in nuclear genes. Among nuclear genes, the 18S and 25S ribosomal RNA genes have been particularly useful, offering the advantage of highly conserved regions useful in assessing ancient divergences among green algal lineages. Studies based on 18S and 25S ribosomal RNA sequences support the concept of close phylogenetic relationship between charophycean algal genera and the land plants (Chapman and Buchheim, 1991; Wilcox et al., 1993). The use of 5S rRNA genes to assess plant and protist phylogenies has been criticized, however, because short lengths and greater variability have led to production of trees that are confounded by high levels of homoplasy (Bremer et al., 1987).

In contrast to the plant nuclear genome, the higher plant chloroplast genome consists almost entirely of single-copy sequences; the exception is the nearly universal "inverted repeat" region wherein rRNA genes are encoded. A summary of recent information on algal and plant chloroplast genomes may be found in a review by Sugiura (1992). Complete sequences for several plant chloroplast DNAs, including that of the liverwort *Marchantia* (Ohyama et al., 1986), are now available. Studies of such phylogenetically disparate plants as the liverwort *Marchantia*, the fern *Angiopteris* (Yoshinaga et al., 1992), *Nicotiana* (tobacco), and *Oryza* (rice) indicate that the chloroplast genome is structurally quite stable in embryophytes. However, a 30-kilobase pair inversion has been found to characterize plastid genomes of all vascular plants except three lycopsids, whose gene order within this region is similar to that of bryophytes examined. This evidence suggests that lycopsids are the earliest divergents from the vascular plant lineage (Raubeson and Jansen, 1992).

In common with the genome of procaryotes, the land plant chloroplast genome is characterized by the occurrence of tRNA Ala and Ile genes in the transcribed spacer between the 16S and 23S rRNA genes. In all land plant chloroplast DNAs so far examined, group II introns are found in the tRNA Ala and tRNA Ile genes, but these genes are typically uninterrupted in algae and eubacteria. However, Manhart and Palmer (1990) discovered that both introns are present in these genes in the charophycean algae *Coleochaete* and *Nitella*, and that the tRNA Ala intron also occurs in *Spirogyra*. These workers obtained a complete nucleotide sequence for a 2.3 k base pair region containing the spacer and part of the contiguous rRNA genes for a species of *Coleochaete* and compared it with the homologous sequence of *Marchantia*. They found that the introns were of similar size and were in exactly the same positions in both genes (Fig. 3-19). Moreover, regions of the introns adjacent to exons had sequence similarities of 66.9 to 75 percent, strongly suggesting a common origin (Manhart and Palmer, 1990). The intron evidence supports the conclusion that *Coleochaete* and *Nitella* are more closely related to land plants than is *Spirogyra*, inasmuch as an intron occurs in only one of the two tRNA genes examined in the latter genus. It is thus postulated that the tRNA Ala intron was acquired first, with the tRNA Ile intron appearing later in the charophycean lineage leading to land plants.

Baldauf et al. (1990) surveyed charophycean algae for the genomic location of the *tuf* A gene, which encodes the chloroplast protein synthesis elongation factor

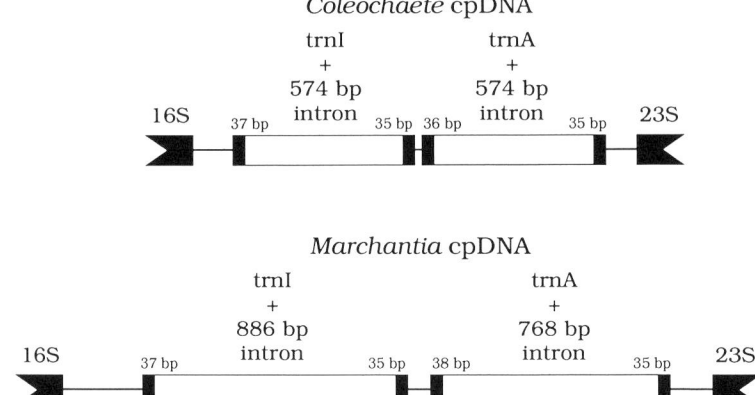

Fig. 3-19. Strong molecular evidence for alliance of modern charophytes with the ancestry of land plants is provided by Group II intron analysis of chloroplast transfer RNA sequences. Uninterrupted cpDNA tRNA genes are characteristic of most algae and eubacteria, but all land plants and charophytes so far examined have at least one intron in these genes. This diagram from Manhart et al. (1990) illustrates similarity in size and position of two introns in the charophyte *Coleochaete* to those of the liverwort *Marchantia*. *Nitella* also possesses two cpDNA t-RNA introns, but only one occurs in *Spirogyra*, suggesting earlier divergence of Zygnematales from the charophyte-embryophyte lineage. Reprinted with permission from Manhart, J. R., and J. D. Palmer, 1990. The gain of two chloroplast tRNA introns marks the green algal ancestors of land plants. *Nature* 345:268. Copyright 1990 Macmillan Magazines Limited.

Tu (EF-Tu), because this gene is located in the nuclear genome of land plants, but in the chloroplast genome of most green algae. They made the dramatic discovery that *tuf* A, ancestrally present in the green algal chloroplast, was apparently transferred to the nucleus in the charophycean lineage before the divergence of land plants. A probe for the *tuf* A gene hybridized with multiple regions of nuclear DNA from *Spirogyra* and *Sirogonium* (Zygnematales), *Coleochaete, Chara,* and *Nitella* (Fig. 3-20). In the higher plant *Arabidopsis, tuf* A is present as a single copy, and is characterized by a typical chloroplast-targeting peptide sequence. It is not known which, if any, of the charophycean nuclear *tuf* A genes are transcribed, or if transit peptide sequences are present. However, Baldauf et al. (1990) found that a heterologous (*Arabidopsis*) *tuf* A probe hybridized strongly to chloroplast DNAs of several noncharophycean green algae, as well as *Chara* and *Nitella*, and gave a weak signal with *Coleochaete* plastid DNA, but no hybridization was obtained for *Spirogyra, Sirogonium,* or land plant plastid DNA. This was interpreted as evidence for variation in the fates of the (presumably) redundant plastid *tuf* A genes of various charophytes. Apparently, the plastid gene has been completely lost from the zygnematalean algae examined, with parallel loss occurring in land plant plastids. Charalean algae apparently have retained the plastid gene in sufficiently conserved form as to be recognizable to the hybridization probe. In

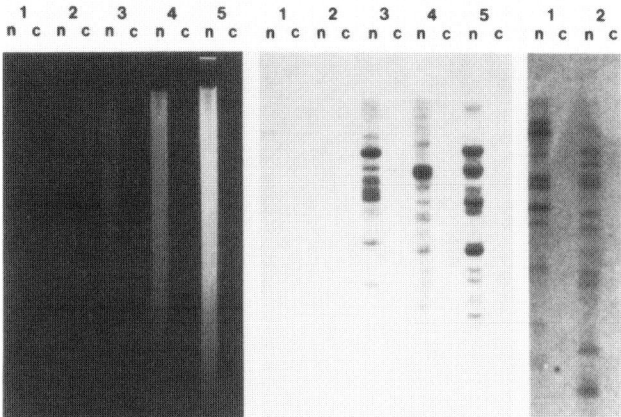

Fig. 3-20. Molecular data (from Baldauf et al., 1990, used with permission) reveals that charophytes, unlike other green algae, share with embryophytes the occurrence of homologous, nuclear-encoded *tuf*A gene sequences. In these hybridization studies, lanes are as follows: 1. *Spirogyra*, 2. *Sirogonium*, 3. *Chara*, 4. *Nitella*, 5. *Coleochaete*. Restriction enzyme digests of nuclear DNA are shown in columns n; columns labeled c contain chloroplast DNA digests. On the left is the original agarose gel separation; at center and right are nylon filter replicas probed with a radiolabeled *tuf*A fragment from the flowering plant *Arabidopsis*. These data indicate that the gene *tuf*A was transferred from the chloroplast to the nuclear genome in the charophycean ancestors of plants. Subsequently, the chloroplast copy of this gene has been lost completely from zygnematalean algae and land plants (probably independently), but retained in charalean algae and highly modified in *Coleochaete*.

Coleochaete, however, sequence analysis of the plastid gene showed less than 55 percent similarity to other *tuf* As, explaining the weak hybridization signal. The *Coleochaete tuf* A gene sequence was found to be sufficiently divergent that active regions of the transcription product would be affected. Baldauf et al. (1990) suggested that the gene may have retained some ancestral functions or has acquired a new role. These exciting results render alliance of charophycean algae with the ancestry of land plants essentially noncontroversial and raise many interesting questions for future research. Which of the charophycean nuclear copies of *tuf* A have acquired transit peptide sequences? Are charophycean nuclear *tuf* A genes properly transcribed and targeted back to the chloroplast? It would indeed be surprising if nuclear *tuf* A genes of the zygnematalean representatives and *Coleochaete* were not, inasmuch as the plastid gene is lacking from the former and probably at least partially nonfunctional in the latter.

Comparison of the sequences encoding transit peptides in charophycean algae and land plants could shed light on the evolution of transit peptide sequences and the process of organellar evolution. The higher plant outer chloroplast envelope apparatus for import of proteins is beginning to be well characterized. A constitutively expressed heat shock protein that is immunologically related to hsp70 and an additional 86 kDa protein appear to form part of the receptor complex in *Pisum*

(Waegemann and Soll, 1991). Comparison of the chloroplast envelope receptor systems of charophyte algae with those of higher plants might be phylogenetically illuminating.

Surprisingly little is known about the evolutionary origin of the unusual plant mitochondrial genome, which is typically large and structurally variable. Plant mitochondrial genomes range in size from 40 to 2500 kilobase pairs and may be linear, circular, or divided into separate subgenomic circles (Li and Grauer, 1991). In contrast, the mitochondrial genomes of green algae range from 20 to 150 kb in size (Gray and Boer 1988; Moore and Coleman, 1989; Misonou et al., 1989), and other algal mitochondrial genomes are also relatively small. Nothing is known about the size or structural complexity of charophycean mitochondrial genomes, nor have any mitochondrial genes been sequenced in this group. Because plant mitochondrial gene sequences are characterized by very low nucleotide substitution rates, comparative mitochondrial genome analyses for charophycean algae and seedless plants have the potential to be very useful in tracing the process of early plant evolution. In this connection, it is interesting that the mtDNA of the liverwort *Marchantia polymorpha* consists of a single DNA species having no large repeated sequences or homologous recombination mechanisms similar to those commonly associated with higher plant mtDNA. This suggests that the larger recombinational capacity of higher plant mitochondrial genomes arose after the divergence of liverworts (Ohyama et al., 1991).

Approaches to Phylogenetic Tree Construction

Whether an evolutionary study involves morphological or molecular characters, or both, decisions must be made regarding the mode of phylogenetic analysis. Most phylogenetic trees related to land plant origins that are based on morphological characters have been generated according to the rules of cladistic analysis as discussed by Hennig (1966) and lucidly explained by Wiley (1981), but also independently derived by Wagner and his students (Mishler 1986). Cladistic analyses attempt to define the branching pattern in evolution by using derived characters that are shared among the members of a group of organisms at the particular hierarchical level of interest (synapomorphies). In contrast, shared primitive characters (symplesiomorphies) and derived characters present only in one operational taxonomic unit (autapomorphies) are not useful in assessing relationships according to cladistic methods. A monophyletic group is defined as all and only organisms descended from a common ancestor, as based on shared, derived characters. Therefore, groups defined only on the basis of symplesiomorphies cannot properly be described as monophyletic. A group that contains some, but not all, descendants of a common ancestor is paraphyletic.

Character states representing morphological, biochemical, amino acid, or nucleotide changes, or presence/absence of deletions or insertions in DNA sequences may be used in construction of phylogenetic trees. Determination of the polarity (direction of change) of character states is determined by comparison with the nearest identifiable taxon, called the out-group or sister taxon. When conflicting trees result, the branching pattern requiring the fewest number of homoplasies (instances of parallel or convergent evolution) necessary to explain character distribution is preferred. In other words, the phylogenetic hypothesis having greatest parsimony is favored over hypotheses of relationship requiring greater numbers of parallel or convergent changes. However, cladistic methods have been criticized for overreliance on the principle of parsimony, inasmuch as evolution does not always take the simplest or shortest route (Crawford, 1990). Further, Caro-Beth Stewart (1993) has pointed out that parsimony analysis is characterized by two major pitfalls. These are failure to find the shortest tree, and the possibility that the shortest tree does not reflect the correct phylogeny. The first pitfall is usually a consequence of including too many taxa, too little phylogenetically-informative data, or a combination of these factors. The second pitfall results from the occurrence of homoplasy (Stewart, 1993).

A common misinterpretation of cladograms is that one extant group gave rise to another. Actually, cladograms imply only that extant groups terminating adjacent branches shared a common ancestor represented by the branch node (Figs. 3-8 and 3-18) (Mishler and Churchill, 1985). Advantages of the cladistic method include rigorous techniques for testing concepts of homology (synapomorphy) and procedures for recognizing characters resulting from convergent or parallel evolution (homoplasies). Another benefit is that cladistic analyses yield testable phylogenies that point out where further research is required on characters or taxa (Mishler, 1986). Cladistic argumentation is regarded as being relatively free of the influences of personal belief and valuation. However, the outcome of a particular analysis may be influenced by an investigator's choice of characters used. For example, different workers may or may not choose to use characters for which a complete matrix of occurrence is not available for all taxa. The effects of character choice can be minimized if all potential characters are listed and reasons are given for inclusion or noninclusion of particular characters. Then, later analysts can revise cladograms when the missing data become available, without overlooking potentially useful sources of phylogenetic information.

In contrast to cladistic methods of phylogenetic data analysis, phenetic approaches seek to classify organisms on the basis of as many characters as possible, whether derived or primitive. Distance matrix methods are phenetic approaches that have been widely used in the construction of gene trees. Branch lengths are directly related to the number of nucleotide substitutions that have occurred within the group analyzed (Li and Grauer, 1991). Many molecular systematists analyze the same data set using both distance and parsimony methods, because when resulting trees are congruent they are regarded with greater confidence. The development of increasingly more sophisticated methods for analysis of phylogenetic data is an extremely active field. Current views may be found in journals such as *Cladistics, Systematic Biology* (formerly *Systematic Zoology*), and *Molecular Evolution* that feature articles discussing innovations in the area of phylogenetic analysis.

4

THE CHAROPHYCEAN ALGAE

As we have seen, there is compelling ultrastructural and biochemical evidence that land plants evolved from ancestors that would likely be classified as charophycean green algae, and this concept is strongly supported by comparative studies of nuclear and chloroplast genomes. In this chapter, we take a closer look at individual charophycean taxa because they are all important as representatives of the ancestors of plants. We begin with examination of proposed relationships among charophyte taxa and end with discussion of the evidence relating to origin of the charophycean lineage. This intensive focus on charophycean algae sets the stage for subsequent consideration of charophycean morphology, reproduction, and ecology in relation to land plant evolution.

In 1985 three cladistic treatments of charophycean algae based on morphological and biochemical data were published independently by Mishler and Churchill, Bremer, and Sluiman. Later modifications of the first two analyses, presented by Bremer et al. (1987) and Graham et al. (1991), supported Mishler and Churchill's (1985) conclusions that extant land plants (embryophytes) are monophyletic and that the closest sister taxa to the embryophyte clade are *Coleochaete* and the Charales. The analysis by Graham et al. (1991) revealed that derivation of embryophytes from a "higher charophyte" clade containing *Coleochaete* and Charales (the algal order containing *Chara, Nitella,* and relatives) was strongly supported by the data. Trees positioning *Coleochaete* as the sister taxon to embryophytes were also supported, but less strongly. Alternate trees showing Charales as the sister group to embryophytes indicated that the common ancestor of Charales and *Coleochaete* would most likely have resembled *Coleochaete* in any event (Graham et al., 1991). In two recent studies, nuclear ribosomal DNA gene trees place both *Coleochaete* and Charales close to land plants, but suggest that Charales represents the sister taxon to embryophytes (Chapman and Buchheim, 1991; Wilcox et al., 1993). These results do not mean that the extant taxa *Coleochaete* or Charales gave rise to modern land plants, but instead suggest that these three taxa were probably linked by a common ancestor.

Sluiman (1985) suggested that the embryophytes may have been polyphyletic, i.e., that various land plant groups were independently derived from different charophycean taxa. However, Graham et al. (1991) cited a group of embryophyte autapomorphies for which independent origin in separate land plant lineages arising from different charophyte ancestors would be rather unlikely. The list includes the highly conserved structure of the land plant chloroplast genome in those embryophytes ranging from bryophytes through higher plants, the occurrence of bicentriolar centrosomes in the development of biflagellate sperm in various groups of seedless plants, the occurrence of preprophase bands in cells entering mitotic division, details of plant sperm MLSs, sporic meiosis (i.e., occurrence of a multicellular sporophyte), and production of spores protected by walls containing sporopollenin. In contrast, the chloroplast genome of charophycean algae shows considerable size variability (Manhart and Palmer, 1988), MLSs of charophycean algae are subtly different from those of embryophytes, meiosis in charophytes is zygotic, meiospores are not enclosed by sporopollenin-containing cell walls, and charophycean algae normally lack bicentriolar centrosomes. There is some evidence for the possible occurrence of an evolutionary precursor to land plant preprophase bands and teratogenic production of bicentriolar centrosomes in particular charophytes; these are discussed in more detail in Chapters 6 and 7. A cladistic analysis of rRNA sequences of various land plants (including bryophytes) (Waters et al., 1992) and an analysis of chloroplast gene order in land plants (Raubeson and Jansen, 1992) both suggest that extant embryophytes are monophyletic. In the absence of specific evidence that various embryophyte groups evolved from different charophytes, for the present it is reasonable to assume that extant embryophytes are monophyletically derived from the charophycean lineage, and that the common ancestor of embryophytes resembled the ancestors of modern *Coleochaete* and Charales.

Corollary to the conclusion that embryophytes are derived from charophytes are the assumptions that the ancestors of land plants were freshwater algae, rather than marine forms, and that the land plant life cycle arose from the delay of meiosis in forms lacking alternation of generation (the antithetic theory). Supporting these assumptions is evidence that modern charophycean algae appear to be fundamentally freshwater algae (with tendencies toward the terrestrial habitat), and that they uniformly lack alternation of generations. Charophycean taxa that exhibit sexual reproduction (Charales, Coleochaetales, and Zygnematales) also have "zygotic" life cycles in which the only diploid cells are zygotes. In the discussion that follows, we first consider the characteristics of these "advanced" charophytes, then review the features of charophytes regarded as more primitive because they lack sexual reproduction.

CHARALES

The modern order Charales includes the well-known stoneworts *Chara* and *Nitella*, and several other less well studied genera (Khan and Sarma, 1984) (Fig. 4-1), with an estimated 400 species (Grant, 1990). Calcified remains of reproductive structures, known as gyrogonites, together with some remains of vegetative parts provide a substantial fossil record for the Charales. The oldest known charophyte species is *Praesycidium siluricum* (Scycidiaceae) described from the upper Silurian period of the Ukraine (Feist and Grambast-Fessard, 1991). The two major subfamilies of modern Characeae are known as separate lineages from as far back as the Triassic period (Grambast, 1974). The fossils *Eochara* and *Paleonitella* indicate that forms structurally comparable to modern charalean algae were present during Devonian times (see Taylor et al., 1992, for example). The relatively late appearance of fossils that can be related to modern Charales has been cited as evidence that charalean algae were not directly involved with the radiation that led to higher plants (Tappan, 1980). However, it is possible that noncalcified charophytes similar to modern forms could have existed earlier than the late Silurian period; such forms would not have been so well preserved in the fossil record.

The large and structurally complex thalli of modern charalean algae are conspicuously similar to those of certain higher plants. The whorled pattern of branching

Fig. 4-1. *Chara zeylanica* illustrates the morphology of charalean algae, which are as large as, or larger than, many aquatic flowering plants.

and nodal organization of *Chara*, for instance, superficially resembles that of the gymnosperm *Ephedra* and the pteridophyte *Equisetum*. The aquatic angiosperm *Ceratophyllum* is commonly confused with *Chara*, because the two are so similar in size and organization (Fig. 4-2). Such striking similarity has been viewed by some as evidence for close relationship to land plants. However, it must be noted that the charalean thallus represents the gamete-producing portion of the life cycle, homologous to the gametophyte generation of embryophytes. The gametophytes of modern bryophytes and pteridophytes are diverse in structure, many having a dorsiventrally flattened, "thallose" morphology with dichotomous branching that probably represents the ancestral form (Fig. 4-3). Among bryophytes, only the comparatively derived mosses and leafy liverworts have nodally organized gametophytes (Mishler and Churchill, 1985). It appears, therefore, that the similarity between thalli of charalean algae and those of various higher plants represents parallel evolution rather than homology.

Another conspicuous feature of charalean algae is the unusually complex female reproductive structure, consisting of an egg cell surrounded by five spirally elongated tube cells that each terminate in one or two coronal cells (Fig. 4-4). These are clearly the most highly derived female gametangia known in the green

Fig. 4-2. The angiosperm *Ceratophyllum* greatly resembles *Chara* in size and production of whorls of branches at nodes. *Ceratophyllum* may be distinguished from *Chara* by presence of bifurcation of branch tips in the former. In addition, *Ceratophyllum* plants are sporophytes, whereas *Chara* thalli represent the gametophyte generation.

60   THE CHAROPHYCEAN ALGAE

Fig. 4-3. The liverwort *Conocephalum* illustrates the thalloid, dichotomously branched form widely regarded as the likely morphology of the earliest land plants. These prostrate organisms are adapted for survival in moist terrestrial habitats. In contrast, erect and radially symmetrical gametophytes such as those of many mosses and the fossil *Lyonophyton* are more likely derived morphologies possibly related to increased light harvesting or dispersal potential.

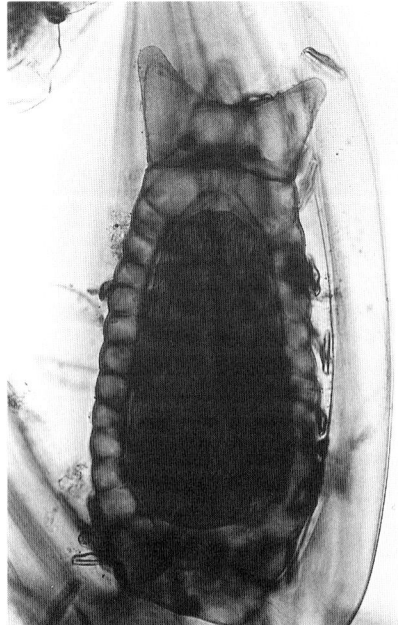

Fig. 4-4. Zygotes, shown here, and oogonia of charalean algae exhibit enclosure of the egg/zygote by five elongated, spirally twisted cells, each capped by one or two coronal cells.

algae and are reminiscent of the jacketed female gametangia (archegonia) of modern seedless plants. However, developmental studies, including ultrastructual work by Pickett-Heaps (1975), have revealed that there is little evidence for homology of the charalean egg cell cortication to the parenchymatous jacket of archegonia. Graham et al. (1991) proposed that egg cortication in Charales may be homologous to intercellular growth responses associated with sexual reproduction that occur in other charophycean algae; these are discussed below in this chapter and again in Chapter 8.

There are three rather more compelling cellular features that have been used to link Charales directly to the ancestry of embryophytes: absence of centrioles from somatic cells; the occurrence of multiple discoid plastids; and the backward-directed flagella of sperm cells. These features commonly occur in embryophytes, but are usually absent from other charophycean algae. There is evidence that similarity between charalean algae and embryophytes in these characteristics represents the occurrence of homoplasy, and that evolutionary changes related to plastid form, the cytoskeletal apparatus (including centrioles), and flagellar orientation were developmentally and phylogenetically linked during the transition to land (Graham and Repavich, 1989; Graham and Kaneko, 1991). These complex developmental and evolutionary interactions are considered in more detail in Chapters 6 and 7.

COLEOCHAETALES

The only genera included in the order Coleochaetales with any degree of certainty are *Coleochaete,* of which there are approximately 15 species (Pringsheim, 1860; Jost, 1895; Printz, 1964; Szymanska, 1989) (Figs. 4-5 and 4-6), and *Chaetosphaeridium* (Fig. 4-7), for which 4 species have been described (Thompson, 1969). A few other genera, such as *Conochaete* and *Radioramus* (Hu and Wei, 1991), may be members of this group but have not been studied ultrastructurally, biochemically, or with modern molecular techniques. *Chaetosphaeridium* has been classified with *Coleochaete* because both genera are characterized by distinctive hair cells (McBride, 1967; Marchant, 1977; Moestrup, 1974) (their names are derived in part from the Greek word *chaetos,* meaning "hair") (Fig. 4-8) and because their zoospores are ultrastructurally quite similar (Moestrup, 1974; Sluiman, 1983). Charalean algae and *Coleochaete* are known to produce land-plantlike phragmoplasts and plasmodesmata during cytokinesis (Pickett-Heaps, 1967; 1975), but cell division in *Chaetosphaeridium* has not yet been studied. Charales, *Coleochaete,* and *Chaetosphaeridium* are like archegoniate land plants in displaying oogamy (larger nonmotile eggs and smaller, flagellate sperm), but *Chaetosphaeridium* does not retain eggs on the parental thallus during fertilization (Thompson, 1969), as do Charales, *Coleochaete,* and embryophytes. This evidence suggests that *Chaetosphaeridium* is less closely related to embryophytes than are *Coleochaete* and Charales, but the hypothesis should be tested by examining molecular characters.

Among charophytes, *Coleochaete* is remarkable because zygotes are retained on the gamete-producing thallus through the process of spore development and

62   THE CHAROPHYCEAN ALGAE

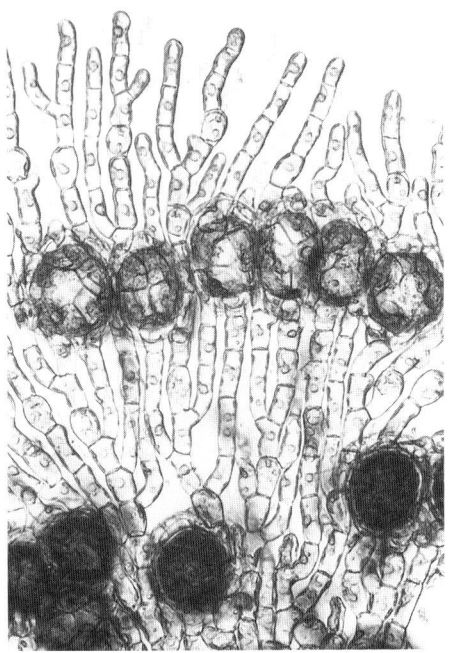

Fig. 4-5. *Coleochaete pulvinata* illustrates the branched, filamentous morphology characteristic of most *Coleochaete* species. The spherical structures are zygotes, enclosed, like zygotes of *Chara*, by other thallus cells.

Fig. 4-6. *Coleochaete orbicularis* illustrates radially organized, thalloid morphology which is also characteristic of *C. scutata*. These species somewhat resemble early, radially organized stages in development of gametophytes of the primitive moss *Sphagnum*.

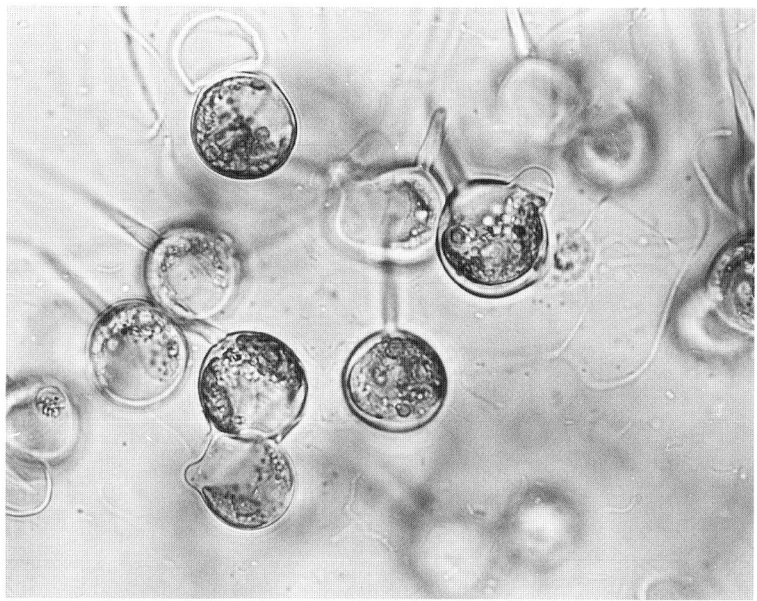

Fig. 4-7. *Chaetosphaeridium* often appears to be unicellular, as does *C. globosum,* shown here, but actually consists of more or less adherent branched filaments in which every cell bears a distinctive hair.

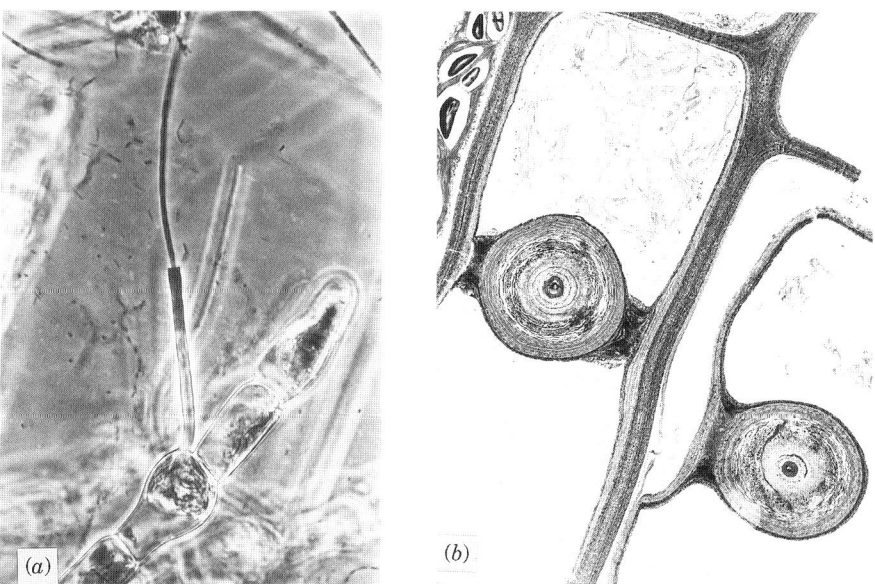

Fig. 4-8. *a. Coleochaete* hair cells (setae) occur throughout the thallus, but in most species, hairs do not extend from every cell. Hairs typically have a basal sheath region and often break off at the sheath tip. Seta cell plastids typically rotate by means of a cytoplasmic streaming mechanism attributed to presence of an actin-myosin system. *b.* Seta cells of *C. scutata* and

63

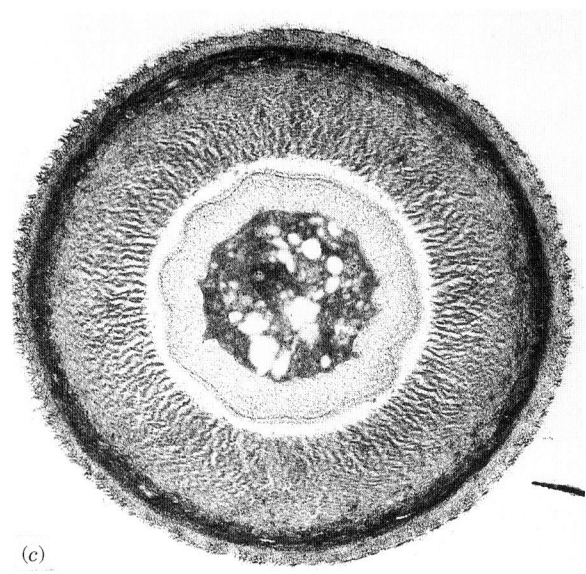

Fig. 4-8. *(Continued) C. orbicularis* have a large internal, concentric wall structure, which is absent from *Chaetosphaeridium*. This, and other differences suggest that comparative hair cell ultrastructure may provide useful characters for systematic analysis of the Coleochaetales. *c.* A cross-section through the sheath region of a *C. pulvinata* hair reveals complex, layered wall structure whose chemistry is not known. The internal region is cytoplasm that has extended into the hair.

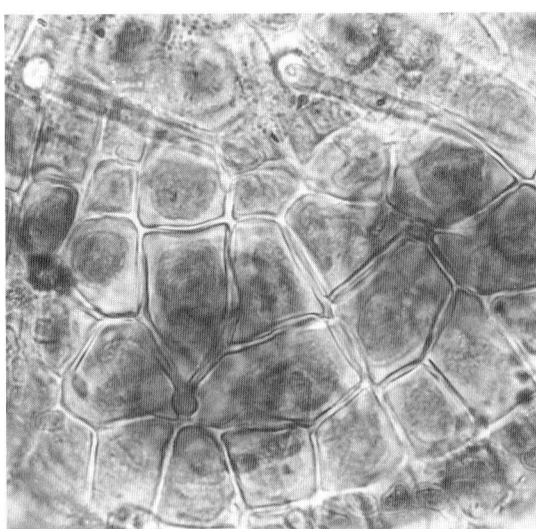

Fig. 4-9. Cortical cells typically enclose zygotes of *Coleochaete* species, such as *C. orbicularis,* shown here. However, the extent of coverage varies among species. Cortical cells probably function to retain zygotes on parental thalli, which usually grow attached to organic or inorganic substrates in shallow fresh waters. This strategy favors subsequent littoral attachment and growth of meiospores which are produced at zygote germination.

release, similar to the retention of the spore-producing generation (sporophyte) on the parental gametophyte generation in hornworts, liverworts, and mosses. The mechanism of retention in *Coleochaete* is the growth of a layer of cortical cells over the zygotes (Oltmanns, 1898) (Fig 4-9). In at least one species, these cortical cells develop localized wall ingrowths that are very similar to those occurring in placental transfer cells located at the junction of sporophyte and gametophyte tissues of land plants (Figs. 4-10 and 4-11). Because *Coleochaete* zygotes accumulate large stores of photosynthate in the form of lipid and starch (Fig. 4-10), the occurrence of adjacent cells that structurally resemble cells of the embryophyte placenta is circumstantial evidence for a nutritional relationship between cells of different generations. The presence of wall ingrowths in one *Coleochaete* species may well represent the occurrence of parallel evolution. However,

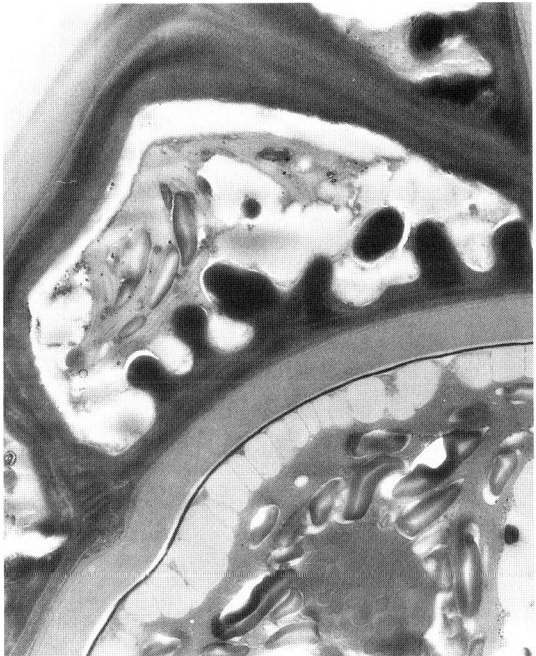

Fig. 4-10. Electron micrograph of the junction between a cortical cell and zygote of *Coleochaete orbicularis* reveals the presence of numerous ingrowths of cortical cell walls localized in the region of immediate contact with zygotes. This suggests response to chemical influences emanating from zygotes. The hypothesis that wall ingrowths function in transport of photosynthate from the haploid thallus to diploid zygotes is supported by accumulation of photosynthetic storage materials (lipid and starch) in greatly enlarged zygotes. The thin layer of electron-dense material just inside zygote walls has been identified by FTIR microscopy as sporopollenin and indicates zygote maturity. Photo from Graham (1985), *American Scientist,* and Graham and Kaneko (1991), *Critical Reviews in Plant Science,* with acknowledgement of the use of the high-voltage electron microscope at the Integrated Microscopy Facility, University of Wisconsin-Madison.

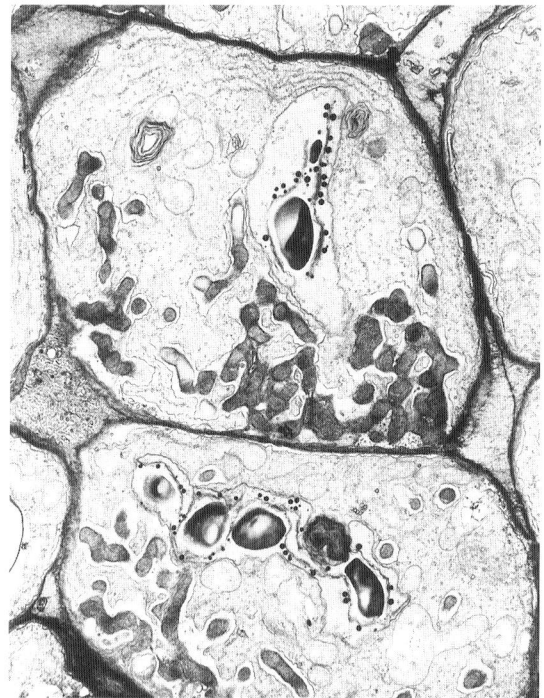

Fig. 4-11. Transmission electron micrograph of cells at the placental region of embryophytes (such as the hornwort *Anthoceros,* shown here) reveal wall ingrowths similar in structure and location to those of *Coleochaete orbicularis.* Wall ingrowths of *Coleochaete* may or may not be homologous to those of embryophytes, but occurrence in both groups suggests analogous nutritional and developmental function. Photo from Graham and Kaneko (1991).

sexual reproduction in *Coleochaete* may illustrate evolutionary stages that occurred during origin of the plant embryo (Graham, 1985). Evidence supporting the hypothesized homology of the *Coleochaete* placenta to that of embryophytes is the presence of distinctive acetolysis-resistant, autofluorescent materials in the cortical cell walls of all *Coleochaete* species examined and in the walls of cells at the placental junction of the hornwort *Anthoceros* (Delwiche et al., 1989). These materials are discussed further in Chapter 9.

Other intriguing similarities between *Coleochaete* and land plants include the involvement of callose in gametogenesis and sporogenesis; preprophase plastid division and migration to tetrahedral positions in germinating zygotes during meiotic divisions leading to production of spores; and asymmetrical, diagonal divisions directly preceding formation of spermatids. These characteristics of *Coleochaete* are discussed in more detail when we later consider the evolutionary origins of fundamental features of land plant developmental and reproductive processes (Chapter 8).

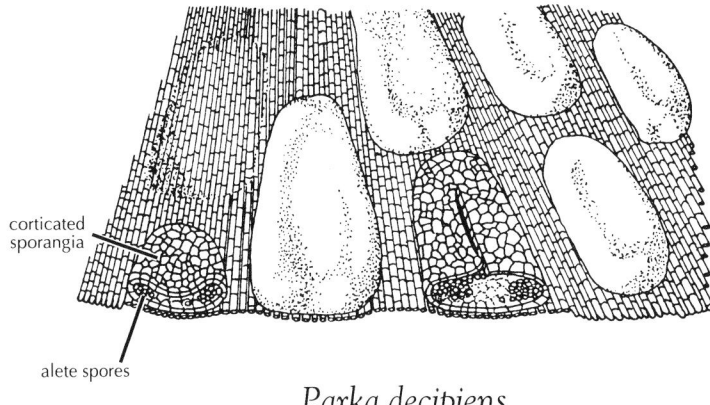

Parka decipiens

Fig. 4-12. The Lower Devonian fossil *Parka* is regarded as similar in overall thallus organization, reproductive mode, and chemistry to green algae such as *Coleochaete*. The flattened, orbicular thallus and polygonal cell layers covering sporangia are especially suggestive of thalli and corticated zygotes of *Coleochaete*. Reproduced by permission of the Royal Society of Edinburgh and K. J. Niklas from *Transactions: Earth Sciences* 69 (1986):487.

Although fossils that can be unambiguously attributed to the Coleochaetales are not known, Niklas (1976a, b) suggested that the lower Devonian impression and compression fossil *Parka* was similar in organization, reproduction, and chemistry to green algae such as *Coleochaete*. At 0.5 to 7 cm in diameter, *Parka* was considerably larger than extant *Coleochaete* (which rarely approach the lower limits of this range), but *Parka* bore large "sporangia" that were corticated, as are zygotes of *Coleochaete* (Fig. 4-12). Various differences between *Coleochaete* and *Parka* led Hemsley (1989) to suggest that the latter may be more representative of modern liverworts (see Chapter 8 for further discussion).

ZYGNEMATALES

Zygnematales, a group of about a thousand species, includes the unbranched filaments *Spirogyra* (Fig. 4-13), *Mougeotia, Zygnema,* and *Sirogonium,* and the unicellular or "filamentous" desmids, which are among the most beautiful of algae (Figs. 4-14 and 4-15). Evidence derived from study of cell division, enzymes, chloroplast DNA, and nuclear ribosomal DNA clearly shows Zygnematales (also known as Conjugales) to be members of the charophycean lineage (Chapman and Buchheim, 1991; Graham et al., 1991). However, no member of this order exhibits flagellate reproductive cells, and sexual reproduction involves a unique process known as conjugation (described in Chapter 8). Therefore, various phylogenetically informative characters related to zoospore or sperm microanatomy and development are not available for comparison with those of other charophytes and

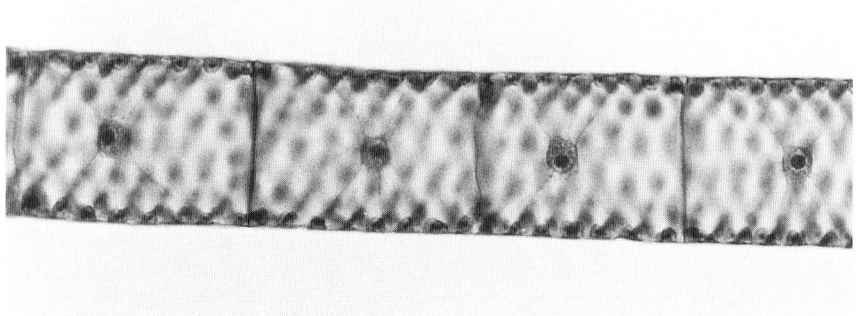

Fig. 4-13. A filament of *Spirogyra*, illustrating central nucleus and helically twisted chloroplasts.

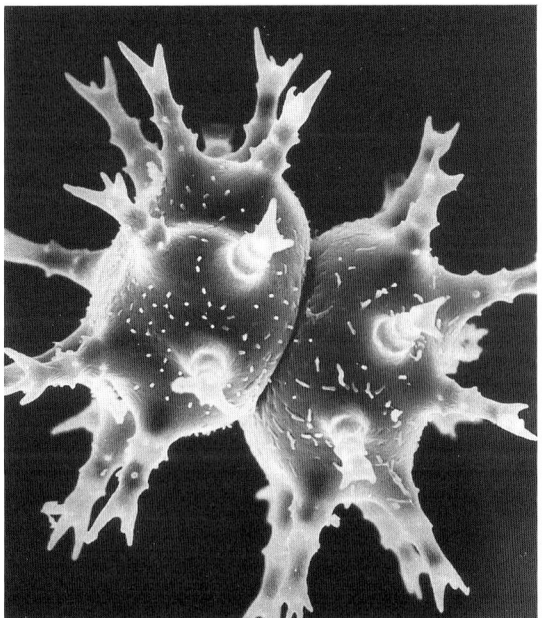

Fig. 4-14. Scanning electron micrograph of the desmid *Staurastrum*, illustrating morphological complexity in a unicellular charophyte. Photo courtesy of Sonia Cook (University of California at Davis).

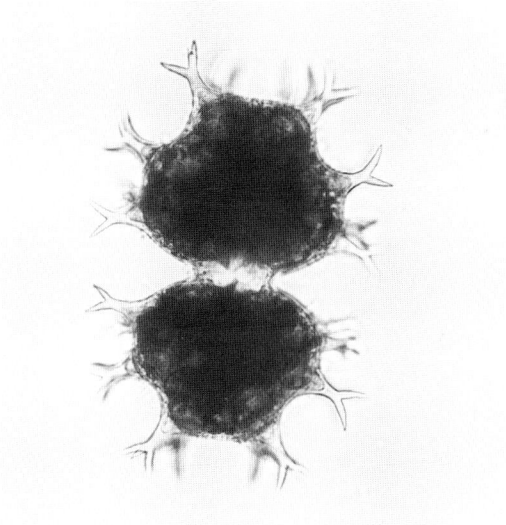

Fig. 4-15. Light micrograph of the desmid *Staurastrum*, a member of the placoderm desmids, which are usually constricted into two semi-cells. Such desmids often exhibit wall pores through which mucilage can be extruded, the basis of a form of motility.

land plants. It has been suggested that zygnematalean conjugation represents the first appearance in the charophycean lineage of cell-to-cell growth responses related to sexual reproduction, and that aspects of conjugation are homologous to growth of cortical cells around eggs of charalean algae and zygotes of *Coleochaete*. These growth responses are postulated to involve excretion and reception of diffusible organic substances, but none have been investigated at the biochemical level (Graham et al., 1991).

It is of considerable importance that zygnematalean algae, specifically *Mougeotia* and *Mesotaenium* (Fig. 4-16), are the most ancient organisms known to exhibit phytochrome-mediated photomodulatory responses (chloroplast reorientation) (Grolig and Wagner, 1988). Unfortunately, aside from documentation of red/far-red reversible control of zygote germination in *Chara* (Takatori and Imahori, 1971; Stross 1979) and *Chara* sporeling growth (Rethy, 1968), there is little other evidence for the occurrence of phytochrome in charophytes. This situation most likely results from lack of study rather than demonstrated absence of phytochrome.

The fossil record of zygnematalean algae extends back to the mid-Devonian *Paleoclosterium leptum* (Tappan, 1980). Cell preservation may be related to the occurrence of resistant materials, such as the "lignin" reported by Gunnison and Alexander (1975) to occur in walls of *Staurastrum*. Zygnematalean zygote walls may contain sporopollenin, but fossil remains of zygotes attributed to

70 THE CHAROPHYCEAN ALGAE

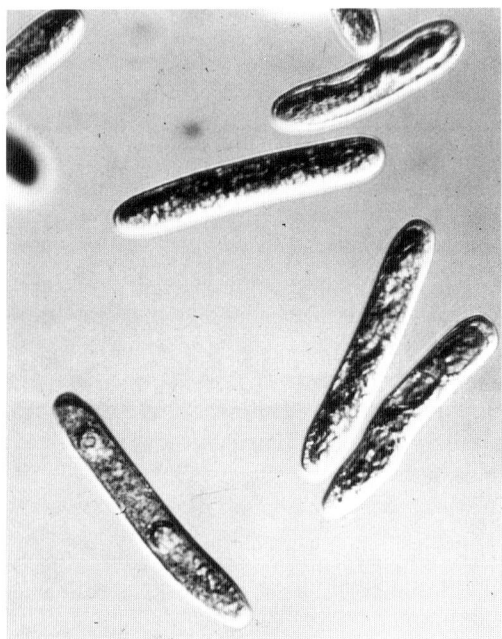

Fig. 4-16. Saccoderm desmids, such as the *Mesotaenium* species illustrated here, are not constricted and lack cell wall pores. Hence, they are regarded as less derived members of the Zygnematales, but branching order in this charophycean group is controversial.

zygnematalean algae are known only as far back as the Carboniferous period (Hoshaw et al., 1990).

KLEBSORMIDIALES

Only three genera, *Klebsormidium* (Fig. 4-17), *Stichococcus* (Fig. 4-18), and *Raphidonema*, having a total of about 40 species, form the group Klebsormidiales (Mattox and Stewart, 1984). Because none of these genera is known to exhibit sexual reproduction, they are regarded as more primitive than Zygnematales, Coleochaetales, or Charales. The zoospores of *Klebsormidium* exhibit MLSs. Ultrastructural features of mitosis (open and persistent spindle) are consistent with inclusion in the Charophyceae, but phragmoplasts and plasmodesmata are absent. *Klebsormidium* is known to produce the enzymes glycolate oxidase (Frederick et al., 1973), Cu/Zn SOD (deJesus et al., 1989), and class I aldolases (Jacobshagen and Schnarrenberger, 1990) as do other charophytes and land plants. In addition, nuclear rRNA sequences of *Klebsormidium* indicate close relationship to charophytes and embryophytes (Chapman and Buchheim, 1991). Nuclear rRNA data obtained by Zechman et al. (1990) suggested that *Klebsormidium* was closer to the

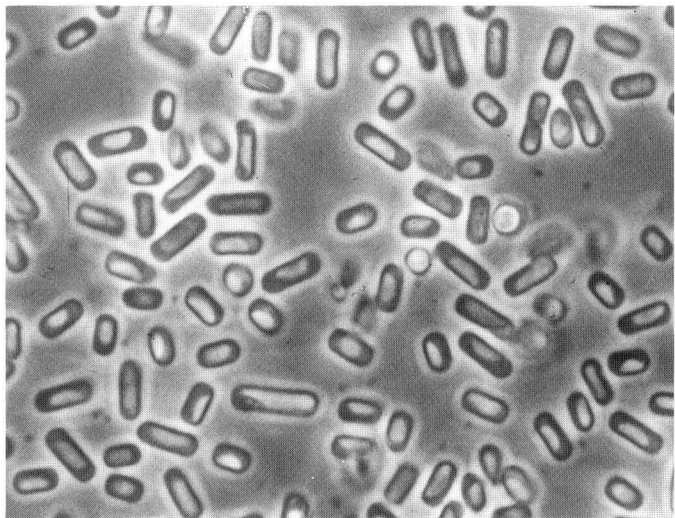

Fig. 4-17. *Stichococcus* is a unicellular charophyte, but short filaments consisting of a few cells may sometimes be observed. This organism is commonly identified in collections made from terrestrial habitats and air.

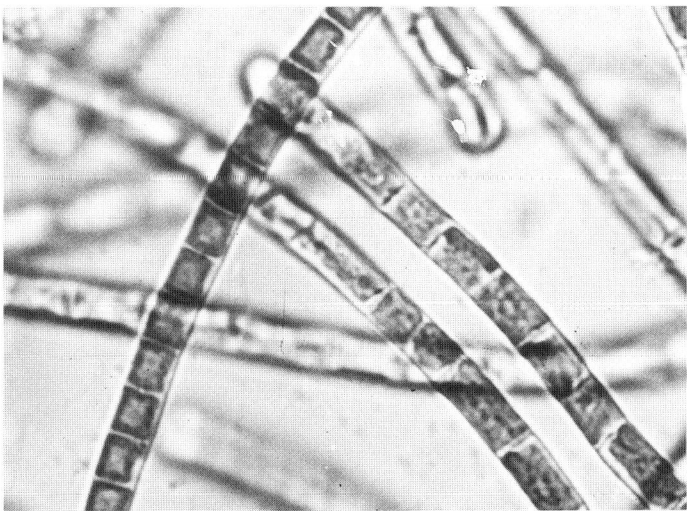

Fig. 4-18. *Klebsormidium* consists of unbranched filaments of varying length and is also encountered in terrestrial habitats. It differs from *Stichococcus* in being able to generate flagellate asexual cells (zoospores), useful in disperal in aquatic or moist terrestrial habitats.

land plants than were zygnematalean algae. It would be interesting to determine whether *Klebsormidium* has one or more introns in chloroplast tRNA genes as do Zygnematalean algae, Charales, *Coleochaete,* and land plants. Inclusion of *Stichococcus* and *Raphidonema* in the group is based entirely on ultrastructural characters related to cell division; no zoospores are produced by these genera and they have not been studied with the use of molecular techniques.

CHLOROKYBALES

Mattox and Stewart (1984) established the order Chlorokybales to include the monospecific genus *Chlorokybus,* which occurs as cubical packets of two to eight cells (Fig. 4-19). It exhibits an open and persistent spindle at mitosis, but cytokinesis occurs by development of a cleavage furrow, and plasmodesmata are not present (Lokhorst et al., 1988). A peroxisome typical of charophycean algae is present, but it is not located between the nucleus and plastid as in most other charophytes. In addition, there are two pyrenoids associated with the single plastid, which is unusual (Lokhorst et al., 1988). The flagellate basal bodies of zoospores are associated with an MLS similar to that observed in flagellate cells

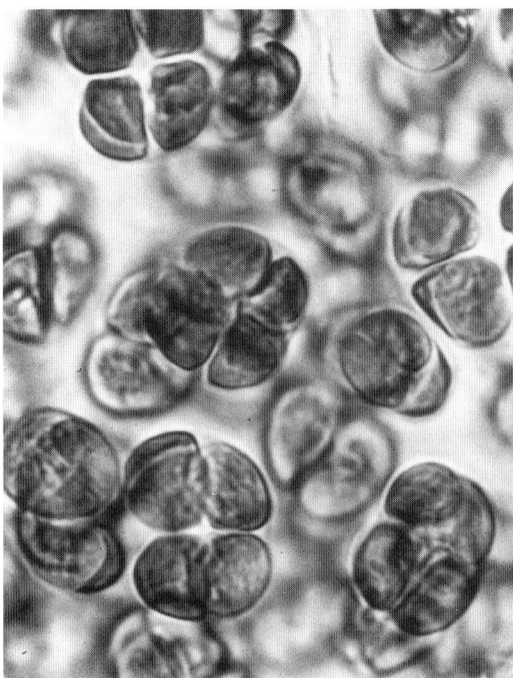

Fig. 4-19. *Chlorokybus* forms packets of cells held together by extracellular matrix material. It has been isolated only from terrestrial sites, and it produces zoospores.

of other charophytes, but some features of the zoospores resemble aspects of primitive flagellates classified among the prasinophytes (Rogers et al., 1980). Sexual reproduction is not known to occur.

Chlorokybus appears to be the most primitive organism known that can be confidently assigned to the Charophyceae and associated with the ancestry of land plants. It is generally assumed that the charophytes arose from among ancestral green flagellates, but despite considerable effort, no unambiguous candidate has yet been identified from among modern protists. The features of *Chlorokybus* zoospores can be used, however, to predict the features of the flagellate ancestral to the charophytes. These characteristics can then be used to evaluate various flagellates that have been proposed as possible ancestors of the charophycean lineage.

FEATURES OF A HYPOTHETICAL ANCESTOR OF THE CHAROPHYCEAE

Generally, the Charophyceae are considered to be a monophyletic group (Mattox and Stewart, 1984), although some nuclear rRNA gene data suggest that this assumption may be questionable (Chapman and Buchheim, 1991). A list of shared primitive features (symplesiomorphies) of the charophycean clade can be used to predict the attributes of an unknown ancestor. However, it should be kept in mind that any of these characters may actually be derived as compared with character states of the ancestral form, if transition in the character state(s) occurred at the same time as divergence of the derivative taxon. In other words, one or more of these characters may represent autapomorphies of the charophycean clade, and are thus absent from ancestral forms.

Putative Characteristics of an Ancestor of the Charophyceae:

1. Unicells biflagellate, flagella emerging from the side of cells (as is characteristic of charophycean flagellate cells, zoospores, or sperm).
2. Flagellar membrane covered with hairs and a single layer of small diamond-shaped or square scales (as is characteristic of charophycean sperm and most zoospores) (Fig. 4-20).
3. Cells wall-less, but covered with a single layer of diamond-shaped or square scales (as occur on sperm of *Coleochaete* and Charales, and on zoospores of *Chlorokybus, Chaetosphaeridium,* and some *Coleochaete* species) (Fig. 4-21).
4. Stellate structure present at the flagellar transition zone (as is characteristic of green algal flagella and those of embryophytes, excepting hornworts) (Fig. 4-22).
5. At least one MLS associated with flagellar basal bodies. MLS lamellae oriented at right angles to the long axis of microtubular layer (as in charophyte cells possessing MLSs) (Fig. 4-23).

74 THE CHAROPHYCEAN ALGAE

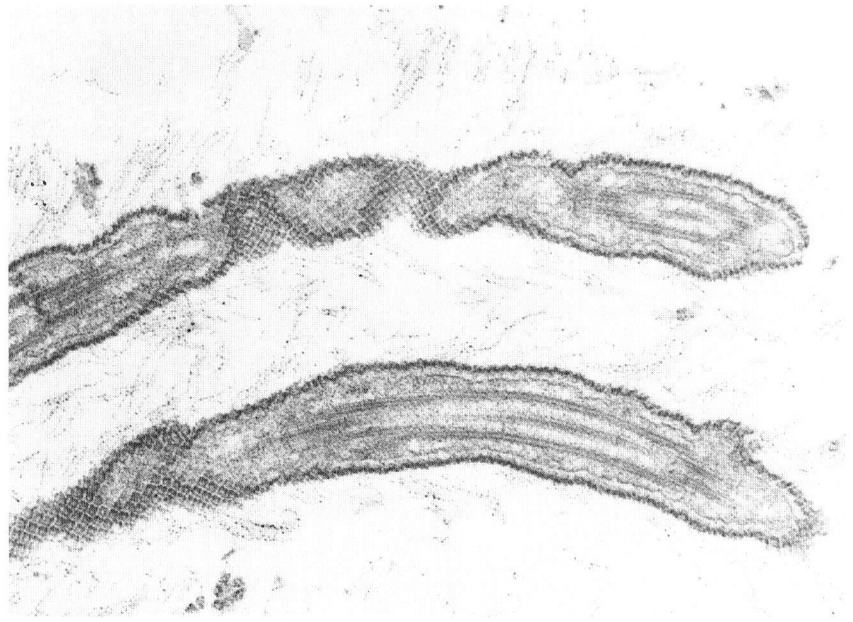

Fig. 4-20. Flat, diamond-shaped scales typically cover surfaces of charophyte flagella, including those of *Coleochaete scutata,* shown here. Photo from Graham and McBride (1979), *American Journal of Botany* 66:891.

6. A striated band connects flagellar basal bodies, but a musclelike rhizoplast connecting basal bodies with the nucleus is absent (characteristic of all flagellate cells of charophycean algae) (Fig. 4-23).
7. At least one peroxisome containing catalase and glycolate oxidase is present (as probably occurs in all charophycean algae and embryophytes) (Fig. 4-24).
8. Peroxisome and mitochondria attached to flagellar basal bodies during mitosis, as a mechanism ensuring equal partitioning of organelles to daughter cells (which occurs in zoospores of *Chlorokybus,* but not other charophytes).
9. Open mitosis (early breakdown of the nuclear envelope, as occurs in all charophytes and land plants) and spindle persistent at cytokinesis (as in less highly derived charophycean algae).
10. Phytochrome (occurs in at least two members of Zygnematales, possibly in Charales, and certainly in embryophytes).
11. Plastid contains chlorophylls a and b, thylakoids in stacks of more than three, a pyrenoid traversed by thylakoids, but no eyespot (all features characteristic of most charophytes and hornworts).
12. Sexual reproduction absent (as in *Chlorokybus* and Klebsormidiales).

FEATURES OF A HYPOTHETICAL ANCESTOR OF THE CHAROPHYCEAE

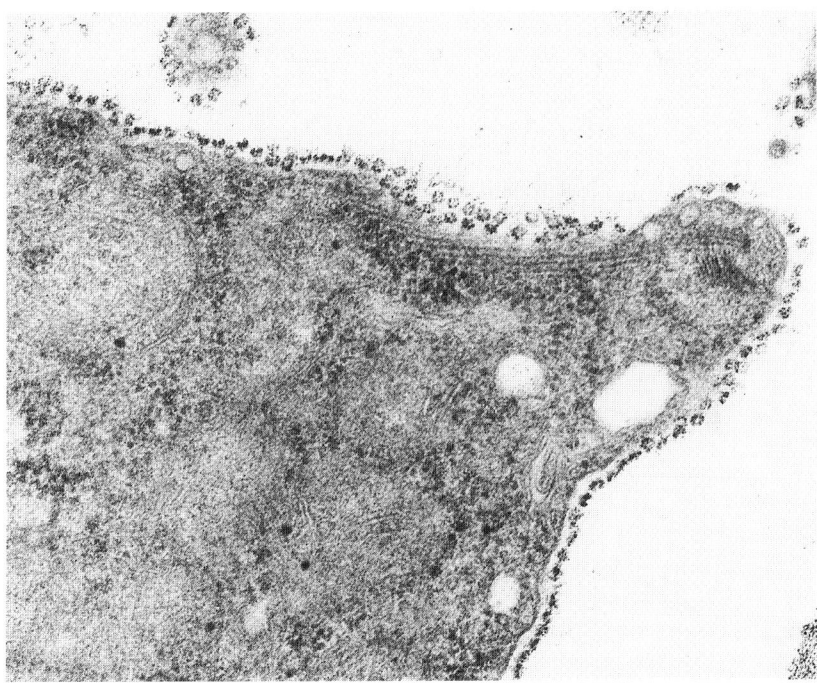

Fig. 4-21. Flat, diamond-shaped scales may also adorn cell membranes of charophyte spermatozoids (sperm of *Coleochaete pulvinata* shown here), as well as zoospores of some charophytes. Body scales are characteristic of most prasinophyte flagellates, but do not occur on the surfaces of land plant sperm.

There is no indication as to whether the ancestral flagellate was freshwater or marine, and a variety of green flagellates (prasinophytes) have been proposed as possible ancestors. These include *Pedinomonas, Nephroselmis* (= *Heteromastix*), *Pyraminomas, Mesostigma, Halosphaera,* and *Pterosperma* (Fig. 4-25). The first three genera are characterized by persistent spindles, but the nuclear envelope remains intact during mitosis, and glycolate dehydrogenase, rather than glycolate oxidase, occurs in all three (Floyd and Salisbury, 1977). Flagellar scales and the lowermost layer of body scales of *Pyramimonas* (Sym and Pienaar, 1991) resemble charophyte scales, for example, scales on *Chara* sperm (Moestrup, 1970). *Pedinomonas* has a conspicuous eyespot in the plastid, and neither *Pedinomonas, Pyraminomas,* nor *Nephroselmis* has a true or "bona fide" MLS. MLSs similar to those of charophytes or plants are found in *Mesostigma* (Rogers et al., 1981; Melkonian, 1989), *Pterosperma* (Inouye et al., 1990), and *Halosphaera* (Hori et al., 1985), but none of these flagellates has been examined for the presence of glycolate oxidase, glycolate dehydrogenase, or Cu/Zn SOD. Fawley and Lee (1990) discovered that the pigments of *Mesostigma* were similar to those of a derived member of the ulvophycean algae (*Bryopsis*).

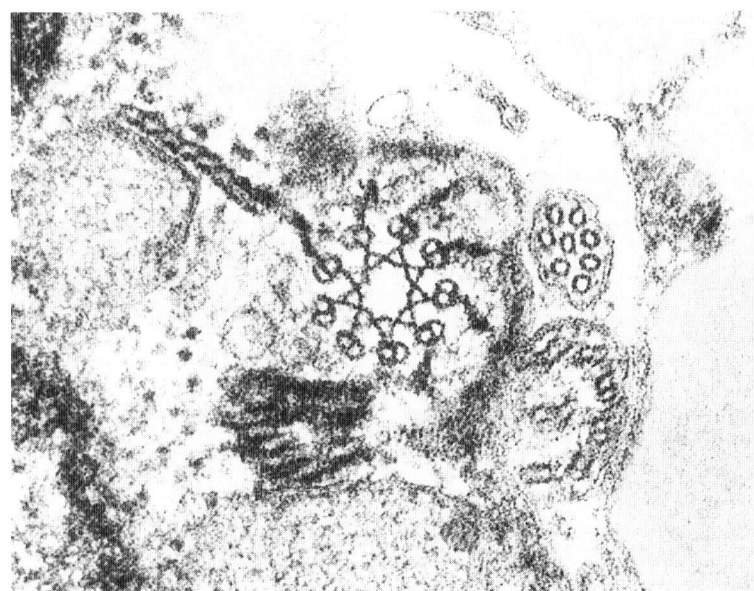

Fig. 4-22. Stellate structures are formed of fibers that interconnect microtubules at the flagellar transition region of green algae (zooids of *Trentepohlia aurea* shown here) and most flagellate land plant sperm.

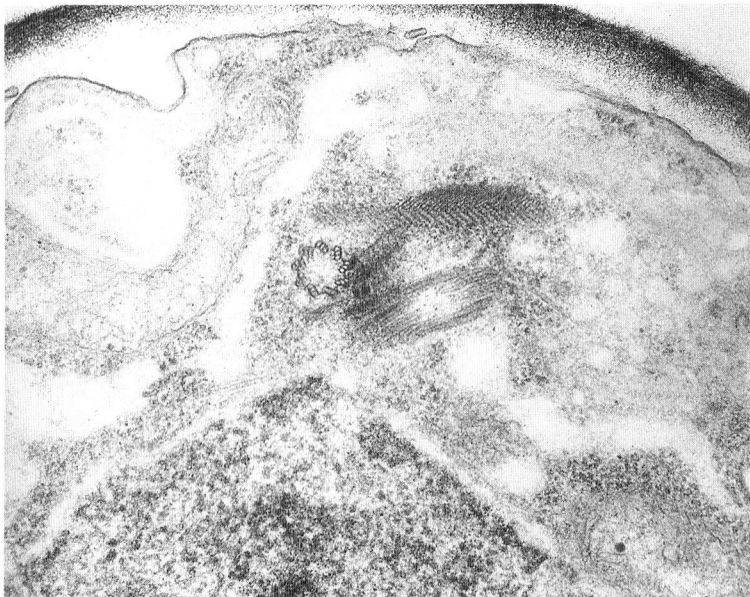

Fig. 4-23. A complex striated connective joins flagellar basal bodies and is also associated with the MLS of charophycean reproductive cells. Vestigial connectives occur in certain land plant sperm as well.

FEATURES OF A HYPOTHETICAL ANCESTOR OF THE CHAROPHYCEAE   77

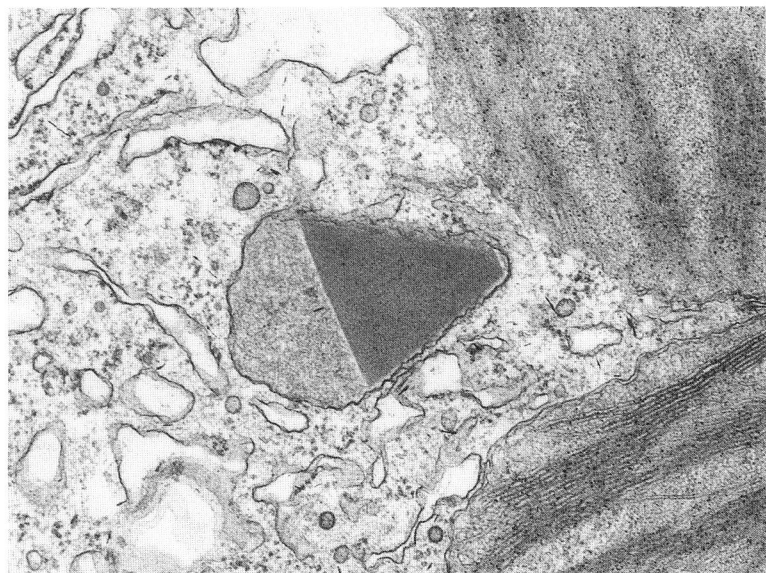

Fig. 4-24. Peroxisomes containing catalase characterize cells of charophycean algae and land plants. A peroxisome of *Chara zeylanica,* shown here, contains a crystal which is probably catalase. Photo from Graham and Kaneko (1991). Crit. Rev. in Plant Sciences 10:329.

Pterosperma cristatum

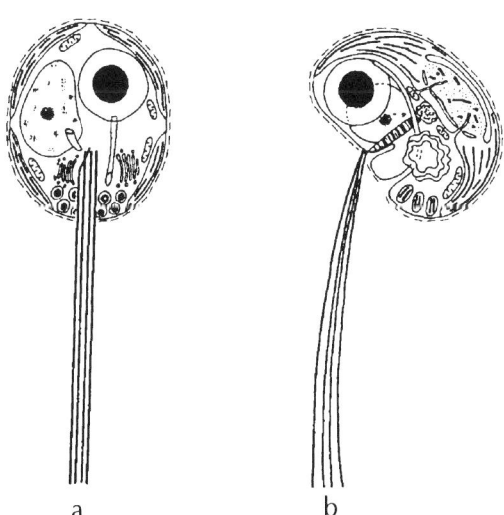

Fig. 4-25. Diagram of the flagellate phase of the prasinophyte *Pterosperma cristatum* reproduced from Inouye et al. (1990), *Journal of Phycology* 26:330, used with permission. These scaly flagellates produce a nonmotile stage known as phycomata, which are characterized by ornamented walls. It is currently believed that ancestors of multicellular green algae may have been related to modern prasinophytes.

The life cycles of *Pterosperma, Halosphaera,* and *Pachysphaera* involve both scaly flagellates and nonmotile, walled cysts called phycomata. The walls of phycomata may contain sporopollenin, and these cells have been allied to fossil leiospheres, especially *Tasmanites,* which appear in the fossil record between 800 and 1,500 million years ago (Tappan, 1980). O'Kelly (1992) has suggested *Halosphaera* as a model charophyte ancestor, but noted that forms intermediate between *Halosphaera* and *Chlorokybus* are unknown. Since *Halosphaera* is quadriflagellate, with flagella emerging from a pit (as is characteristic of prasinophytes), the transition to charophytes would have to include "halving" of the flagellar apparatus and a change in flagellar location to a lateral groove. Scales of neither *Halosphaera* nor *Pterosperma* resemble those of charophycean flagellate cells. It is intriguing that *Pterosperma* flagella possess both a stellate structure (very similar, but not identical, to that of green algae) and a transitional helix (characteristic of the chromophyte algae).

The green algae, including prasinophytes, are thought to have arisen through a primary endosymbiotic event involving an oxygenic phototrophic procaryote and a colorless eucaryotic flagellate (Stewart and Mattox, 1980). Although the early origins of green algae are not understood, the recent discovery that the heterotrophic deep-sea flagellate *Jacoba* has a plantlike MLS (O'Kelly, 1992) suggests that further study of poorly known protists may clarify primary events in the origin of green algal lineages. For example, phagocytosis (particle ingestion) and digestion occur in the chlorophyte *Cymbomonas tetramitiformis,* and further study of this interesting protist is indicated (O'Kelly, 1992). Until considerably more ultrastructural and molecular survey work on protists has been accomplished, it is not possible to speculate further regarding the origins of chlorophytes. However, it is worthwhile to consider further the shared features of green algae with respect to the origin of a charophycean clade. These shared features include the presence of chlorophylls a and b, but not phycobilins; a stellate structure in the flagellar transition region; and nuclear encoding of the small subunit of ribulose bisphosphate carboxylase (Rubisco).

The green algae are characterized by plastids delimited by two membranes that contain the photosynthetic pigments chlorophyll a and b, differing only in the nature of the R1 substituent ($-CH_3$ in the case of chlorophyll a, and -CHO in the case of chlorophyll b) (Fig. 4-26). The prochlorophytes (*Prochloron, Prochlorococcus,* and *Prochlorothrix*) have been cited as possible models for the green algal chloroplast precursor, inasmuch as these procaryotes contain chlorophylls a and b, but not phycobilins. However, there are substantial differences between the chlorophyll-associated proteins (apoproteins) of prochlorophytes and those of land plants, and because the monophyletic origin of chlorophyll b has not been established, prochlorophytes are not thought to have been directly ancestral to green algal plastids (Stackebrandt, 1989). The 16S rRNA data suggests that *Prochloron* falls within the realm of the cyanobacteria, and there has been some controversy as to whether *Prochlorothrix* is more closely related to cyanophytes or to green plastids. However, a tree based on 16S rRNA data places it closer to cyanophytes (Turner et al., 1989). Sequence data for the *rpo* C1 gene, which

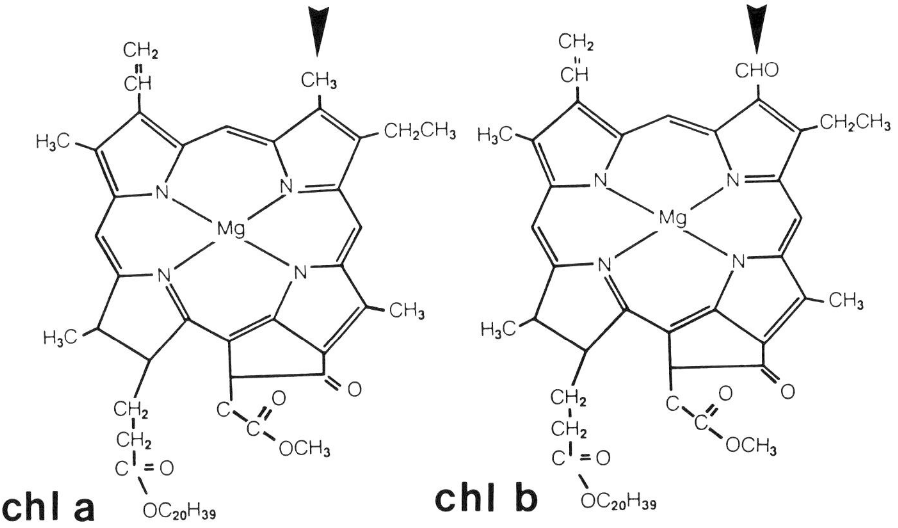

Fig. 4-26. Chlorophylls a and b differ only slightly in structure. Thus, it is possible that chlorophyll b diverged from chlorophyll a more than once during the evolution of chlorophyll b-containing organisms.

encodes a subunit of DNA-dependent RNA polymerase (Palenik and Haselkorn, 1992), and additional 16S rRNA gene sequencing studies (Urbach et al., 1992) indicate that prochlorophytes are a highly divergent, polyphyletic group that did not give rise to green plastids. Lokhart et al. (1992) cloned and sequenced *atp* BE genes from *Prochloron didemni* and determined that the evidence supported a relationship to cyanobacterial rather than plastid lineages. These authors suggested that sequences of oxygenic taxa may tend to group because of shared substitution processes rather than (necessarily) common ancestry.

Ribosomal RNA data (Urbach et al., 1992) substantiated earlier findings (Giovannoni et al., 1988) that the cyanelle of *Cyanophora paradoxa* is the closest known prokaryotic relative of the *Marchantia* chloroplast. This suggests that loss of phycobilin pigments and gain of chlorophyll b may have occurred at least twice, i.e., separately in prochlorophytes and in the direct ancestor of green algal plastids. It is interesting that members of the Glaucocystophyta (including *Cyanophora*) typically possess MLSs, but glaucophyte MLSs are sufficiently different from those of charophytes and plants that homology is questionable. More definitive means are needed for testing the homology of MLS-like structures. These could include specific antibody localizations (O'Kelly, 1992), but have not yet been developed.

Another shared derived character widely cited as evidence that the green algae are of common ancestry is the stellate structure that occurs at the transition zone between basal bodies (centrioles) and the flagellar axoneme (Fig. 4-22). Stellate structures also occur in most flagella produced by land plants (they are, however, absent from hornworts). Flagellar transition regions of some other protists may also exhibit stellate structures, but these can be differentiated from those of green algae and plants (O'Kelly, 1992). The biochemical composition of the stellate structure has not been completely defined. However, the stellate structure of *Chlamydomonas* contains the contractile protein centrin (Salisbury et al., 1988; Salisbury, 1989, 1992), and centrin has also been localized within the transition region of other green flagellates (Melkonian et al., 1992). Centrin (= caltractin) is a calcium-modulated "EF-hand" protein having considerable sequence similarity to other calcium-binding proteins, including calmodulin. Contraction of centrin is based on supercoiling, rather than sliding as in actin-myosin systems, and is independent of ATP. Extension of the protein requires ATP, however. Centrin has been found in almost all eukaryotic cells examined. It is most prominent in flagellate cells, as a component of the flagellar base apparatus. Possible functions in flagellates include rhizoplast contraction, closure of thecal slits, and flagellar abscission (Melkonian et al., 1992). Although the presence of centrin in stellate structures of charophytes and embryophytes has not been documented, Vaughn (1992) reported that MLSs in developing bryophyte spermatids contain centrin, and that the MLS develops from an amorphous centrin-containing structure at the spindle poles. This discovery will no doubt stimulate further examination of centrin localization in charophytes and embryophytes.

Nuclear coding of the gene for the small subunit of ribulose bisphosphate carboxylase (Rubisco) is a third character shared by green algae and land plants. The *rbc* S gene is encoded in the chloroplast genome of all other algal groups studied (Douglas and Durnford, 1989; Newman and Cattolico, 1990), indicating that this is the ancestral condition and that the gene has been transferred to the nucleus in the ancestors of green algae and plants. The small subunit is required for maximum catalysis, but not for activation of Rubisco, and its role is poorly understood. The degree of small subunit gene sequence variation among taxa suggests that there are weaker functional constraints on the small subunit than on the large subunit, which harbors the active catalysis sites (Newman and Cattolico, 1990). The evolutionary advantage acquired by modern green algae and plants in segregation of the genes for small and large Rubisco subunits is not understood (Glover, 1989).

For a gene that functions in an organelle, movement to the nucleus requires all of the following: physical relocation of a complete gene, insertion into a favorable nuclear locus, and acquisition of expression signals and a transit peptide coding sequence located within the reading frame, all within a relatively short period of time to prevent loss of function resulting from mutation (Baldauf and Palmer, 1990). Transit sequences provide the structural information for targeting cytoplasmically synthesized proteins to the appropriate cellular compartment. An organellar receptor system capable of recognizing the transit peptide must also be functional (Fig. 4-27). After importation of the targeted protein, the 3-7 kD transit peptide is proteolytically cleaved at the N-terminus. In view of these complexities, it does not seem likely that *rbc* S could have been transferred from the plastid to the nuclear genome more than once in the evolutionary history of green algae and plants.

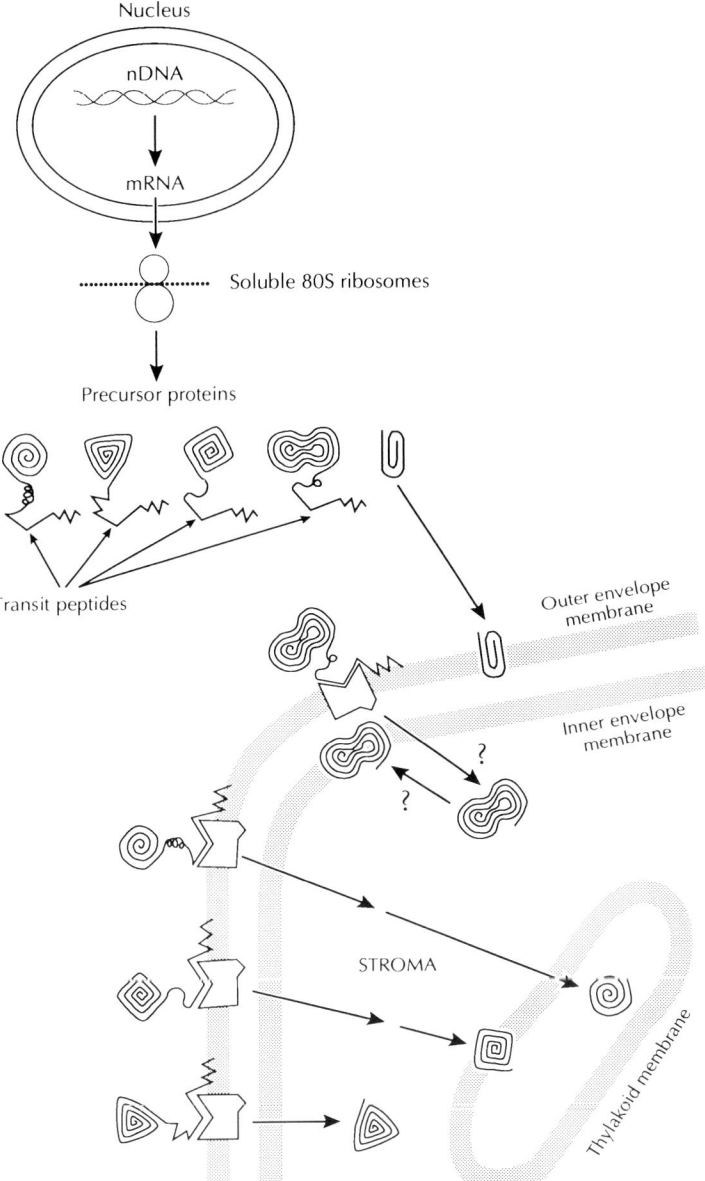

Fig. 4-27. Protein import into chloroplasts is incompletely understood. However, this preliminary model (courtesy of Kenneth Keegstra) suggests that various outer and inner envelope proteins, including heat shock protein cognates (or chaperonins), are required for uptake of cytosolically synthesized polypeptides bearing appropriate targeting sequences. Once proteins have been imported, presequences are removed by proteases. Evolution of signal presequences and protein import systems has been integral to the formation of eukaryotic cells and organelles, in general, and plant cells in particular.

Surprisingly, some data indicate that transfer of the sequence for a small Rubisco subunit has occurred more than once. Franzen et al. (1990) compared the sequences of 12 chloroplast transit peptides (including that for the small subunit of Rubisco) from *Chlamydomonas reinhardtii* with those of higher plants and the mitochondrial targeting peptides of yeast. They found that in terms of length and composition, the chloroplast transit peptides of *Chlamydomonas* are more similar to mitochondrial targeting peptides than to chloroplast targeting peptides of higher plants. These authors speculated that the chloroplasts of *Chlamydomonas* and those of higher plants are of different endosymbiotic origin, such that transfer of genes to the nucleus and evolution of transit peptides occurred independently in the ancestors of *Chlamydomonas* and in those of plants.

In contrast to transit peptides that appear to mediate entry into plastids, the signal sequences for protein translocation through the thylakoid membrane (the C-terminal domain of lumen-targeting chloroplast transit peptides) of *Chlamydomonas* and land plants were quite similar, indicating common origin. These sequences are most likely older than signals for chloroplast protein import, because they also occur on proteins targeted to the thylakoid lumen in cyanobacteria and were probably already present in protein coding regions of the procaryote that became the chloroplast in both *Chlamydomonas* and land plants (Franzen et al., 1990). This evidence suggests that the ancestors of *Chlamydomonas* plastids and higher plant chloroplasts were quite similar, but opens the possibility that host cells were different with respect to the nuclear genome, the mitochondrial genome, or both. Presumably, both host cells were flagellates having stellate structures, inasmuch as putatively homologous structures occur in *Chlamydomonas,* other green algae, and the charophycean relatives of higher plants. However, the host cells that eventually gave rise to charophytes and embryophytes could have differed from those leading to other green lineages (such as the Chlorophyceae and *Chlamydomonas*) in various features relating to the cytoskeleton, mitosis, and biochemistry that were nuclear encoded. Postulation of a single green flagellate ancestor for all of the green algae may not be appropriate. The transit peptide evidence indicates that neither possession of green plastids nor the presence of nuclear-encoded Rubisco small subunits are evidence for monophyly of the green algae. These observations suggest that homology of green algal nuclei and mitochondria should not be assumed, but examined with the use of sequence data from multiple genes.

The results of a comparative study of a partial sequence of the nuclear encoded 28S rRNA gene in 24 protists led Perasso et al. (1989) to propose that rhodophyte, chromophyte, and chlorophyte algae emerged as three distinct groups relatively late in protist radiation, close to the time of origin of embryophytes and animals. Representing green algae in this study were *Pyramimonas parkeae* and *Chlorogonium elongatum,* neither of which is associated with the charophycean lineage. Nuclear rRNA gene data obtained by Gray et al. (1989) also suggest that the higher plant lineage branched off at about the same time as animals and fungi. Nuclear rRNA gene sequences do not indicate that the nuclear genome of charophycean algae is dramatically divergent from that of other green algal lineages (Chapman and Buchheim, 1991), and 16S rRNA sequence data also suggest that the nuclear genomes of *Chlorella* and *Chlamydomonas* are related to those of *Zea mays* and *Zamia* (Ariztia et al., 1991). In contrast, analysis of mitochondrial rRNA sequences results in trees showing divergence of higher plants close to the root of purple bacteria and separate from fungi, noncharophyte green algae, and animals (Gray et al., 1989). In other words, rRNA genes of plant mitochondria appear to be more similar to those of purple bacteria, the likely ancestors of mitochondria than they are to the mitochondrial rRNA genes of other green algae. Other differences observed between plant mitochondrial DNA and that of other organisms include the finding that plant mtDNA uses the universal genetic code, whereas a modified code operates in yeast and humans (Ohyama et al., 1991). In addition, the amino acid sequence of liverwort cytochrome oxidase subunit II is deduced to be at least 82 percent similar to that of angiosperms, whereas it shows only 62 percent similarity with *cox* II from nonplant species (Ohyama et al., 1991). Additional *cox* II sequence information from green algae would be useful in tracing the evolutionary origin of plant mitochondria. It would be particularly helpful to determine whether mitochondrial transit sequences of nuclear-encoded mitochondrial proteins and mitochondrial membrane receptors in charophycean algae resemble those of higher plants, because such data shed light on the process of endosymbiotic origin of organelles (Franzen et al., 1990) and can be used to accept or reject hypotheses of organelle homology.

With regard to the origin of plant peroxisomes, Keller et al. (1991) found that antibodies to a microbody-targeting signal localize in peroxisomes of plants as well as microbodies of animals, suggesting that these sequences (and their receptors) are evolutionarily related and highly conserved. A recent survey of the occurrence of glycolate oxidase and glycolate dehydrogenase among algal protists revealed that most chromophyte algae (and *Cyanidium*) possess glycolate oxidase coupled with high levels of catalase activity, as do charophytes and embryophytes. It is not known whether chromophyte and charophyte glycolate oxidases are homologous, but the enzymes are similarly insensitive to cyanide and exhibit similar substrate stereospecificity (Suzuki et al., 1991). In contrast to apparent conservation of ancient peroxisomal targeting mechanisms in plants and animals, protein targeting through the secretory system to vacuoles differs in plants on the one hand, and yeast and animals on the other (Chrispeels and Raikhel, 1992). However, little or nothing is known of the molecular nature of secretory/vacuolar targeting in green algae, including charophytes.

In summary, although additional data are needed, at present it appears that there may be some reason to doubt the monophyly of chlorophytes. It is possible that charophycean algae arose through endosymbiotic events separate from those giving rise to other green algal lineages. Perhaps host flagellates had similar nuclear genomes, but acquired genetically different mitochondrial symbionts and similar plastid endosymbionts separately. Such a possibility may suggest that the eucaryotic lineage culminating in the land plants was distinct since its origin, and that green algae related to the ancestry of plants will appear to be variously related to other green algae, depending on which genome is used for comparison. This has some important implications for classification schemes based on the

Hennigian philosophy that only monophyletic groups should be named. The name Viridiplantae has been formally proposed for the green plants (green algae and higher plants) (Cavalier-Smith, 1981), but until monophyly of the green algae has been firmly established, the taxon Viridiplantae should be only provisionally accepted. Jeffrey (1982) proposed the term Streptophyta for the land plant lineage (charophytes and land plants); this taxon is dependent on proof that the Charophyceae is a monophyletic group. Testing hypotheses of monophyly for these groups requires considerably more research focus on protist diversity and elucidation of the evolutionary history encoded in protist genomes and cellular structure.

5

MORPHOLOGY, ECOLOGY, AND PHYSIOLOGY OF CHAROPHYTES

Molecular, biochemical, and ultrastructural evidence provides compelling support for the theory that modern land plants and charophytes share a common ancestor. Therefore, consideration of relationships between morphology, ecology, and physiology of charophycean algae may shed light on the origin of adaptations important in the conquest of land. Adaptation has been defined as "a character state or suite of character states within a set of heritable character variation whose presence in an individual results in an increase in birth and/or a decrease in mortality rates under a given set of environmental conditions" (Knoll and Niklas, 1987). To be adaptive, a character must provide current utility to an organism and "have been generated through the action of natural selection for its current biological role" (Baum and Larson, 1991). In extrapolating the morphological and physiological adaptations of modern charophytes to the past, it is important to remain cognizant of the fact that the characteristics of extant organisms may reflect modern selection pressures that may not have been relevant some 450 to 470 years ago. Justification for this approach is, however, provided by evidence that aquatic plants may exhibit very slow rates of evolution and extremely long species duration, attributable to the uniformity of aquatic environments and widespread occurrence of vegetative reproduction in hydrophytes. Unequivocal fossils of the aquatic angiosperm *Ceratophyllum*, for example, are known from Tertiary (Eocene) deposits (Herendeen et al., 1990). Data derived from sequencing of the Rubisco large subunit gene indicate that *Ceratophyllum* is a descendent of some of the earliest angiosperms. It is not believed that the earliest flowering plants were aquatic, but that *Ceratophyllum*, having diverged from the ancestral lineage, secondarily migrated into the aquatic habitat and in so doing entered upon a phase of evolutionary stasis (Les et al., 1991). Similarly, as we have previously noted, fossil evidence suggests that charophycean algae have maintained similar morphologies over hundreds of millions of years.

Morphological, physiological, and ecological features of modern, but archaic, animals have been useful in deducing ancient transitions from aquatic to terrestrial habitats. For example, among invertebrates, various adaptations present in marine eurypterid ancestors are thought to have provided scorpions with the ability to resist the effects of desiccation in early terrestrial ecosystems. Based on study of extant lungfish and hagfish, the marine fish regarded as ancestors of the earliest terrestrial vertebrates (tetrapods) are thought to have possessed air-breathing organs, urea production via the ornithine cycle, and water and urea retention via absorption in the glomerular kidney, adaptations to aquatic life which were also useful in the terrestrial habitat (Seldon, 1985; Rolfe, 1985). The evidence indicates that various invertebrate groups invaded the land directly from the marine habitat, with other invertebrates, and vertebrates, passing through an intermediate freshwater aquatic phase (Bray, 1985; Halstead, 1985; Selden and Edwards, 1989).

In contrast, all of the available structural, biochemical, and molecular data indicate that derivation of land plants from multicellular marine algae, represented by modern seaweeds, is extremely unlikely. Land plants are now generally regarded as derived from algal inhabitants of transitory freshwater bodies, such as shallow pools, backwater lagoons, or lakes (Selden and Edwards, 1989). It is quite possible, however, that marine flagellates gave rise to a freshwater charophyte lineage. Some have argued that land plants evolved from unicellular terrestrial charophytes in direct response to terrestrial selection pressures (Stebbins and Hill, 1980). Others contend that land plants were more probably derived from multicellular aquatic charophytes that had become preadapted for invasion of land in response to selection pressures operating in shallow freshwater environments (Graham, 1985). A survey of the morphological, physiological, and ecological attributes of charophycean algae is relevant to this controversy and useful in evaluating the alternative hypotheses.

MORPHOLOGICAL CHARACTERS IN RELATION TO ENVIRONMENTAL VARIATION

Chlorokybus and the Klebsormidiales are especially interesting because they may occupy terrestrial habitats. *Chlorokybus* was described by Geitler (1942, 1955) from two terrestrial habitats in Austria. In common with many other terrestrial algae, *Chlorokybus* readily produces flagellate zoospores (Rogers et al., 1980), and therefore is presumably derived from an aquatic ancestor. It forms packets of 2 to 8 cells surrounded by a thick mucilage, and thus resembles various noncharophycean soil algae classified in the Chlorosarcinales (Bold and Wynne, 1985). *Klebsormidium* (= *Hormidium*) is an unbranched filament, whereas *Stichococcus* and *Raphidonema* form short filaments or occur as unicells. *Klebsormidium* and *Stichococcus* may occur in lakes or bogs (Prescott, 1951) but have also been found in various terrestrial habitats, inhabiting limestone walls or growing on bark or wood (Graham et al., 1981), rooftops (Brook, 1968), or rock surfaces (Allen, 1971). *Stichococcus* may occur as a phycobiont in lichens (Tschermak-Woess, 1988). *Raphidonema* is most commonly found in permanent snowfields (Smith, 1950). The presumed loss of zoospore production from some charophytes (*Stichococcus, Raphidonema,* and some *Klebsormidium* species) has been interpreted as adaptation to terrestrial life (Stebbins and Hill, 1980). However, many other terrestrial eucaryotic algae produce zoospores that are released when conditions become moist. Examples include soil chlorophytes such as *Chlorococcum* and numerous relatives, more complex green algae such as *Trentepohlia* and its relatives and *Fritschiella* (Figs. 5-1 and 5-2), and the chromophytes *Botrydium* and *Vaucheria* (Bold and Wynne, 1985). Among the algae, production or absence of zoospores does not seem to be highly correlated with habitat, but this hypothesis should be tested further.

A wide diversity of zygnematalean algae, including desmids, may be found in fresh waters of low pH, such as *Sphagnum* bogs. In recent years, unbranched, filamentous zygnematalean algae such as *Mougeotia* and *Temnogametum* have been observed to form nuisance growths in lakes suffering from the effects of

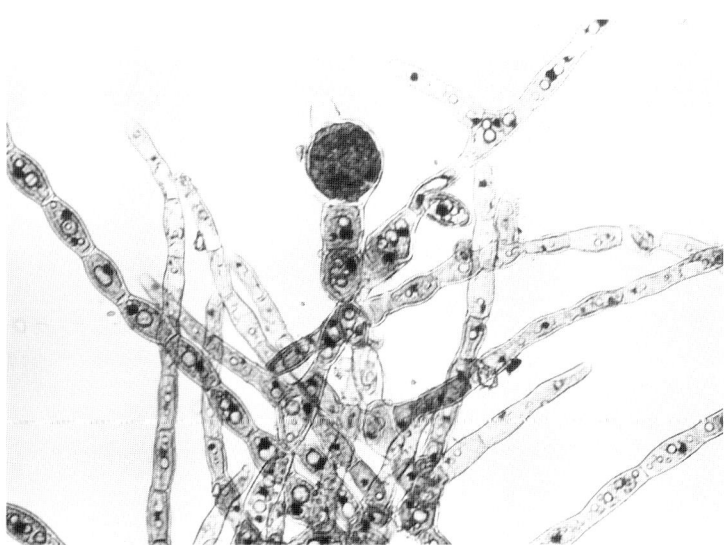

Fig. 5-1. *Trentepohlia aurea* is a member of the terrestrial green algal group Chroolepidaceae (=Trentepohliales). It commonly occurs on rocks, tree trunks, or wood in moist areas. The highly branched filaments reproduce by means of flagellate zoospores and gametes generated within distinctive urn-shaped sporangia/gametangia. Bright orange carotenoids accumulate in lipid droplets (the dark structures within cells). Chroolepidacean algae produce MLSs; these are different in structure from MLSs of charophytes or land plants, but similar to MLSs in the glaucophytes (a group of algae that contain blue-green cyanelles). Although phragmoplasts, cell plates, and plasmodesmata occur in the Chroolepidaceae, the nuclear envelope persists during mitosis (in contrast to open mitosis of charophytes and embryophytes).

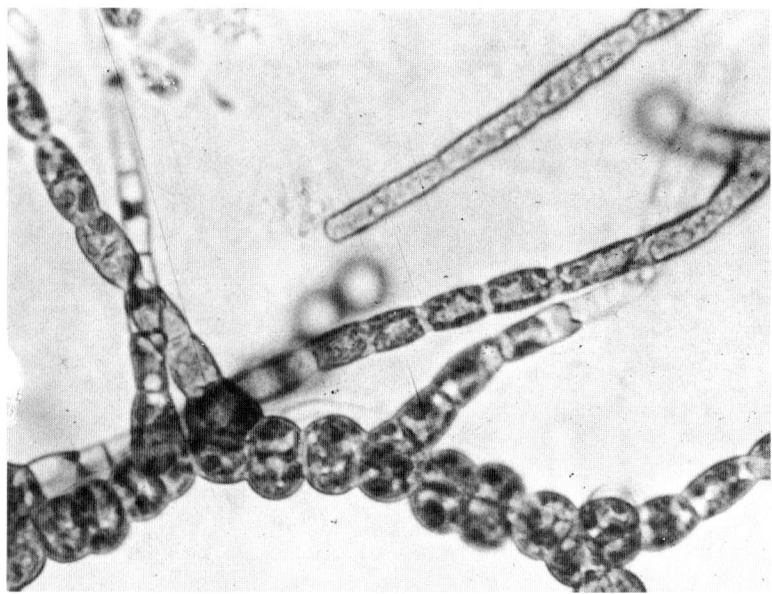

Fig. 5-2. *Fritschiella tuberosa* is a relatively complex terrestrial green alga, which has aerial photosynthetic branches, multiseriate basal regions, and colorless branches that penetrate moist soil as do bryophyte rhizoids. Although *Fritschiella* exhibits advanced cytokinetic mechanisms (there is a cell plate, but no phragmoplast), mitosis and reproductive cell structure indicate absence of close relationship to charophycean algae or the ancestry of land plants. However, *Fritschiella, Trentepohlia,* and other terrestrial algae show that there have probably been multiple colonizations of the terrestrial environment by green algae.

acid rain, or in aquatic systems that have been experimentally acidified (Schindler et al., 1980; Turner et al., 1987; Webster et al., 1992). Although some zygnematalean taxa occur in more alkaline waters, there is a tendency for the group as a whole to prefer acidic habitats (Hoshaw et al., 1990). Certain taxa are found in moist terrestrial environments (Prescott, 1951; Allen 1971). Some desmids (particularly *Cosmarium* and *Staurastrum*) may be important components of the phytoplankton, but other unicellular forms more typically occur as mucilaginous aggregates associated with submerged shallow water substrates. Filamentous forms may aggregate into large metaphytic masses (*Mougeotia*) or surface mats (*Spirogyra*) that are buoyed by oxygen bubbles trapped in the mucilaginous matrix formed by the sheaths of individual filaments. Some authors have found it difficult to understand why aquatic algae would adopt conjugation as a method for bringing gametes into proximity (Stebbins and Hill, 1980). However, it is possible that production of copious mucilage and the resulting tendency of zygnematalean algae to form large aglomerated populations may have preadapted them toward conjugation. Because putatively more primitive charophytes (*Chlorokybus* and *Klebsormidium*) produce flagellate reproductive cells (asexual zoospores), it is generally assumed that the ancestors of the Zygnematales had the ability to produce flagellate cells, but that these were subsequently lost. Sexual reproduction does not occur in *Chlorokybus* or Klebsormidiales, so it is not known whether zygnematalean conjugation might have derived from a preexisting form of sexual reproduction involving flagellate gametes, or whether conjugation arose after the loss of flagellate cells in the ancestors of modern Zygnematales (see Chapter 8 for further discussion). Sexual reproduction results in resistant zygotes whose walls, in *Spirogyra* at least, are impregnated with sporopollenin (Fig. 5-3). Presumably, this adaptation provides for survival should the freshwater habitat dry up and allows long-distance dispersal. Because many other freshwater green algae, such as the noncharophytes *Chlamydomonas, Volvox, Oedogonium,* and *Sphaeroplea* also produce resistant, readily-dispersed zygotes (Figs. 5-4 and 5-5) as the result of sexual reproduction (Bold and Wynne, 1985), it can be hypothesized that resistant zygotes represent adaptation to the seasonally unstable, relatively temporary nature of freshwater habitats. Shallow freshwater ponds may dry up; lakes may undergo successional transition to dry land. It cannot be assumed that the resistant features of zygnematalean zygotes necessarily arose in response to selective pressures experienced during a terrestrial phase of evolution.

The fact that zoospores are not produced by zygnematalean algae has been interpreted as reflecting adaptation by ancestral forms, perhaps resembling modern Klebsormidiales, to terrestrial environments (Stebbins and Hill, 1980). However, there are alternative explanations for loss of zoospores by filamentous algae. Comparative studies of the reproductive ecology of other freshwater green algae have

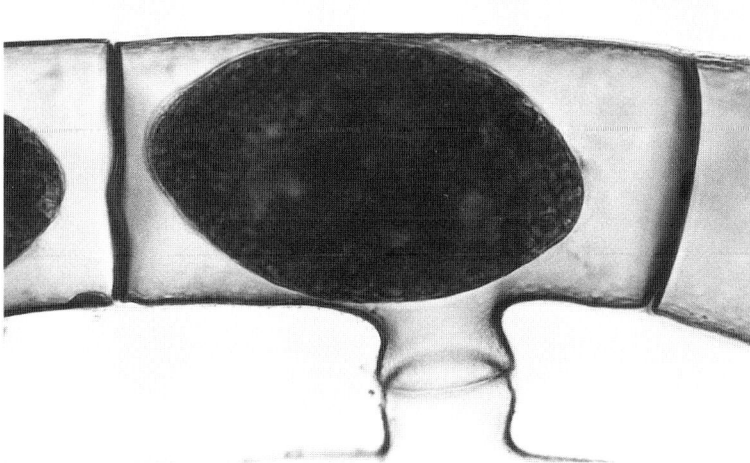

Fig. 5-3. *Spirogyra* zygote walls are impregnated with resistant sporopollenin, which contributes to their ability to withstand desiccation. This feature most likely illustrates evolutionary preadaptation leading to the origin of sporopollenin-enclosed spore walls of land plants.

90 MORPHOLOGY, ECOLOGY, AND PHYSIOLOGY OF CHAROPHYTES

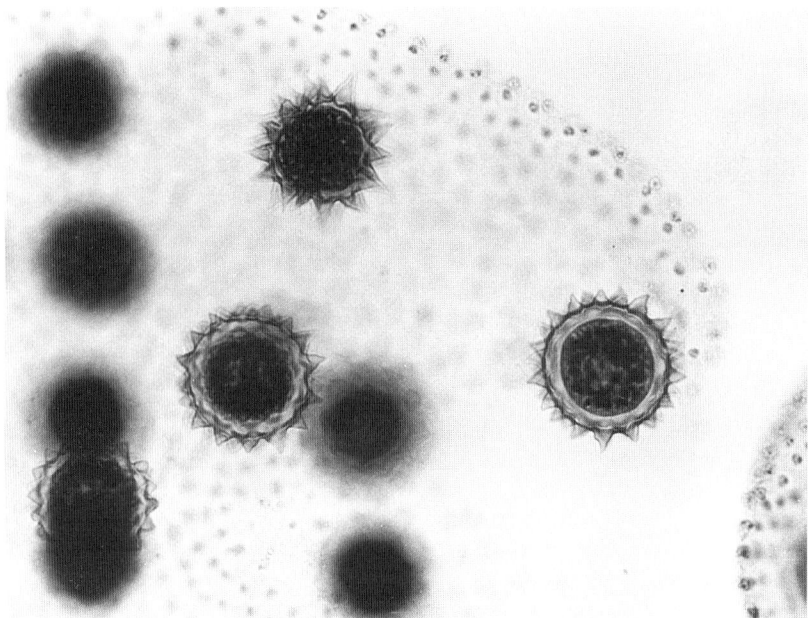

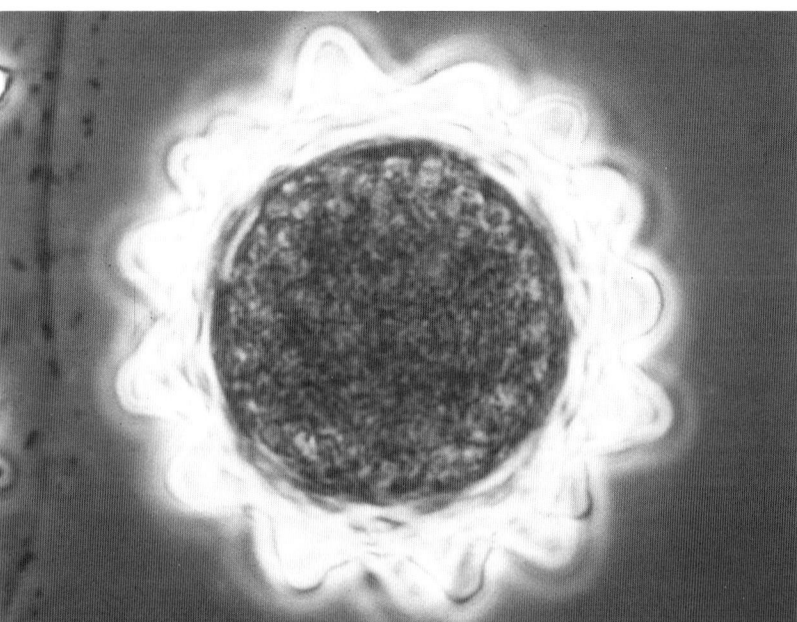

Figs. 5-4 (top) and 5-5 (bottom). *Volvox* and various other green algae also produce resistant, often highly ornamented zygotes, shown here in bright-field (**5-4, top**), and dark-field (**5-5, bottom**) microscopy. Evolution of resistant zygotes in green algae is likely to be correlated with seasonal or longer-period perturbations of freshwater habitats, but the adaptive function of elaborate wall ornamentation has not been established.

MORPHOLOGICAL CHARACTERS/ENVIRONMENTAL VARIATION   91

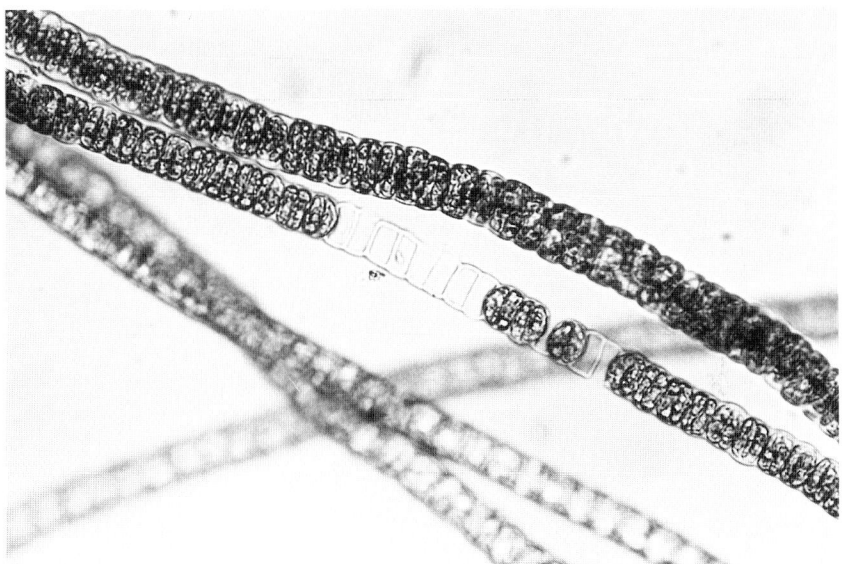

Fig. 5-6. Zoospore production by unbranched filamentous algae such as the attached ulvophyte *Ulothrix zonata* results in wholesale filament disintegration and disappearance of most algal biomass from the periphyton.

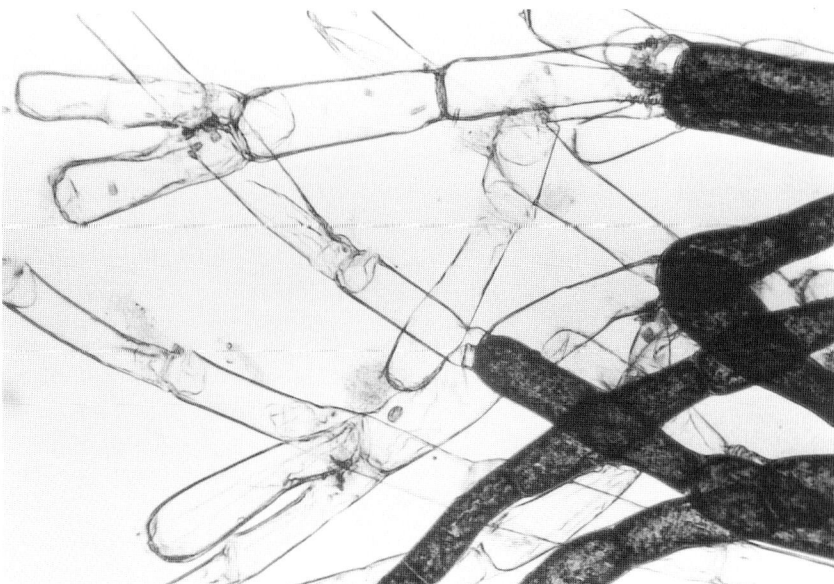

Fig. 5-7. Zoospore production by terminal regions of branched filaments such as the attached ulvophyte *Cladophora glomerata* allows basal thallus regions to remain intact and to persist in the periphytic environment for relatively long periods.

demonstrated that zoospore production by unbranched filaments may result in extensive thallus disintegration, whereas branched filamentous forms can generate zoospores from localized regions without experiencing loss of thallus integrity (Auer et al., 1983) (Figs. 5-6 and 5-7). Zoospore production by filamentous zygnematalean algae, which are unbranched, would result in thallus disruption and perhaps also render them more vulnerable to loss of photosynthetic biomass by predation of zooplankton upon the zoospores. Many unicellular desmids can move along surfaces by mucilage extrusion through wall pores and thus do not require conversion into flagellate forms in order to accomplish local dispersal in aquatic habitats. In view of these considerations, loss of zoospores by the Zygnematales cannot be unambiguously interpreted as adaptation to terrestrial life.

The Coleochaetales (*Coleochaete* and *Chaetosphaeridium*) produce branched thalli and are therefore considered to be more advanced than *Chlorokybus,* Klebsormidiales, or Zygnematales. However, the Coleochaetales generally grow attached to macrophytes or inorganic substrates in shallow lake waters; none are known to be terrestrial in habitat. Zoospore production, useful in dispersing these forms to substrates throughout the littoral zone, is primarily controlled by temperature, with maximum zoospore production by *Coleochaete scutata* occurring between 20° and 25°C. (Graham et al., 1986) (Figs. 5-8 and 5-9). Because thalli are branched or thalloid and attached to substrates, retention of the primitive capability for production of flagellate stages does not result in disintegration of parental thalli (Wesley, 1928) (Figs. 5-10 and 5-11). The morphologically complex terrestrial green alga *Fritschiella* (Fig. 5-2) was once thought to represent a possible progenitor of land plants, but is now known to be unrelated to the charophytes or embryophytes, on the basis of ultrastructural characteristics (Melkonian, 1975). The Chroolepidaceae (= Trentepohliales) are also branched filamentous green algae of terrestrial environments (Fig. 5-1), whose phylogenetic affinities are as yet undetermined. Given the occurrence of forms such as *Fritschiella* and Chroolepidaceae, there is no obvious reason that terrestrial charophytes having morphologies intermediate between those of *Chlorokybus* or *Klebsormidium* and land plants, if they existed, should not have survived. There is also no evidence that the parenchymatous tissue organization characteristic of land plants arose directly from unicells, cell aggregates such as those of *Chlorokybus,* or unbranched filaments. In contrast, the higher charophytes (Coleochaetales and Charales) exhibit variations in the morphology of branched filaments that can be arranged in a hypothetical transformation series, which could explain the origin of more complex morphologies attained by the simplest extant land plants (Graham, 1982b) (Fig. 5-12). Charophycean morphological variations can also be correlated with the environmental variation that exists in shallow freshwater habitats, and thus most likely represent adaptations to freshwater selective pressures. In view of their potential phylogenetic importance, relationships between morphological variation occurring in higher charophytes and freshwater littoral environments are explored in greater detail.

Coleochaete species fall into three main morphological groups: branched filaments radiating horizontally (in the plane of the substrate) and vertically (ex. *C. pulvinata,* Fig. 5-13a), branched filaments radiating in the horizontal plane only (ex.

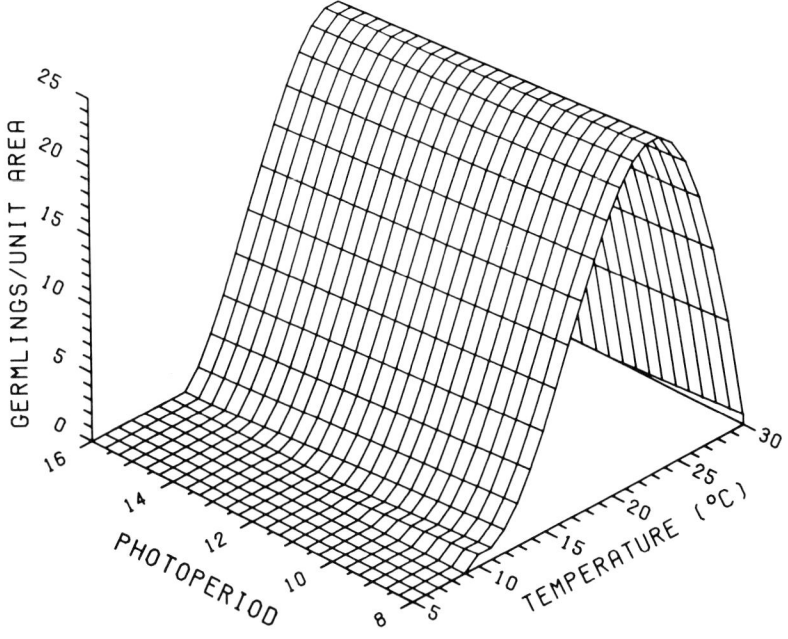

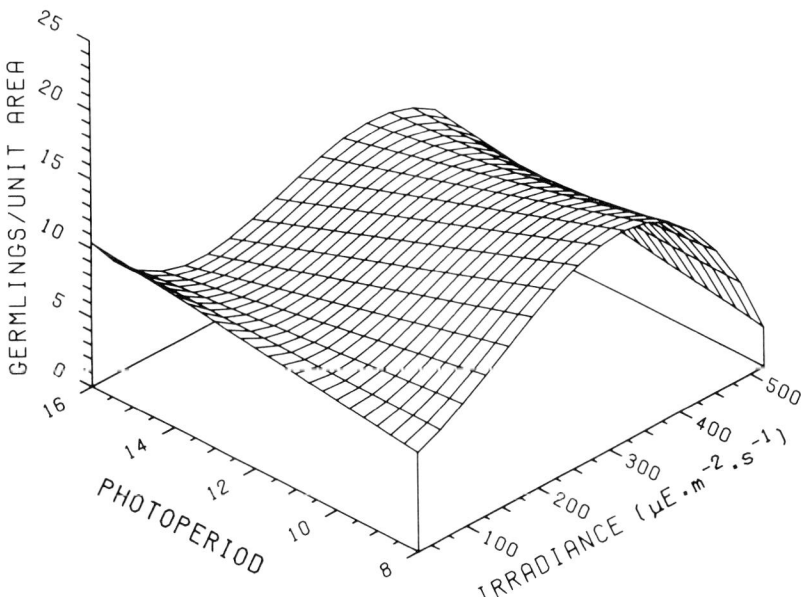

Figs. 5-8 (top) and 5-9 (bottom). Factorial experiments indicate that temperature is more important than photoperiod and irradiance in controlling asexual reproduction (zoospore production) in *Coleochaete scutata* (graphs from Graham et al., 1986, used with permission of the *Journal of Phycology*). This kind of information is useful in interpreting seasonal population behavior (phenology) of *Coleochaete* in nature, and in devising methods for isolation and growth of laboratory cultures.

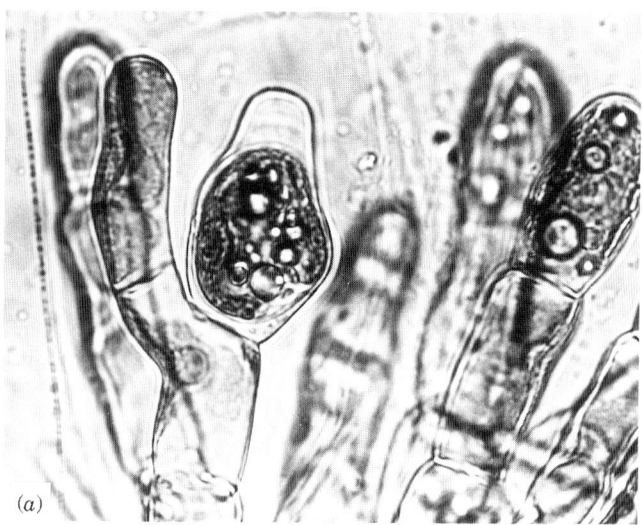

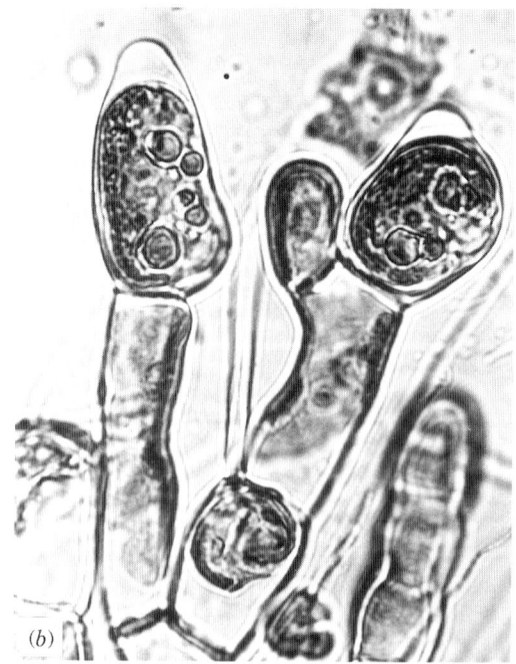

Fig. 5-10. *a, b.* Zoospores are produced from terminal cells in the attached, filamentous, branched *Coleochaete pulvinata*. Production of zoospores on branch termini is advantageous because it does not contribute to thallus disruption. However, once development of terminal cells has been diverted toward zoosporogenesis, these cells no longer function meristematically, that is, to increase thallus size.

Fig. 5-11. Zoospore production occurs throughout the body of thalloid *Coleochaete* species, such as *C. scutata,* shown here. The holes represent places where zoospores have escaped by wall hydrolysis. Because the thallus is more highly consolidated than that of related filamentous species, internal zoospore production does not result in thallus disintegration. The advantage of internal zoospore production is that peripheral growth is not limited by loss of terminal meristematic cells in the production of zoospores. Scanning electron micrograph by Sonia Cook (University of California-Davis).

C. soluta, Fig. 5-13b), and circular monostromatic thalli composed of closely adherent cells (thalloid species). Peripheral cells of the thalloid species divide parallel to the thallus periphery or perpendicular to it, increasing the thallus circumference (ex. *C. scutata* and *C. orbicularis,* Fig. 5-14). It has been suggested that the radially symmetrical species *Coleochaete pulvinata* models an ancestral form from which either horizontally branched filaments (other *Coleochaete* species) or vertically branched axes (Charales) might have been derived (Graham et al., 1991). Thalloid *Coleochaete* species and charalean algae are capable of precisely controlling division planes during gametangial formation to create relatively complex gametangia. This suggests the evolution of developmental mechanisms in charophytes, which may underlie land plant histogenetic and reproductive processes (Graham and Kaneko, 1991) and will be discussed in more detail in Chapters 6, 7, and 8.

Species of *Coleochaete* exhibiting a range of morphological variation often occur in the same habitat, allowing analysis of relationships between morphological variation and environmental factors. Such studies have indicated that morphological variation in thallus organization is correlated with microhabitat variation,

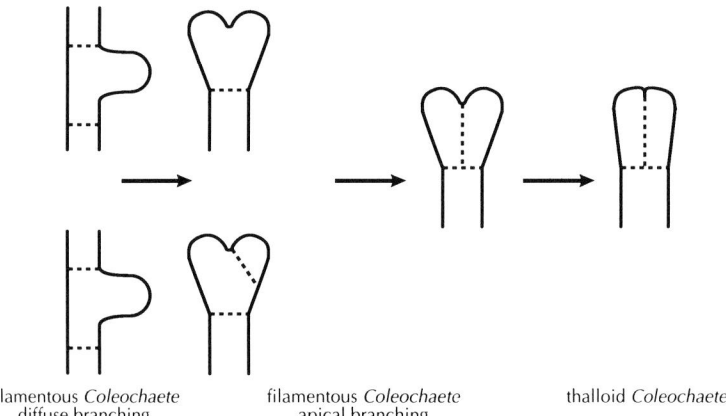

filamentous *Coleochaete* diffuse branching   filamentous *Coleochaete* apical branching   thalloid *Coleochaete*

Fig. 5-12. *Coleochaete* species can be arranged in a transformation series that illustrates how the charophycean ancestors of land plants could have made the change from branched filaments to an early form of parenchyma (tissuelike organization). This does not mean that modern thalloid forms of *Coleochaete* are necessarily homologues of the earliest plants, merely that their thallus structure may be analogous to that of early forms of plant tissue. Thalloid species of *Coleochaete* may have been influenced by the same kinds of littoral selection pressures that preadapted charophycean ancestors of plants for survival in the land habitat. Transition to thalloid morphology confers reduction in surface area/volume ratio useful in reducing desiccation stress.

and that in modern littoral environments, biotic factors (interspecific competition and herbivory) are less important than physical parameters in controlling the occurrence of *Coleochaete* (Graham et al., 1992). However, there is some evidence that the hair cells (setae) of *Coleochaete* (Fig. 5-15) (and presumably those of *Chaetosphaeridium*) function to deter herbivores such as snails (Marchant, 1977).

Thalloid species of *Coleochaete* may occur in very shallow water and in a transect study were found to have a depth optimum of 0.25 meters (the shallowest depth examined) to 0.75 meters (Fig. 5-16). In contrast, the same study revealed that filamentous species, represented by *C. pulvinata* and *C. soluta*, were more commonly present at greater depths (Fig. 5-17). The filamentous forms also prefer leaves of macrophytes as substrates (Fig. 5-18), whereas thalloid species are more abundant on inorganic substrates (glass slides as analogues of shoreline rocks) (Fig. 5-19) (Graham et al., 1992).

These results suggest that thalloid species may be particularly well adapted to the conditions of high irradiance and high turbulence of nearshore waters. Morphological variation in extant *Coleochaete* may thus illustrate the kinds of morphological transformations that occurred when charophytes invaded the terrestrial environment some 450 to 470 million years ago. The reduced surface area to volume ratio characteristic of thalloid morphology might have been useful to the earliest land plants in resisting desiccation, and ability to tolerate irradiance conditions

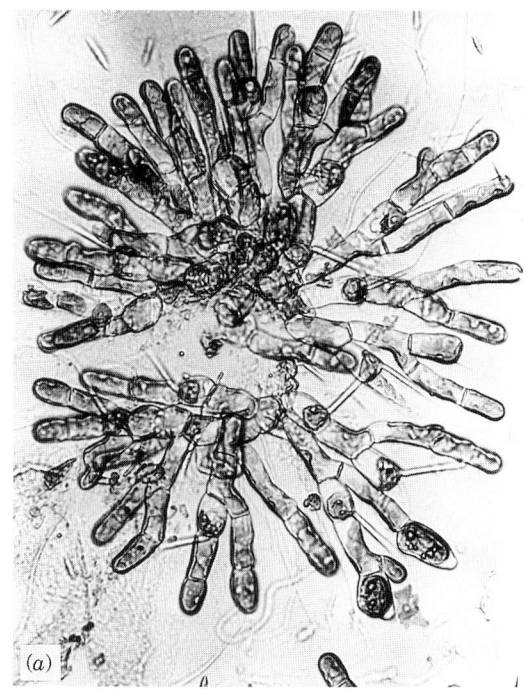

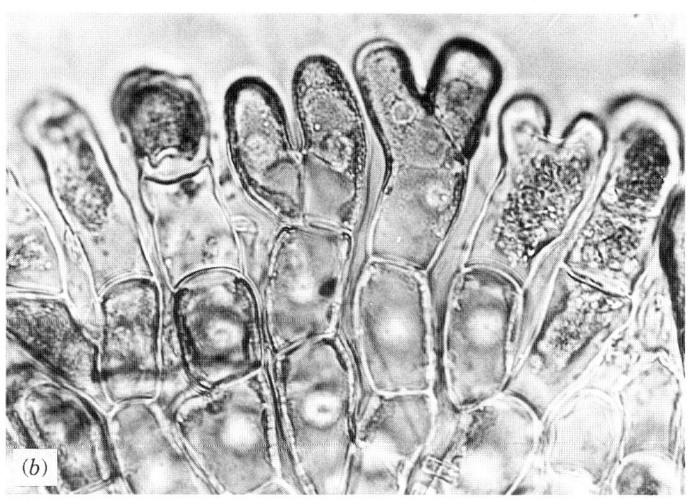

Fig. 5-13. Filamentous species of *Coleochaete* may vary considerably in morphology. *a. Coleochaete pulvinata* has highly branched, radially symmetrical thalli with both erect and flattened (prostrate) systems, which may have been similar to the morphology of the common ancestor of *Coleochaete,* Charales, and embryophytes. *b. Coleochaete soluta* resembles the flattened basal system of *C. pulvinata,* except that the angle of branching is more acute and, occasionally, cells divide radially with bisection as in thalloid species of *Coleochaete.* Morphology of *C. soluta* thus illustrates a hypothetical evolutionary intermediate between branched filaments and thalloid construction.

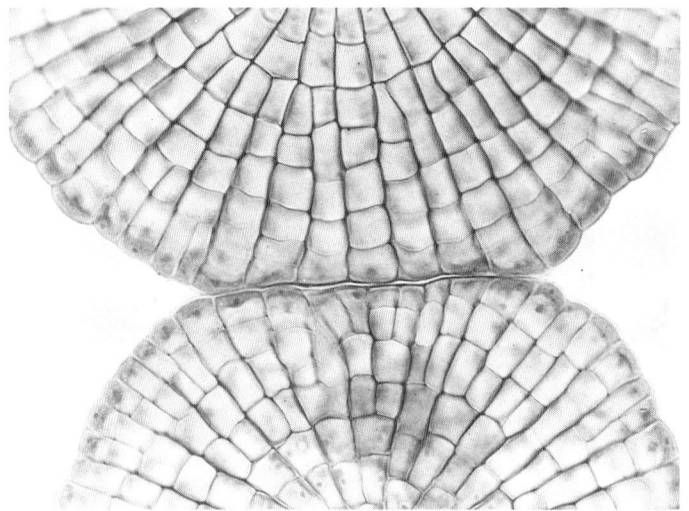

Fig. 5-14. Thalli of *Coleochaete orbicularis* represent a derived morphology within the genus. All peripheral cells may divide, extending the thallus radially in all directions, unless an obstacle is encountered, whereupon contact inhibition is exhibited (as shown here). Peripheral cells tend to divide radially if there is room to expand laterally, or circumferentially, if not. These observations suggest that mechanical constraint, or lack thereof, may be a critical determinant of meristematic capability and division polarity.

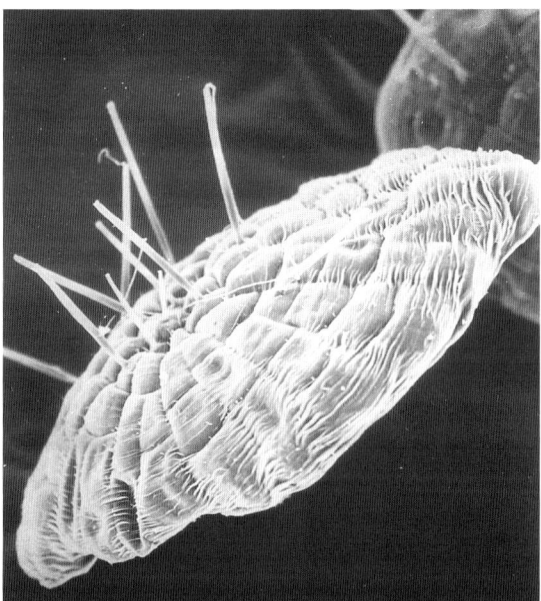

Fig. 5-15. An SEM view of *Coleochaete scutata* shows the position of hair cell sheaths from which hairs have broken off. There is some evidence to suggest that hair cells deter herbivores, such as snails, but it is possible that they also function in nutrient absorption, as is postulated for hair cells of other algae. The wrinkled appearance of the thallus results from the occurrence of dorsal "cuticular" material that may serve a protective function. Photo courtesy of Sonia Cook (University of California-Davis).

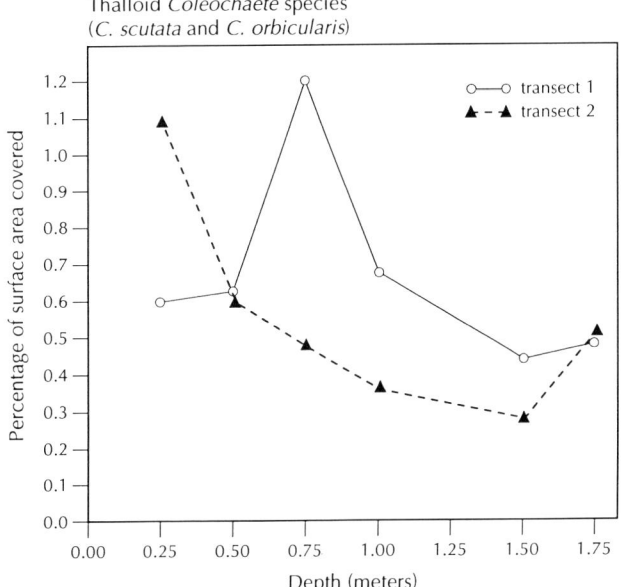

Fig. 5-16. Depth transect analysis indicates that thalloid *Coleochaete* species (*C. orbicularis* and *C. scutata*) prefer shallow water, with depth optima of 0.75 meters or less. The depth optimum differences between transects reflect greater near-shore shading at Transect 1. Eight sampling dates are averaged ($N = 8$ for each point on the graph). It is postulated that thalloid morphology confers greater resistance to drag effects of high turbulence shallow littoral habitats, allowing these forms to take advantage of higher irradiance and increased levels of dissolved carbon dioxide. Unpublished results by L. Graham, J. Graham, J. Downs, and J. Gerwing.

in shallow water might also have been useful on land. Hypotheses of morphological adaptation and preadaptation can be tested using the phylogenetic protocol recommended by Baum and Larson (1991). These authors suggest that one approach to evaluating adaptive utility of a character is to compare the success of two taxa—one having, and the other lacking, the character in question—in the same selective environment. Although filamentous and thalloid *Coleochaete* species co-occur, the putatively plesiomorphic filamentous habit seems more successful in deeper waters, whereas the evidence suggests that presumably derived thalloid morphology confers an advantage in dealing with selective pressures operating in very shallow water. The hypothesis that thalloid morphology is derived can be effectively tested when comparative gene sequence data are available for thalloid and filamentous species of *Coleochaete*. If such data indicate that thalloid species diverged later than filamentous species, the concept that thalloid morphology in *Coleochaete* is derived will have been supported.

Charalean algae primarily occupy aquatic habitats characterized by low phosphorus levels, in the pH range of 5 to 10, and prefer muddy substrates. A few

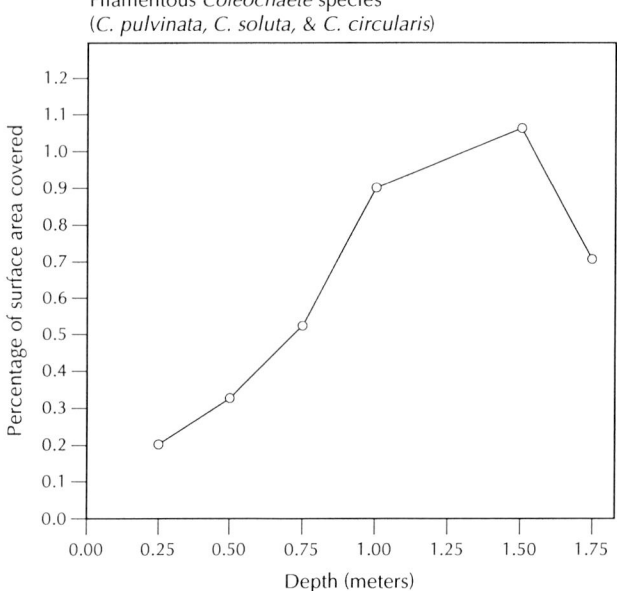

Fig. 5-17. Depth transect assessment of various co-occurring *Coleochaete* species shows that filamentous species (*C. pulvinata, C. soluta,* and *C. circularis*) have relatively greater depth optima than do thalloid species. Whether they are actually adapted to lower-light environments or are merely excluded from high-light, high-turbulence environments is not known. The data for sampling dates and duplicate transects were averaged (N = 10 for each point on the graph). Unpublished results by L. Graham, J. Graham, J. Downs, and J. Gerwing.

species are adapted to brackish water habitats (Grant, 1990), and one species is reported to grow on moist soil. Members of the Charales appear to be fundamentally branched filaments that have achieved a high degree of morphological complexity (Pickett-Heaps, 1975). Alternating large internodal cells and smaller nodal cells are generated by the division of apical cells located at the apex of a main vertical axis and the ends of branches that emerge in whorls (Fig. 5-20). Some forms are characterized by more robust thalli that are based on the growth of corticating filaments from nodal cells over the surface of internodal cells (Fig. 5-21). If zoospores were produced, their dispersal could disrupt the integrity of the main axis of uncorticated species, unless zoospore production could be developmentally restricted to branches. Loss of the capacity to produce zoospores by charalean algae could be interpreted as a necessary adjustment for uniaxial growth.

The morphology of Charales has been interpreted in terms of adaptation to the irradiance environment of the freshwater benthos (Andrews et al., 1984). Although most charalean algae grow at depths of 1 to 8 meters, they have been found in 18 meters of water in the Finger Lakes of New York state, at 30 meters depth in Japanese and German lakes (Tappan, 1980) and, amazingly, in waters more than

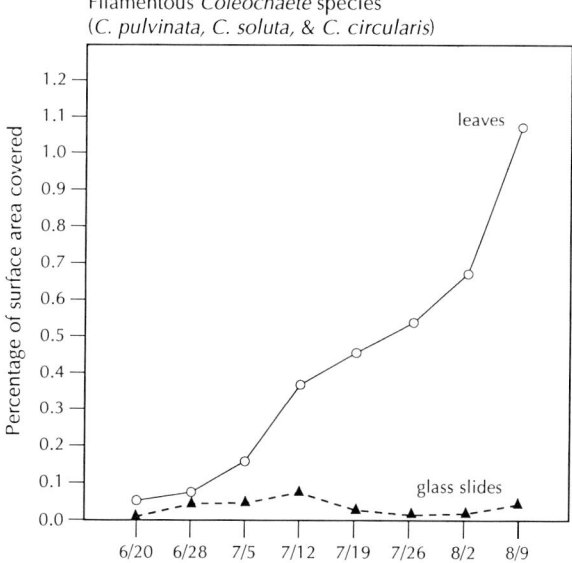

Fig. 5-18. Filamentous species of *Coleochaete* exhibit greater percentage cover on leaves of macrophytes, such as *Potomogeton,* than on inorganic substrates, such as glass slides. Data for seven depth stations and two transects have been averaged. Unpublished results by L. Graham, J. Graham, J. Downs, and J. Gerwing.

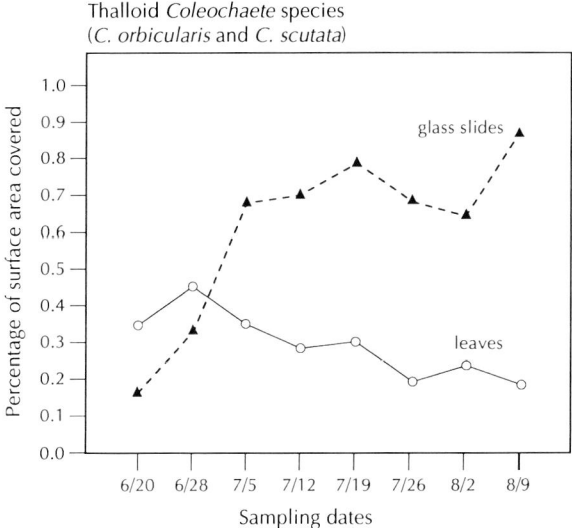

Fig. 5-19. Field observations indicate that thalloid *Coleochaete* species occur more frequently on inorganic substrates (glass slides as analogues of rock substrates), which may be more abundant in very shallow water of freshwater lakes than are organic substrates (aquatic macrophyte leaves or stems). Data for the two thalloid species were added together, and data for depth and two transects were averaged (N = 12 for each graphed data point). Unpublished results by L. Graham, J. Graham, J. Downs, and J. Gerwing.

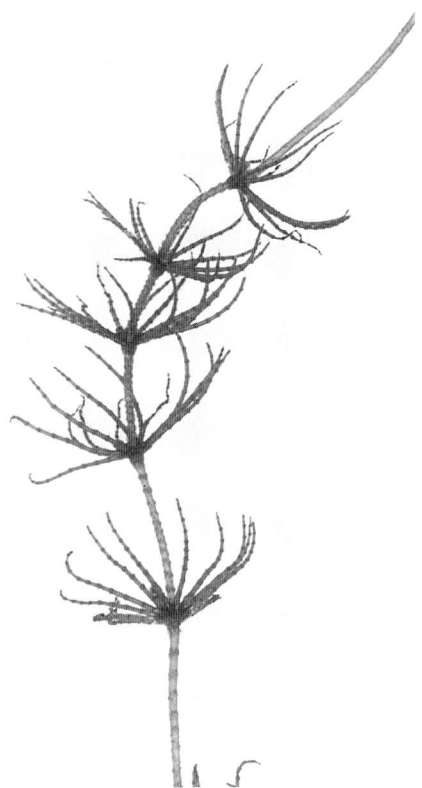

Fig. 5-20. Thalli of Charales (such as *Chara zeylanica,* shown here) are characterized by elongate internodal cells which do not divide, alternated with smaller nodal cells, which retain meristematic function. Nodal cells generate whorls of branches, which themselves are organized into alternating branched nodal and nondividing internodal regions—the whole having a somewhat fractal organization.

150 meters deep in Lake Tahoe (Frantz and Cardone, 1967). Experimental evidence suggests that giant-celled algae, such as members of the Charales, are "shade adapted" in contrast to small-celled algae, which may be either "sun" or "shade" adapted (Raven et al., 1979). In contrast to other charophytes, which have a single, platelike plastid containing a pyrenoid, the plastids of charalean algae are numerous, discoidal, and lacking pyrenoids (Fig. 5-22). It has been suggested that the evolution of discoidal plastids in charophytes is a shade adaptation for increased flexibility in light harvesting, and that occurrence of numerous discoidal plastids in both Charales and most land plants is an example of parallel evolution (Jeffrey, 1968). However, empirical evidence in support of this hypothesis is lacking.

Stebbins and Hill (1980) suggested that the charophytes that were ancestral to the relatively complex Charales and *Coleochaete* evolved in the terrestrial

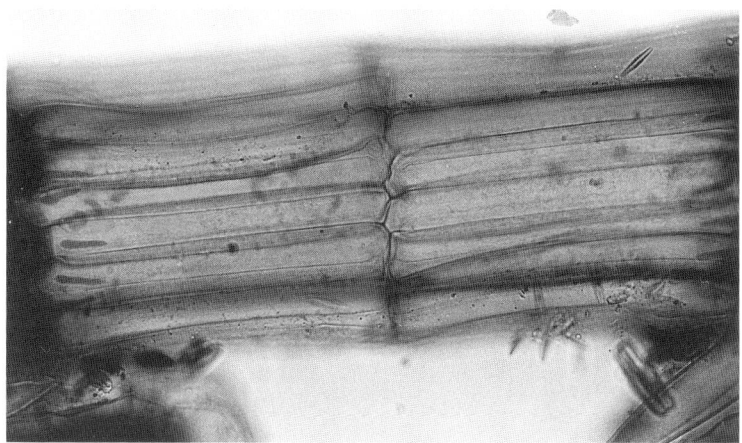

Fig. 5-21. Internodal cells of various *Chara* species are buttressed by the appressed growth of cortical cells upward and downward from each nodal region. In cross-sectional view, internodal regions resemble the pericentral organization of ceramialean red algal filaments. Other charophytes, including *Nitella,* lack cortication and consequently are relatively delicate.

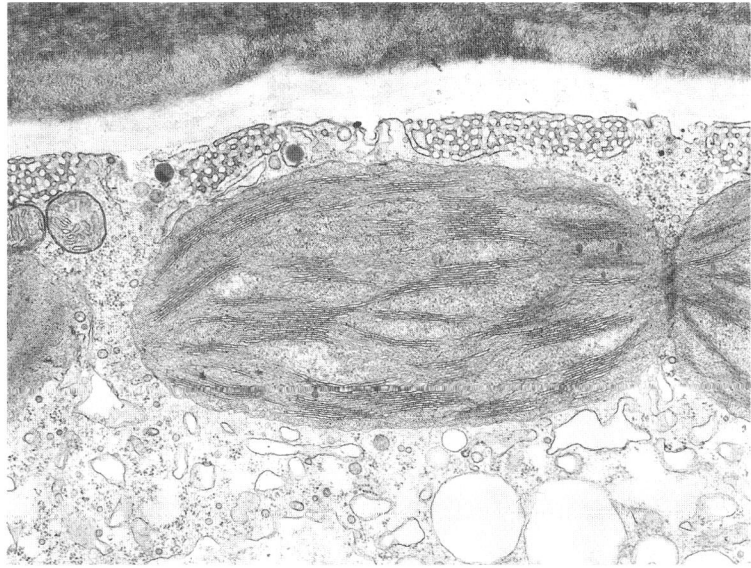

Fig. 5-22. Cells of *Chara zeylanica* (and those of other charalean algae) are polyplastidic. In discoidal shape and absence of pyrenoids, plastids of Charales resemble those of some other green algae (such as *Trentepohlia*) and most photosynthetic cells of embryophytes. Derivation of polyplastidy from ancestral monoplastidy seems to have occurred independently in charophytes and embryophytes. However, the origin of polyplastidy in algae has been poorly studied, and the adaptive value of the transition is not understood. This photograph (from Graham and Kaneko, 1991) also illustrates charasomes, highly organized membrane arrays at the cell surface (see also **Fig. 5-28**).

environment, then became secondarily adapted to fresh water. They hypothesized that the "three-dimensional" plant bodies exhibited by modern *Coleochaete* and Charales would have greater adaptive value in a terrestrial rather than an aquatic habitat. However, the complex charalean thallus appears to have resulted from acquisition of shade-tolerance adaptations useful in offshore aquatic habitats (Raven et al., 1979). Inhabitants of early terrestrial ecosystems would most likely not have been light limited, inasmuch as competitors in the form of taller vascular plants were not present. Furthermore, most species of *Coleochaete,* as well as species of *Chaetosphaeridium,* are actually quite delicate filamentous forms, suggesting that this is the plesiomorphic (primitive) condition for *Coleochaete*. Only two species exhibit more robust thalli whose morphology appears to be correlated with occurrence in shallow water, suggesting that thalloid morphology may represent adaptation to the turbulence characteristic of the lake edge. It would not be unreasonable, however, to conclude that the reduced surface area to volume ratio exhibited by thalloid *Coleochaete* species might model a kind of preadaptive morphology useful in resisting desiccation on land. Experimental evidence shows that variation in shallow freshwater environments correlates well with morphological variation in higher charophytes. Thus, evolution within the aquatic habitat represents the most parsimonious explanation for the morphological characteristics of *Coleochaete* and Charales. Proposals that parenchymatous land plants evolved directly from unicellular algae or forms consisting of short, unbranched filaments in response to terrestrial selection pressures, although theoretically reasonable, are not supported by empirical data.

PHYSIOLOGICAL ADAPTATIONS OF CHAROPHYTES

With few exceptions, charophycean algae as a group prefer low-nutrient (oligotrophic) aquatic environments. This is relevant to Beerbower's (1985) view that oligotrophic lakes were more common in Ordovician and early Silurian times than at present. Modern charophytes, however, vary substantially in optimal pH environment, with many zygnematalean algae preferring low pH habitats, charalean algae often adapted to higher pH waters, and Coleochaetales favoring more neutral conditions. These differences in pH range are probably related in large part to variations in the form of carbon dioxide taken up during photosynthesis. Equilibria governing transformation of dissolved inorganic carbon (DIC) species dictate that at low pH, DIC will primarily occur in the form of dissolved CO_2, whereas the bicarbonate ion (HCO_3^-) becomes increasingly more abundant as pH rises. Carbon acquisition adaptations are also related to photorespiration and calcification processes in green algae. Photosynthetically linked deposition of calcium carbonate in aquatic environments plays a critical role in global carbon cycling and is also a major factor in the preservation of algal fossil remains. Some authors have pointed out that aspects of carbon metabolism inherited from algal ancestors were probably of great significance in the success of the earliest land plants. Therefore, problems encountered by algae and embryophytes in carbon acquisition and retention, as well as variations in adaptive responses, are discussed in some detail.

Land plants take in carbon dioxide from the air, via stomata or other epidermal pores and moist intercellular air spaces (Raven, 1984; Espie and Coleman, 1985). CO_2 then readily enters cells and chloroplasts, where it is fixed into organic products of photosynthesis. Photosynthetic cells of all aquatic algae and submerged aquatic macrophytes are capable of using dissolved carbon dioxide in photosynthesis, and the concentration of CO_2 in water is similar to that in air. However, CO_2 is much less available in water because it diffuses 10^4 times more slowly in water than in air. In addition, the pH environment may limit the amount of dissolved carbon dioxide available for photosynthesis. Some algae utilize a pump located at the cell membrane to facilitate entry of bicarbonate ions into cells. Many aquatic algae and macrophytes use a cell wall-associated carbonic anhydrase to convert the bicarbonate ion into carbon dioxide for uptake by cells (Beardall, 1985). Such photosynthesizers are often referred to as "bicarbonate users," (Spence and Maberly, 1985) as opposed to "nonusers," which lack extracellular carbonic anhydrase or bicarbonate pumps and whose use of available carbon species is limited to dissolved CO_2. Aquatic algae and macrophytes capable of using the bicarbonate ion as a source of carbon for photosynthesis are at a great advantage in waters of pH greater than 7, because carbon availability is less likely to be limiting to photosynthesis. However, aquatic algae and macrophytes that cannot use the bicarbonate ion as a source of carbon in photosynthesis are typically limited to neutral or acidic waters.

Nonusers face a related carbon economy problem in the form of photorespiration arising from the bifunctional nature of ribulose bisphosphate carboxylase (Rubisco). This most abundant enzyme on earth can function as a reductase, catalyzing the reaction of ribulose 1,5-bisphosphate (RuBP) with CO_2 (carbon fixation). However, O_2 competes with CO_2 for the Rubisco active site, resulting in oxidative production of P-glycolate and, ultimately, loss of fixed carbon from the cell. This undesirable oxidase activity increases under conditions of low CO_2 or high O_2 levels (Tolbert et al., 1985). The enzyme Rubisco originated when the CO_2 concentration in the atmosphere was much higher than it is today and the O_2 level was much lower. Photosynthetic organisms have had to compensate for losses of fixed carbon associated with increasing photorespiration rates as the O_2 level in the atmosphere increased to present levels. Adaptations for concentrating CO_2 inside cells, including cell membrane bicarbonate pumps, production of extracellular carbonic anhydrase, and presence of C4 or CAM metabolism, characterized by the presence of the enzyme phosphoenol pyruvate carboxylase kinase (PEPCK), found in various algae and macrophytes, can reduce the effects of photorespiration. Some algae, notably certain *Chlorella* species and the green procaryote *Prochloron,* excrete glycolate from cells, thereby losing as much as 30 percent of fixed carbon. In these cases, excretion of fixed carbon may be related to the maintenance of host-symbiont relationships, inasmuch as these green phototrophs occur within invertebrate animals. Most algae and aquatic macrophytes,

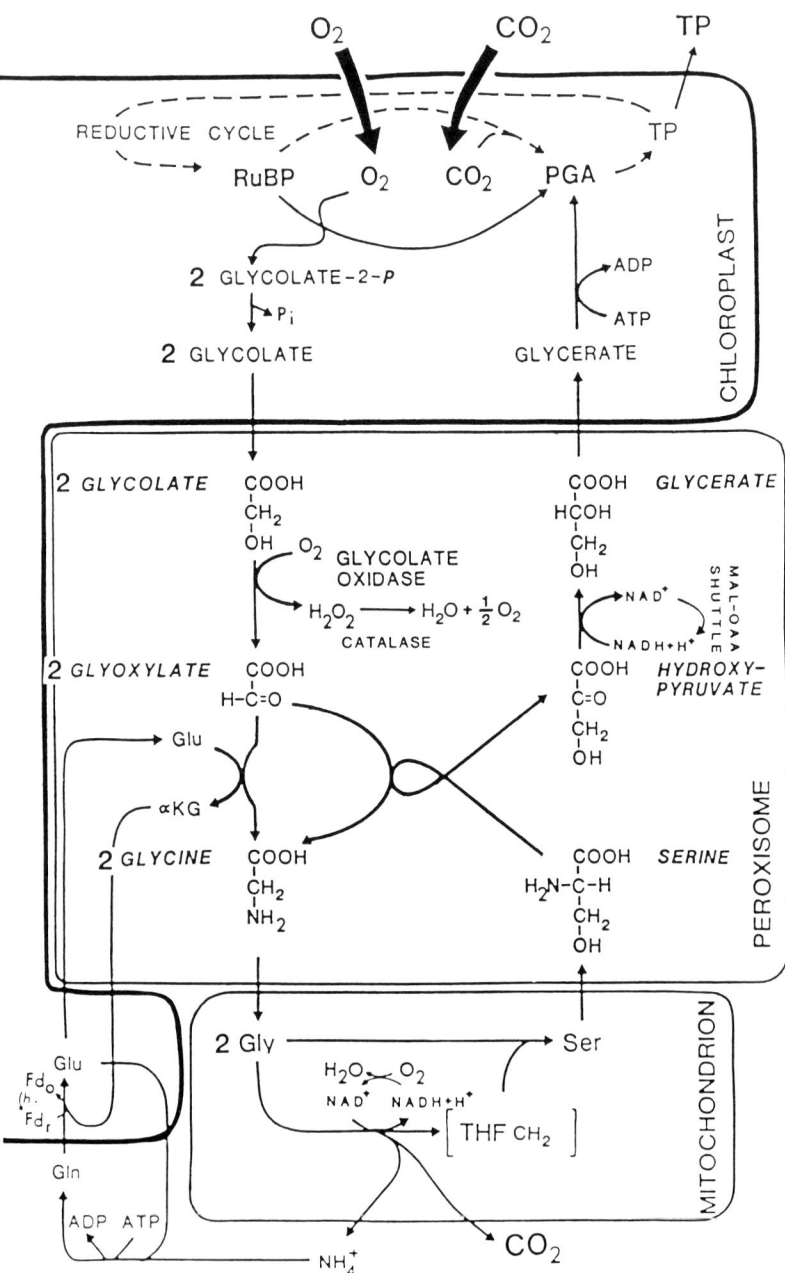

Fig. 5-23. Diagram illustrating the relationship between photosynthetic and photorespiratory pathways in plants (and probably also charophycean algae), from Zelitch (1992) copyright 1992, American Institute of Biological Sciences, used with permission of author. Note the role of peroxisomal enzymes, including glycolate oxidase and catalase, in retrieving carbon which would otherwise be lost from cells in the form of glycolate. The integral role of peroxisomal enzymes in reducing the effects of photorespiration, which may result in loss of as much as 40 percent of photosynthetically produced fixed carbon, explains the physical proximity of peroxisomes to plastids in plant and algal cells.

however, retrieve most of the carbon in glycolate by metabolizing it through the C_2-cycle, also known as the photosynthetic carbon oxidation (PCO) cycle (Tolbert et al., 1985) (Fig. 5-23). In this process, embryophytes and charophytes (and most chromophyte algae) utilize a glycolate oxidase that is coupled to molecular oxygen (Zelitch, 1992), whereas most green algae (and some other algal groups) use glycolate dehydrogenase, which does not react with molecular oxygen. For algae and macrophytes that do not accumulate CO_2, especially inhabitants of neutral and acidic waters, operation of the C_2-cycle is especially important in conserving fixed carbon. In addition, relatively CO_2-rich habitats such as the shallow, wave-agitated littoral zone and moist soil may be particularly favorable habitats for nonbicarbonate users. The neutral to low pH aquatic habitats of Coleochaetales and Zygnematales, and the terrestrial tendencies of *Chlorokybus* and Klebsormidiales, strongly suggest that these algae are nonbicarbonate users. This conclusion is also supported by the tendency of cultured *Coleochaete*, for instance, to grow most abundantly under conditions of increased turbulence, and to grow at the air-water interface (Fig. 5-24).

Monoplastidic charophytes are characterized by single, large plastids containing pyrenoids (Graham and Kaneko, 1991) (Fig. 5-25). There is considerable

Fig. 5-24. Cultures of *Coleochaete* (growing in recycled plastic soft drink bottles, with aeration) tend to exhibit significant growth at the air-water interface and above. *Coleochaete* is thus capable of growth in moisture-saturated air and does not require submersion in an aqueous medium.

immunolocalization evidence that pyrenoids are the major location of Rubisco in algae and hornworts (Vaughn et al., 1990) (Fig. 5-26). Furthermore, the colocalization of Rubisco activase indicates that pyrenoids are the active site of Rubisco activation (and hence, carbon fixation) in *Coleochaete* and other algae (McKay et al., 1991). It has been suggested that pyrenoids function to sequester Rubisco from the oxygen-evolving reactions of photosystem II, thus possibly reducing competition between CO_2 and O_2 for the Rubisco active site (Vaughn et al., 1990). This hypothesis is supported by evidence for the absence in the hornwort *Notothylas,* of PSII activity associated with the chloroplast thylakoids that typically traverse pyrenoids. It is thought that thylakoids embedded in pyrenoids might, however, provide ATP generated in PSI (McKay et al., 1991).

There is substantial evidence that charalean algae are capable of utilizing bicarbonate in addition to CO_2 in photosynthesis, explaining their ability to grow well in alkaline waters (Mimura et al., 1991). This also explains the abundant growth of charalean algae in closed, nonturbulent culture systems (Fig. 5-27). However, there is considerable debate as to whether the Charales primarily utilize cell membrane or wall-associated carbonic anhydrase to generate CO_2 that subsequently diffuses into cells, or whether they operate a proton-bicarbonate cotransport system to move the bicarbonate ion across the cell membrane (Lucas, 1985). *Chara* is known for production of elaborate systems of anastomosing membranes,

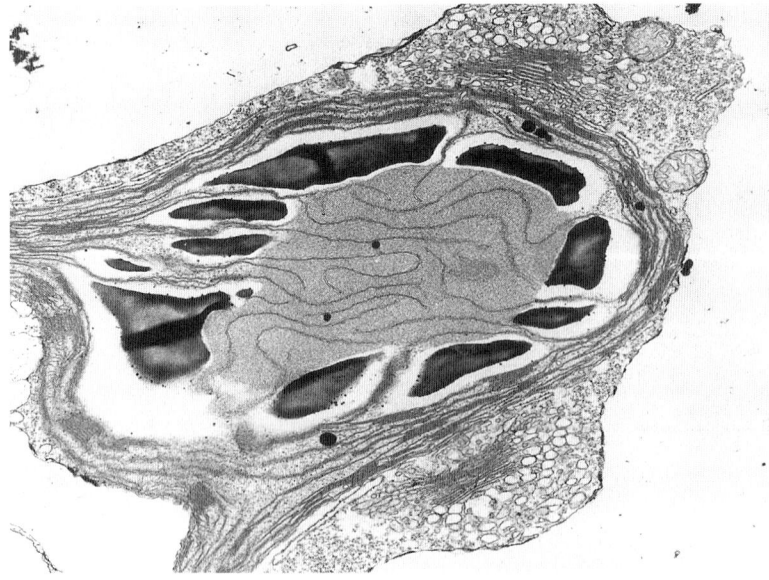

Fig. 5-25. Plastids of most charophycean algae, such as *Closterium,* shown here, contain a differentiated region known as a pyrenoid, most likely derived from the carboxysome of a cyanobacterial plastid precursor. Starch production is associated with pyrenoids of charophytes, as in other algae. Pyrenoids are traversed by thylakoid membranes. From Graham and Kaneko (1991).

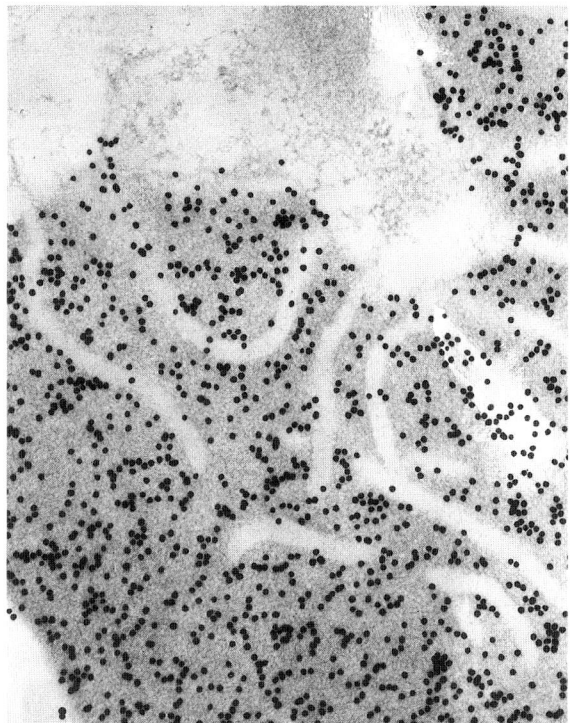

Fig. 5-26. Pyrenoids include concentrations of the enzyme Rubisco, as is demonstrated by application of Rubisco antibodies to the pyrenoid of *Coleochaete orbicularis*. Gold particles bound to a secondary antibody localize Rubisco to the pyrenoid, but not to plastid stroma or thylakoids. Note that intrapyrenoidal thylakoids are generally devoid of the label. Immunolocalization by S. Kroken, L. Graham, and W. Russin.

known as charasomes, adjacent to cell walls (Fig. 5-28). It has been proposed that charasomes are the site of carbonic anhydrase activity, but bicarbonate utilization also occurs in their absence (Lucas et al., 1989). The question could be resolved by measuring cytoplasmic levels of bicarbonate, but this has not yet been accomplished. Even if charasomes are not required for carbon uptake, extracellular carbonic anhydrase could be effective in reducing the loss of carbon at the wall region of charalean algae (Lucas, 1985). The acquisition of a carbon accumulation system may explain why charalean algae have been able to make the transition from single pyrenoid-containing plastids to the multiple plastid condition, which is associated with the loss of pyrenoids. As inhabitants of the benthos, charalean algae may also be able to make use of CO_2 that has been generated by decomposition in the sediments, as is the case with some submerged aquatic angiosperms.

Acquisition of carbon in the form of bicarbonate is related to the tendency of some charalean algae to become covered by a layer of calcium carbonate. In

Fig. 5-27. Charalean algae, including *Chara zeylanica,* shown here, can be maintained in large containers with a bottom soil layer covered by deionized water for long periods of time without supplemental aeration or addition of other exogenous inorganic carbon. This suggests that Charales are adept at acquisition of inorganic carbon in stagnant water conditions, where utilization of bicarbonate would be useful in overcoming carbon limitation of photosynthesis.

Chara the incrustation may amount to 60 percent of the dry weight (Pentecost, 1985). Calcification increases in waters of high pH and high irradiance levels, and with the age of the algae. It is also greater in still water as related to the greater extent of the unstirred boundary layers of water adjacent to cells (Smith, 1985). Not all charalean algae become calcified (*Nitella,* for example, does not), and the possible function of calcification in Charales is unknown. It has been posited that carbonate deposition may be an unavoidable consequence of photosynthesis when bicarbonate is utilized in algae that have no means of inhibiting calcification (Pentecost, 1985). This suggests that some charalean algae do not utilize bicarbonate (one explanation for the absence of calcification), or that some forms use bicarbonate while actively inhibiting calcification. Clearly, there is room for additional study of carbon acquisition by charalean algae and its relationship to calcification. Such studies might be useful in interpreting the fossil record of Charales (which is based on calcification).

In contrast to Charales, aquatic bryophytes (such as *Fontinalis* and *Drepanocladus*) and aquatic lycophytes (e.g., *Isoetes*) are not known to use bicarbonate in photosynthesis. This suggests that ancestors of early land plants may have been nonusers of bicarbonate that inhabited waters of neutral or lower pH. Carbon availability may thus have been limiting to growth unless algae inhabited

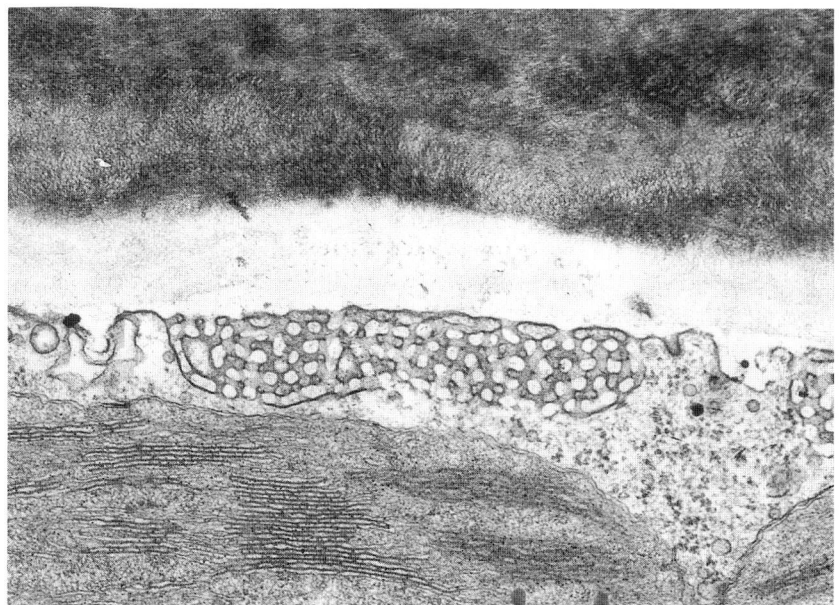

Fig. 5-28. Charasomes are densely staining, highly organized plasmalemma arrays which may occur in charalean cells. They are thought to be somehow involved in acquisition of inorganic carbon, particularly bicarbonate ions. However, the biochemical attributes of charasomes are not well understood, and their precise role has not yet been elucidated.

wave-agitated, shallow waters where boundary layer resistance to diffusion of CO_2 is reduced. Bowes (1985) has pointed out that the aquatic habitat is often depicted as benign in comparison with a harsh terrestrial environment, and although this might be true in terms of water availability, it is certainly not the case with respect to CO_2 availability. Not only is CO_2 diffusion much more rapid in terrestrial habitats than in water, but, as previously discussed, atmospheric CO_2 levels were probably higher during Ordovician and early Silurian times than at present. It is difficult to imagine why the ancestors of land plants made the transition from water to the relatively dry terrestrial environment unless the aquatic habitat was somehow limiting. Therefore, it is likely that the immediate charophycean ancestors of plants were probably nonusers of bicarbonate that benefited first from habitation of the turbulent, CO_2-rich littoral zone, then migration to an even more CO_2-rich habitat, the land. Bicarbonate use by modern Charales may represent a derived characteristic, analogous to bicarbonate usage by some aquatic macrophytes, which is also derived. The Coleochaetales, most of which are inhabitants of the high-energy shallows in waters of neutral pH, may represent good models for the ecophysiology of land plant ancestors (Fig. 5-29). One exception, *Coleochaete nitellarum,* commonly grows within the walls of characean algae, possibly parasitizing the charalean CO_2-accumulation system.

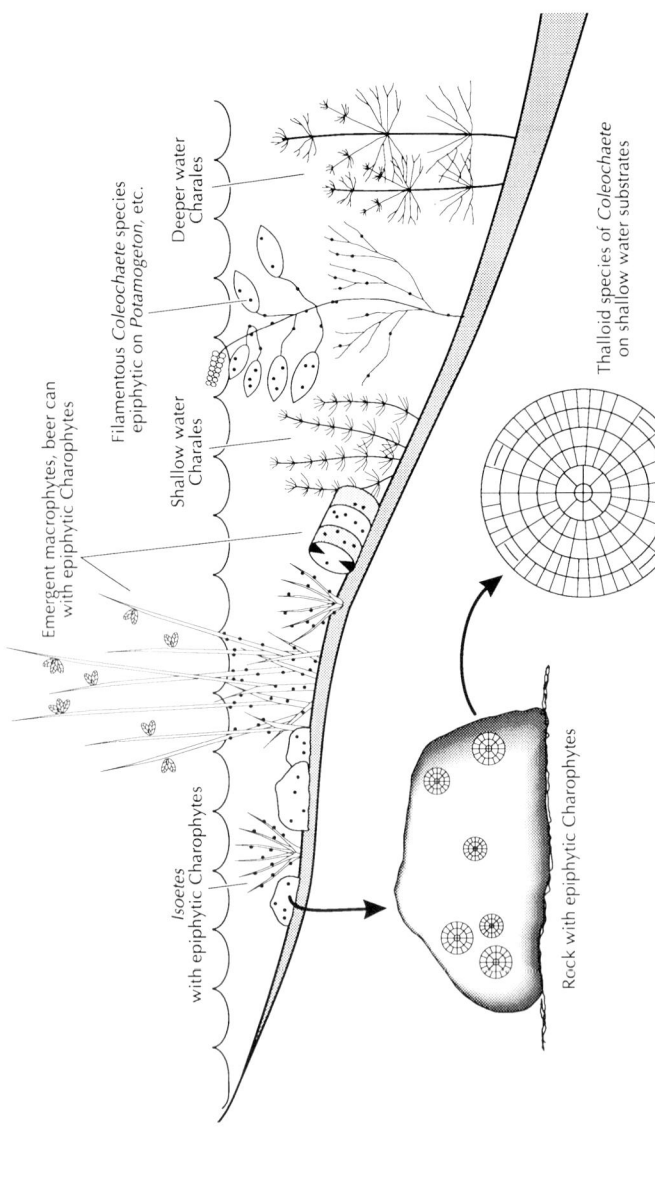

Fig. 5-29. Various strategies are used by aquatic macrophytes to obtain inorganic carbon for use in photosynthesis. Modern charophycean algae similarly use various approaches to solve carbon-limitation problems. Some, like Charales, are able to take up bicarbonate ion and thus occupy habitats that are relatively low in dissolved carbon dioxide. Alternatively, other charophytes such as thalloid, epiphytic *Coleochaete* species occupy shallow, turbulent waters where photosynthetic limitation by carbon dioxide availability (and light) is less likely to occur. The charophycean ancestors of plants might also have been faced with carbon limitation problems that could have stimulated a move to land, where carbon dioxide diffusivity in air is 10,000 times higher than in water.

For the earliest land plants, the much greater availability of carbon dioxide in air as compared with water may have outweighed the problems associated with drought. However, increased levels of O_2 in air, as compared with water, would no doubt have resulted in increased rates of photorespiration. The successful transmigrants, descendants of charophytes, possessed the advantage of glycolate oxidase. This enzyme, in contrast to glycolate dehydrogenase, has the useful property of removing molecular oxygen in the process of retrieving some of the fixed carbon that would otherwise be lost from the cell.

Photoautotrophs also appear to have evolved toward increased carboxylation efficiency and greater affinity of Rubisco for CO_2. Before the active site of Rubisco is catalytically competent, it must contain an activator CO_2 that is covalently bound to the epsilon amino group of a specific lysine residue lying in the cleft of the site. Then Mg^{2+} is complexed onto the site, which is then capable of RuBP carboxylation. These activation reactions are catalyzed by an activase (or carbamylase) (Butz and Sharkey, 1989). For algae such as *Chlamydomonas* that have a CO_2 concentrating system, CO_2 is thought to be sufficiently high for self-activation. In these forms, the activase is thought to prevent inactivation of the site by RuBP or other phosphorylated sugars. Interestingly, *Chlamydomonas* has only a single activase, whose gene is uninterrupted by noncoding sequences. In contrast, higher plants produce two nuclear-encoded activase polypeptides, whose RNA is processed, indicating the occurrence of introns. The *Chlamydomonas* activase has only 60 to 65 percent amino acid sequence similarity to the higher plant activases (McKay et al., 1991). This evidence indicates that evolutionary changes have occurred that may have affected the efficiency of Rubisco activation, but the evolutionary history of land plant activases has not been elucidated.

There is considerable evidence for evolutionary increase in the affinity of Rubisco for CO_2. *Prochloron* and other photosynthetic procaryotes have relatively inefficient Rubisco, with $Km(CO_2)$ = 80 to 330 uM (McKay and Gibbs, 1991). Other marine algal Rubisco $Km(CO_2)$s are lower, indicating higher affinity of Rubisco for CO_2. The $Km(CO_2)$ for the chromophyte *Olisthodiscus* is 45 uM, that of the rhodophyte *Griffithsia* is 41 uM, and the value for the green seaweed *Ulva* is 70 uM (Newman and Cattolico, 1990; McKay and Gibbs, 1991). In comparison, the $Km(CO_2)$ for C3 terrestrial plants ranges from 10 to 25 uM, indicating even greater Rubisco affinity for CO_2 (McKay and Gibbs, 1991). Significantly, the affinity of charalean Rubisco appears similar to that of land plants.

In higher plants and algae, heat shock proteins (chaperonins) are involved in correct assembly of Rubisco subunit aggregations consisting of eight each small and large subunits. Chaperonins are a family of unrelated proteins that mediate the correct assembly of other polypeptides but are not themselves part of the functional assembly. Interestingly, higher plant Rubisco, unlike bacterial Rubisco, fails to assemble correctly in *E. coli*. This suggests a fundamental difference between plant and bacterial chaperonin systems for assembly of Rubisco (Ellis, 1991). It would be illuminating to trace the evolutionary history of Rubisco chaperonin systems. However, no work has yet been done on chloroplast heat shock proteins of charophycean algae, and there are few comparative data for

Km(CO$_2$) in this group . Perhaps future work in this area will illuminate the phylogenetic origin of the comparatively greater carboxylation efficiency exhibited by land plants and will suggest strategies for genetic improvement.

Last, but of no small importance, it is possible that nonbicarbonate-using algae subject to carbon limitation can make use of dissolved organic carbon. Nutritional studies of axenic *Chara zeylanica* sporelings suggested that growth was not stimulated by addition of organic carbon in the form of glucose, sucrose, and a variety of other sugars (Forsberg, 1965). On the other hand, axenic cultures of the desmid *Mesotaenium* grow more abundantly when glucose is added to the growth medium (Taylor and Bonner, 1967), and several desmids that grow in media containing organic carbon cannot be cultured in commonly used inorganic algal media. This suggests that at least some charophycean algae may have hitherto unsuspected abilities to transport organic carbon into their cells (i.e., cell membrane sugar transporters), which may have implications for the evolution of mechanisms for carbon translocation in land plants.

Over 50 hexose (and other) transporters of bacteria, protists, fungi, plants, and animals are similarly characterized by 12 transmembrane-spanning alpha-helices, and therefore are regarded as homologous members of a major superfamily of transporter proteins (Silverman, 1991; Marger and Saier, 1993). These proteins include a glucose-H$^+$ symporter reported for *Arabidopsis thaliana* that shares significant sequence similarity to an inducible hexose transporter in *Chlorella kessleri* (Sauer and Tanner, 1989; Sauer et al., 1990). Genomic and full length cDNA clones of the *Arabidopsis* hexose carrier were obtained by using cDNA of the *Chlorella* H$^+$/hexose cotransporter as a hybridization probe (Sauer et al., 1990). On the basis of sequence comparisons, the *Arabidopsis* and *Chlorella* hexose transporter proteins were regarded as homologous to human and rat transporters that catalyze facilitated diffusion, but non-homologous to a mammalian Na$^+$/glucose cotransporter (Sauer and Tanner, 1989; Sauer et al., 1990). Molecular evidence strongly suggests that H$^+$/hexose transporters are ancient and conserved proteins that were inherited by eucaryotes from bacterial ancestors, and that land plants inherited hexose transport capabilities from ancestral algal protists. The capacity to move hexose sugars across cell membranes is critical to placental function in embryophytes (Renault et al., 1992) (see Chapter 8 for further discussion of this point), and may also underlie phloem and leptom loading in vascular plants and bryophytes. The latter possibility is supported by Sauer et al.'s (1990) finding that sugar transporter protein (STP1) mRNA occurs at high levels in phloem-loading leaves, but at relatively low levels in sink tissues (roots and flowers) of *Arabidopsis*. Since archaic features of plants that have been inherited from bacterial ancestors must have been filtered through charophycean algae, it would be quite surprising if H$^+$/hexose cotransporters were not also present in modern charophytes. Molecular techniques used by Sauer and Tanner (1989) and Sauer et al. (1990) suggest strategies for conducting searches for hexose transporter proteins and their genes in charophytes.

Clearly, there is need for much additional work related to the carbon metabolism of charophecean algae, and the potential for exciting results.

6

GAPS BETWEEN CHAROPHYCEAN ALGAE AND LAND PLANTS

What were the first land plants like? What were the critical adaptations that allowed charophycean algae to colonize the land? In the absence of a substantial fossil record these questions are difficult, but not impossible, to answer. Assuming that modern charophycean algae and embryophytes share a common ancestor, comparative studies of extant charophytes and land plants can shed considerable light on the transition process. Modern embryophytes share a suite of derived characters (autapomorphies) that are absent from charophytes, and in this chapter these apparent discontinuities are defined. This discussion will then serve as a basis for later development of hypotheses that attempt to explain how the gaps may have been bridged.

The comparative approach, however, is complicated by the antiquity of the divergence and the absence of transitional forms from the fossil record. As we have previously noted, the available fossil evidence suggests that land plants emerged perhaps 450 to 470 million years ago. In the intervening time, modern charophytes and "lower" embryophytes have no doubt undergone numerous changes in morphology, biochemistry, and nucleotide sequences, which could confound attempts to use comparative character analyses to deduce the transition process. In addition, variation in the rate of evolution of different characters has occurred within plant and algal groups. Stebbins (1992) notes the common occurrence of "mosaic" evolution, in which taxa exhibit some characters that have changed very little over long time periods along with other characters that are actively evolving. Among the seedless land plants, some hornworts demonstrate such character variation, having comparatively derived sporophytes, but algalike chloroplasts. A compelling example of the co-occurrence in a single taxon of "primitive" and derived traits is the presence of a derived chloroplast DNA inversion character in the morphologically and reproductively simple pteridophyte *Psilotum* (Raubeson and Jansen, 1992). Similarly, zygnematalean conjugation may represent a derived mode of sexual reproduction among charophytes, but cell division and cpDNA tRNA gene structure in these algae are clearly plesiomorphic as compared with those of *Coleochaete* and Charales. Thus, although comparisons most often involve the more advanced charophytes and the simplest land plants, it is not possible to identify a single, most advanced charophyte or most primitive embryophyte taxon for comparative study of all differences between the two taxa. For each embryophyte autapomorphy, the plesiomorphic character state will have to be sought from among variations exhibited by the hornworts, liverworts, mosses, and pteridophytes, for which there is molecular evidence of relationship (Waters et al., 1992). Examination of various charophytes will be necessary to identify possible ancestral character states for embryophyte autapomorphies. This means that even if molecular evidence identifies the extant charophyte that is most closely related to embryophytes, and identifies the modern embryophytes of earliest divergence, it cannot be assumed that these organisms accurately reflect the traits of transitional forms. An understanding of the changes that occurred in various genomes, enzyme pathways, and morphology during early plant evolution will require data from as many charophytes and seedless plants as possible.

Further complicating the comparative process are effects resulting from the human observation process. Psychologists have found that people have trouble noticing and using information related to gaps, and this creates problems in the realm of human judgment and reasoning (Hearst, 1991). A common human response to temporal or spatial absences and nonoccurrences is the urge to fill the gap, reflecting a fundamental need for closure. The presence of unexplained gaps generates a psychological state of tension, which in its positive form operates as a motivating force. However, in attempting to forge explanations for the discontinuities, people, including scientists, tend to select evidence confirming favored hypotheses and weight it more heavily than nonsupportive evidence. People tend to overestimate the number of positive correlations and overemphasize coincidences, thereby possibly coming to unwarranted conclusions and making inaccurate predictions (Hearst, 1991). In examining the evidence related to transition between charophytes and embryophytes, it is necessary to recognize such biases in order to avoid overconfidence in what may be long-held or cherished hypotheses. It is also important to emphasize approaches that allow disproof, rather than confirmation, of one or more alternative explanations.

In the past, issues related to the transition of plants to land have primarily been examined by experts in the structure and reproduction of higher plants. Such workers, including Bower (1908; 1935), have generally viewed the subject from the perspective of the more derived system. However, modern psychological studies show that people tend to recognize changes from simple to complex more readily than changes from complex to simple (Hearst, 1991). Thus, it seems appropriate to take an alternate approach, to consider the origin of complex plant features from the perspective of character state changes occurring in charophycean algae. Concepts of relationship between the emergent properties of modern charophytes and selection pressures common to their modern and ancient environments, which were addressed in the previous chapter, form the basis of this approach.

In taking the comparative approach, it is necessary to consider whether various characters present in modern charophytes are archaic, i.e., were also present in the common ancestors of charophycean algae and land plants, or whether they were more likely to have appeared after divergence of the embryophyte lineage. Adaptations acquired relatively recently are unlikely to illuminate the origin of land plants and, indeed, can confound the issues. In contrast, archaic features present in modern charophytes and the ancestors of land plants may have formed the basis for evolution of complex, derived characters useful in coping with the pressures of terrestrial life. Ancestral characteristics (preadaptations or preaptations) may be later modified and serve new functions (as exaptations) in derivative organisms (Gould and Vrba, 1982). One of the most exciting examples of the role of exaptation in plant evolution has been revealed in the recent discovery, in the gymnosperm *Ephedra,* of double fertilization and development of tissue homologous to endosperm of angiospermous plants (Friedman, 1990a, b; 1992). Previously, double fertilization and endosperm were thought to be unique and defining characteristics (autapomorphies) of the flowering plants. In angiosperms, double fertilization leads to production of nutritive endosperm, regarded as less wasteful of resources than gymnosperm prefertilization storage of nutrients in female gametophytes, whose eggs may not be fertilized. Double fertilization in the common ancestor of *Ephedra* and angiosperms may well have served as the preadaptation allowing endosperm evolution in flowering plants. This evidence leads us closer to solution of Darwin's "abominable mystery," the origin of flowering plants. Similarly, seemingly unbridgeable gaps between algae and land plants are closing as various charophyte features are revealed to be ancestral characters upon which fundamental structures and processes of land plants have been built.

MOLECULAR GAPS

Comparison of chloroplast DNAs (cpDNA) from more than 250 embryophyte species, including bryophytes (Ohyama et al., 1986), ferns, gymnosperms, and angiosperms, indicates that the land plant chloroplast genome architecture is highly conserved (Palmer, 1985). In general, land plant plastid genomes consist of homogeneous circles about 150 kilobases (kb) in size. As is also true of various algal plastid genomes, the DNA molecules are divided into small and large coding regions by an inverted repeat. The consistency of chloroplast genome structure among embryophyte taxa is evidence for monophyletic origin of the land plant chloroplast genome and the embryophytes. Alterations in this highly conserved genome reflect significant phylogenetic divergence. For example, Raubeson and Jansen (1992) showed that a 30 kb pair cpDNA inversion characterizes all vascular plant groups except lycophytes, which share cpDNA gene order with bryophytes.

In contrast to the uniformity of chloroplast genome size exhibited by embryophytes, the chloroplast genomes of charophycean algae reveal considerable size diversity, most likely reflecting the antiquity of the group. The cpDNA of the zygnematalean forms *Spirogyra* and *Sirogonium* are about 130 kb in size, whereas that of *Chara* is 160 to 180 kb. *Nitella*'s relatively large chloroplast genome (350 kb) is presumably derived. *Coleochaete*'s cpDNA, at 100 kb, is the smallest so far encountered among charophytes and is partially explained by loss of the inverted repeat region (Manhart and Palmer, 1988). It is possible that *Coleochaete*'s inverted repeat was lost after divergence of embryophytes from charophytes, inasmuch as similar loss has occurred occasionally among seed plants (pine and eight genera of Leguminosae), as well as other algae (i.e., the seaweed *Codium*) (Manhart et al., 1989). Occurrence of an inverted repeat region coding rRNA genes in plastids of chromophyte and rhodophyte algae (Newman and Cattolico, 1990) and in the cyanelles of *Cyanophora* (Trench, 1991), suggests that the repeat is ancient; however, the reason for its occasional loss is unknown. Because mitochondrial genomes of charophycean algae have not, as yet, been investigated, it is not known whether discontinuities occur between charophytes and embryophytes with respect to mtDNA.

BIOCHEMICAL GAPS

So far as is known, there do not seem to be significant differences between charophycean algae and embryophytes in terms of primary metabolism. As previously discussed (see Chapter 3), charophytes are more similar to land plants than other algae with regard to enzymes associated with photorespiration (glycolate oxidase and Cu/Zn SOD). However, substantial gaps may occur between charophytes and land plants with respect to the enzymes and products of secondary metabolism. Most land plants produce the complex compounds flavonoids, cutin and/or cutan, and lignin. Although there are reports of the occurrence of flavonoids, lignin, and ligninlike compounds (defined on the basis of some functional and chemical similarities to lignin) in charophytes, unambiguous chemical evidence for the production of these materials by charophycean algae is lacking. Further, development and reproduction of land plants (including bryophytes) are commonly regulated by plant growth substances or hormones. There is some evidence for the occurrence of a cytokinin in *Chara* (Zhang et al., 1989), but plant hormones are otherwise not known to occur in charophytes. This situation most likely represents lack of investigation rather than demonstrated absence of plant growth substances from charophytes.

MORPHOLOGICAL GAPS

Because comparative approaches have been applied to studies of plant and algal structure for well over a century, discontinuities that occur in structure at the interface between charophytes and embryophytes are better understood than differences related to genome architecture and enzyme systems of secondary metabolism. In the rest of this chapter, an overview of these morphological discontinuities is provided as a basis for subsequent discussions incorporating hypotheses of transition from the charophycean condition to more derived embryophyte structures. Structural differences separating seedless embryophytes from charophytes can occur in vegetative organization or reproductive morphology.

Differences in Vegetative Structure

Probably the most conspicuous difference between charophycean algae, on one hand, and even the simplest of embryophytes, on the other, occurs in thallus (plant body) organization and development. Most advanced charophytes are filamentous, and even the complex Charales increase in length by division of an apical cell having a single cutting face. In contrast, most embryophytes are composed of three-dimensional tissues (parenchyma) generated by apical meristematic cells that cut off derivatives on more than one face. The occurrence of a wedge-shaped apical cell with four cutting faces is probably a synapomorphy for hornworts and liverworts (Mishler and Churchill, 1985) and most likely represents the primitive condition among extant embryophytes. The evolution of parenchyma, which results in reduced surface area/volume ratio, has often been cited as a critical adaptation for moisture retention in terrestrial habitats. In contrast, the greater surface area/volume ratios provided by the filamentous organization of many charophytes may provide adaptive advantage in nutrient absorption, especially in aquatic habitats where nutrients, including carbon, may be limiting to growth.

Most mosses have an algalike filamentous stage, the protonema (Fig. 6-1) that grows by division of an apical cell in one direction (tip growth). The moss protonema is regarded as an efficient means by which a germinated spore can capture an ecological niche (Stebbins, 1992). Immunofluorescence studies of the developmental transition from filamentous to parenchymatous structure in mosses (protonema to "leafy" shoots) have yielded important insight into the cytoskeletal basis of this change in morphology. For example, preprophase microtubule bands (PPBs), which commonly occur in meristematic tissues of plants, are absent from charophycean algae and usually moss protonemata as well. However, PPBs appear during the switch from tip growth to intercalary growth in the production of the "leafy" stage of the moss gametophyte (Doonan et al., 1987). This evidence has led to the suggestion that evolution of PPBs was closely associated with the evolution of parenchymatous growth (Brown and Lemmon, 1990c), which is discussed in more detail in Chapter 7.

Vegetative (i.e., nonreproductive) cells of embryophytes also differ from those of many charophytes in lacking centrioles. Among charophycean algae, species that produce flagellate zoospores typically exhibit centrioles in vegetative cells, but centrioles are lacking from vegetative cells of taxa that do not produce zoospores (Pickett-Heaps, 1975). Conversely, cortical microtubule arrays are lacking from the former group, but present in the latter. Zoospore-producing charophytes such as *Klebsormidium* and *Coleochaete* produce pointed mitotic spindles that are focused on centrioles, and have astral microtubules (Fig. 6-2). Mitotic spindles in dividing cells of charophytes that lack zoospores may be either spindle-shaped with astral microtubules, as in *Mougeotia* (Galway and

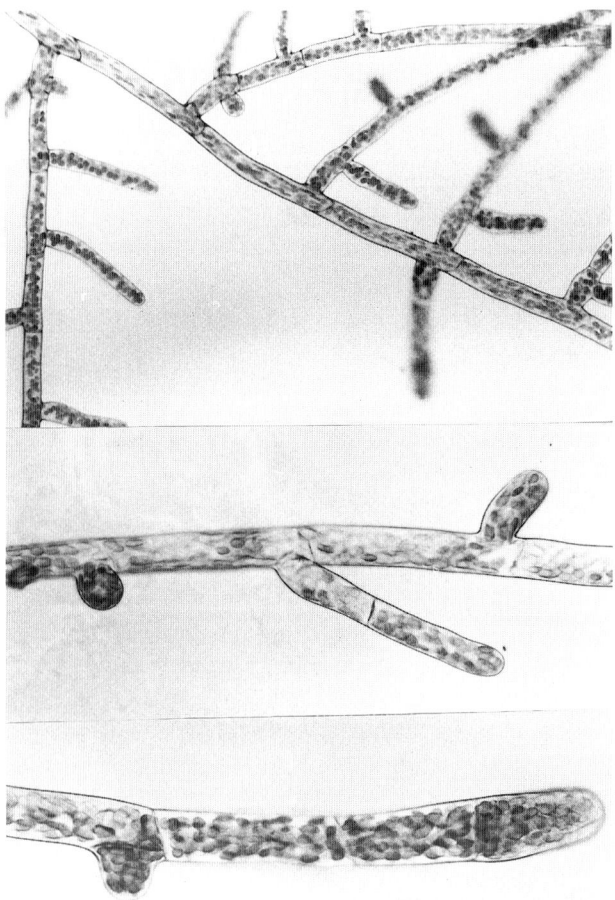

Fig. 6-1. Moss spores typically germinate to form an extensive, algalike, filamentous protonema that exhibits tip growth. Successively higher magnifications of a typical protonema are shown here. This type of protonema is probably derived from ancestral protonemal types that were much less extensively filamentous, because the latter are characteristic of basally placed bryophyte taxa. The protonemal stage of moss development is an adaptation conferring great increase in surface area for absorption of water and inorganic nutrients, which also allows production of multiple "leafy" gametophytic axes from "buds." Elaboration of moss protonemata is analogous to growth of fungal hyphae, which are similarly adapted for maximal nutrient absorption, and which also exhibit tip growth. Fern spores may also produce a filamentous protonemal stage prior to achieving two-dimensional gametophyte growth. Mosses undergo developmental transition from (a type of) mitosis which does not involve production of preprophase microtubule bands (as do charophyte algae) in the protonemata stage to cell division involving preprophase bands (as in most plant meristems) in "buds." In contrast, preprophase bands are present in both unidimensional and bidimensional gametophyte growth of ferns. Thus, it is not clear whether cytoskeletal changes occurring in moss development illustrate ontogenetic recapitulation of phylogenetic history, wherein histogenesis is somehow tied to appearance of preprophase bands, or mitotic specialization related to extensive tip growth.

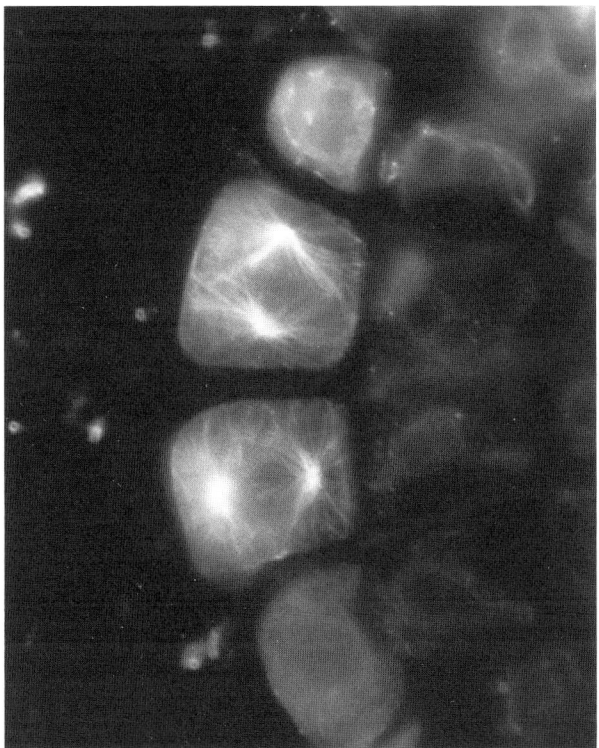

Fig. 6-2. Indirect immunofluorescence localization of microtubules in peripheral dividing cells of *Coleochaete orbicularis* (from Brown and Lemmon [1990], *American Journal of Botany*). The cell below will divide parallel to the thallus circumference, whereas the adjacent cell (above) will divide radially. Centrioles have undergone replication and migrated to polar locations, predicting the plane of cytokinesis. The cytokinetic plane is also predicted by changes in plastid location. Preprophase bands are absent, and microtubules radiating from centrioles will generate the spindle apparatus. Photos generously supplied by R. Brown and B. Lemmon (Southwestern Louisiana State University). Used with permission of the *American Journal of Botany*.

Hardham, 1991), or barrel-shaped and lacking astral microtubules, like those of most land plants, as in Charales. This evidence suggests that microtubule-organizing centers (MTOCs) are localized with centrioles (or the region formerly occupied by centrioles) in some charophytes, as is typical of protists, but when centrioles are absent, MTOC material may be otherwise distributed, contributing to formation of cortical interphase arrays and diffuse spindle poles (Graham and Kaneko, 1991). Interestingly, mitotic spindle poles of certain liverworts are not only pointed during early development, but also exhibit astral microtubules emanating from spherical structures, one at each pole, that do not stain for tubulin (Fig. 6-3). During prometaphase the polar structures and astral microtubules disappear, and the spindle becomes barrel-shaped. These data suggest that the early spindle MTOC is still localized to the site formerly occupied by centrioles in

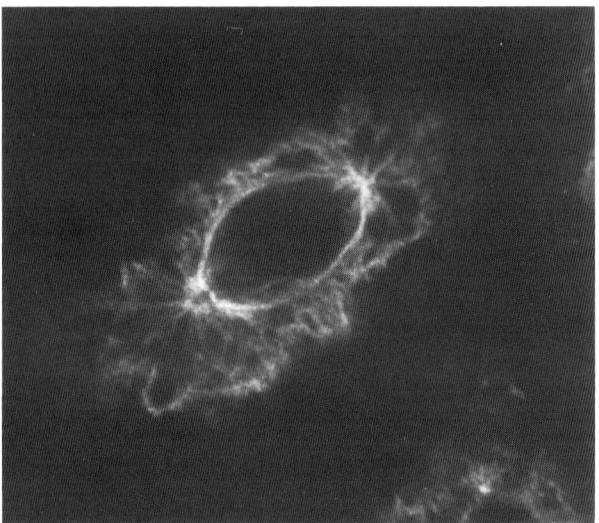

Fig. 6-3. A preprophase band predicts the plane of cytokinesis in dividing cells of the liverwort *Reboulia*. Here, however, in contrast to most other land plants, microtubules also radiate from distinct noncentriolar polar organizers in the same way that microtubules are associated with nuclear poles of *Coleochaete* and other protists. This suggests that the preprophase cytoskeleton of some liverworts represents an evolutionary intermediate between those of charophytes and other embryophytes. Photo generously provided by R. Brown and B. Lemmon was previously published in the *American Journal of Botany* (1990) and is used with permission.

ancestral forms, and that liverwort mitosis is transitional between charophytes and higher plants (Brown and Lemmon, 1990a; 1992). This in turn implies that the charophycean ancestors of liverworts had pointed centric spindles (as in *Coleochaete*), rather than barrel-shaped, acentric spindles as in Charales.

Another difference between charophytes and embryophytes that has attracted much attention is the common (but not exclusive) occurrence of single plastids in cells of algae, whereas most (but not all) land plant cells have multiple plastids (Fig. 6-4). Recent studies have revealed that monoplastidy actually occurs more widely in land plants than is generally recognized and most likely represents the plastid status of ancestral embryophytes (Brown and Lemmon, 1990c). Most hornworts, for example, are monoplastidic throughout the life cycle (Fig. 6-5). Although certain hornworts may exhibit polyplastidy, in these taxa the apical cell, very young cells, and cells of the sporophyte basal meristem have only one plastid (Renzaglia, 1978). Monoplastidic mitosis is also reported for the moss *Timmiella* (Gambardella and Alfano, 1990). In addition, dividing cells of the otherwise polyplastidic pteridophytes *Isoetes* and *Selaginella* are typically monoplastidic, as are cells leading to the production of spores and sperm of mosses. Monoplastidic cells of other seedless plants may also occur in regions undergoing cell division, but are less well studied (Brown and Lemmon, 1990c). The frequent association

MORPHOLOGICAL GAPS 123

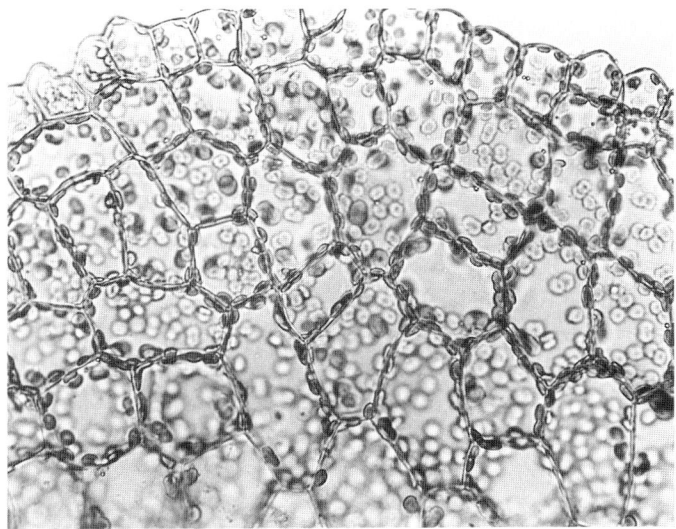

Fig. 6-4. Cells of the liverwort *Blasia*, like those of most embryophytes, are polyplastidic and lack pyrenoids.

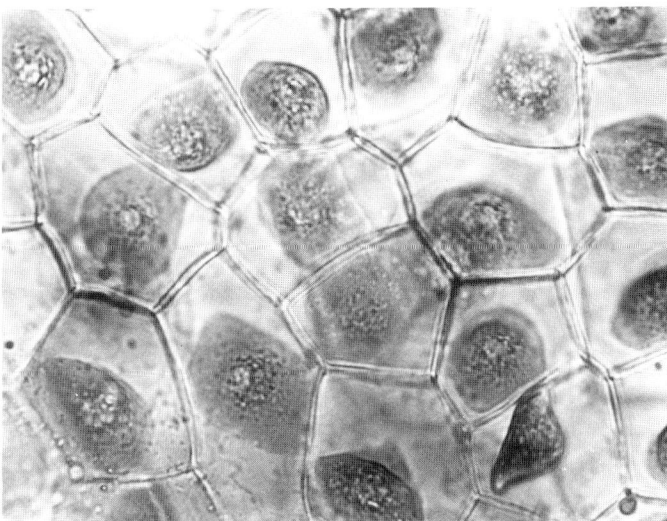

Fig. 6-5. Cells of several hornwort genera (including *Anthoceros*, shown here) are typically monoplastidic. Furthermore, meristematic and sporogenous tissues of various bryophytes and pteridophytes exhibit monoplastidy. This strongly suggests that monoplastidy is plesiomorphic for embryophytes, and that embryophytes are descended monophyletically from a monoplastidic charophycean ancestor.

of monoplastidy with regions of cell division in seedless plants is strong evidence that plastids played an important role in cell division in early land plants. Immunofluorescence studies of bryophytes and pteridophytes have revealed that plastids influence the polarity (or plane) of cell division as a result of predictive plastid migration (Brown and Lemmon, 1990). There is evidence that *Coleochaete* also exhibits predictive plastid migration in mitosis (Brown et al., 1993) and meiosis (Graham, 1990); therefore, similar studies of monoplastidic charophytes may illuminate the origin of predictive chloroplast migration and other important plant developmental processes.

In hornworts and charophytes monoplastidy is generally associated with the presence of pyrenoids in plastids. An exception to this rule is the absence of pyrenoids from two monoplastidic hornworts (Renzaglia, 1978). Immunogold localizations of Rubisco and Rubisco activase in pyrenoids of hornworts (Vaughn et al., 1990) and charophytes, represented by *Coleochaete* (McKay et al., 1991), support the hypothesis that charophyte and hornwort pyrenoids are homologous. As previously noted in Chapter 5, aggregation of Rubisco into pyrenoids may reduce photorespiration by physically separating carbon fixation from the oxygen-producing system of photosynthesis. Immunogold labeling has established that Rubisco is located in the stroma of most land plant plastids, including those of pyrenoidless hornworts (Vaughn et al., 1990). It has been proposed that loss of pyrenoids is associated with the transition from monoplastidic to polyplastidic cells, and that localization of Rubisco to the stroma facilitated chloroplast division (McKay and Gibbs, 1991). However, nothing is known about the system by which Rubisco is targeted to pyrenoid regions in charophytes and hornworts or how the targeting mechanism changed during the diversification of early plants.

Reproductive Differences

Most botanical texts emphasize the presence in all land plants of multicellular sporophytes (embryos, or sporic meiosis) and meiospores coated with protective walls containing sporopollenin. In addition, seedless plants characteristically exhibit "jacketed gametangia," known as archegonia and antheridia (Figs. 6-6 and 6-7), which are believed to function in protection of fragile egg and sperm cells. Embryos, walled meiospores, and specialized gametangia are all regarded as adaptations that facilitate reproduction in the comparatively dry terrestrial habitat. Charophycean algae lack multicellular sporophytes, walled spores, and "jacketed" gametangia, but possess features that are likely to represent ancestral character states.

The multicellular sporophyte generation is commonly regarded as an adaptation that allows plants to amplify the genetic benefits of sexual reproduction. In terrestrial environments, fertilization rates of seedless plants are thought to be limited by the availability of water for transport of flagellate sperm to eggs. The potential for low fertilization rates on land could be offset by the production of many meiotic products (spores) by multicellular sporophytes resulting from successful fertilizations (Searles, 1980). If genetic recombination resulting from outcrossing or crossing-over occurs, the spore products of larger sporophytes

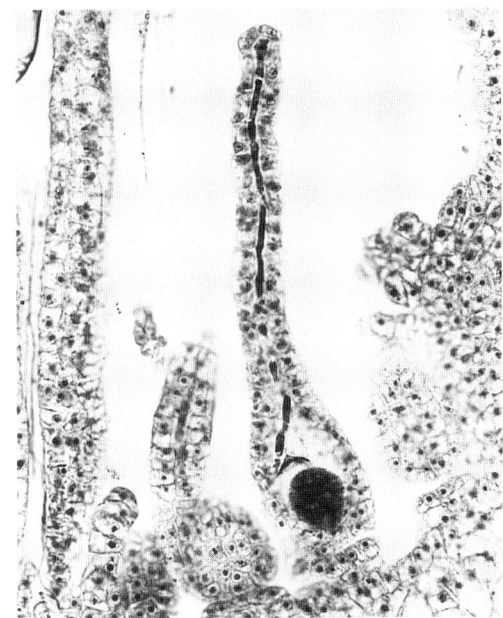

Fig. 6-6. Archegonia of the liverwort *Marchantia* are stalked structures in which the egg cell is protected by a jacket of nonreproductive tissues. These illustrate the stereotypical view of "jacketed" land plant gametangia, widely touted as essential to colonization of land. However, they may actually exhibit derived features that do not accurately reflect the structure of the earliest land plant gametangia.

confer increased genetic diversity. In addition, there is a direct relationship between sporophyte size and the number of spores that can be produced as a result of a single fertilization event. This means that larger sporophytes provide increased dispersal potential. Amplification of spore number and genetic diversity would have been extremely advantageous to the earliest land plants. The origin of the embryo and multicellular sporophyte must have been a critical step in the evolution of land plants.

In all extant land plants, growth and development of the sporophyte generation always begin while the zygote and young embryo are in close association with parental gametophyte tissues. This association continues throughout the life of the sporophytes of mosses, liverworts, and hornworts, but sporophytes of pteridophytes and seed plants become independent of parental gametophytes later in life. Although the prolonged association of sporophyte and gametophyte observed in bryophytes has sometimes been interpreted as derivative, cladistic analyses of structural and molecular characters (Waters et al., 1992; Raubeson and Jansen, 1992) suggest that this relationship probably represents the ancestral condition for embryophytes (Mishler and Churchill, 1985). Comparisons of sporophytes of extant seedless plants has led to the suggestion that liverworts were the earliest

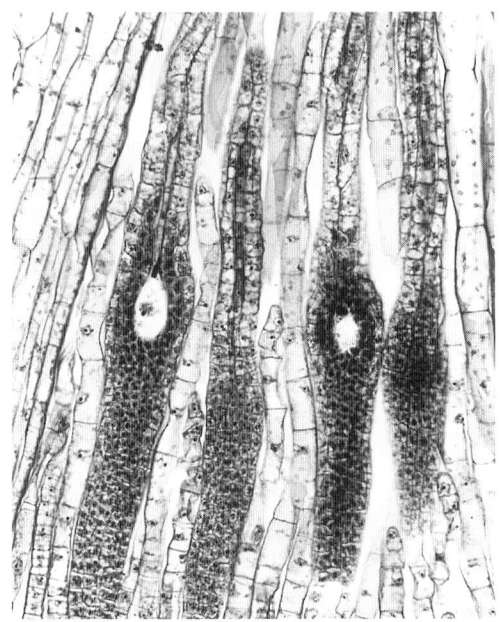

Fig. 6-7. Archegonia of the moss *Funaria* are also emergent, stalked structures with extremely elongated necks and distinct cellular jackets. Gametangia exhibiting this level of complexity are unknown among the green algae, and the nature of transitional morphologies is poorly understood.

diverging group of embryophytes (Mishler and Churchill, 1985), and this conclusion is supported by rRNA sequence analysis (Waters et al., 1992). Liverworts are characterized by extremely small, globose, and poorly differentiated sporophytes (Fig. 6-8) that may represent the plesiomorphic condition. Sporophytes of hornworts, mosses, and pteridophytes are generally regarded as more derived. Mishler and Churchill (1985) suggested that among liverworts, the Sphaerocarpales (*Sphaerocarpos* and *Riella*) may have been one of the earliest diverging groups, on the basis of relatively undifferentiated sporophytes and simple elater structure. Liverwort elaters are sterile cells that are believed to function in spore dispersal. Elaters of most liverworts are elongated cells with hygroscopic, spiral wall thickenings (Fig. 6-9), but homologous cells of Sphaerocarpales are small, green, and lacking wall thickenings, and presumably have reduced spore dispersal effectiveness. This character state difference is a very weak basis upon which to identify sporophytes of Sphaerocarpales as the least derived among those of modern liverworts. It is, of course, possible that sporophytes of liverworts in general, and Sphaerocarpales in particular, may actually represent the results of reduction, as argued by Schuster (1984a). Additional comparative developmental and molecular studies of liverworts are necessary in order to develop a concept of the primitive land plant sporophyte.

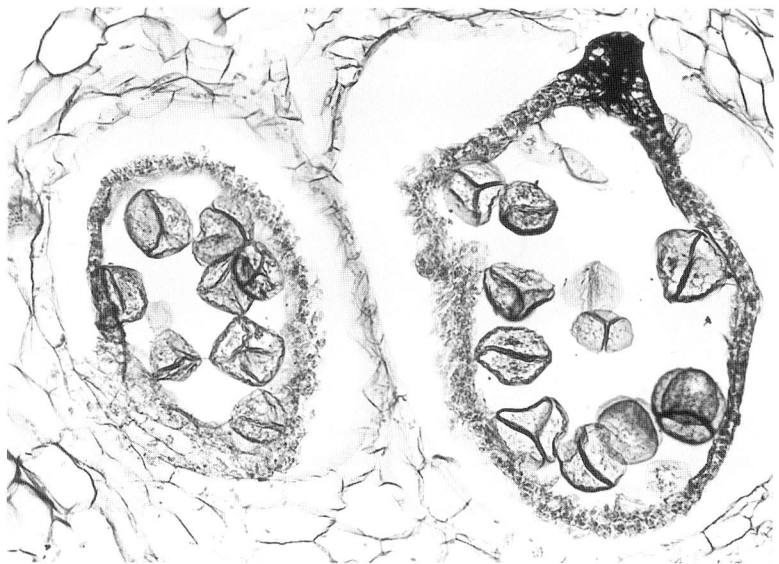

Fig. 6-8. Liverwort sporophytes are poorly differentiated, primarily consisting of sporogenous cells, or trilete spores enclosed within a sporangial wall. Whether such simple sporophytes illustrate the structure of the earliest sporophytes, or are reduced from more complex precursors, is controversial.

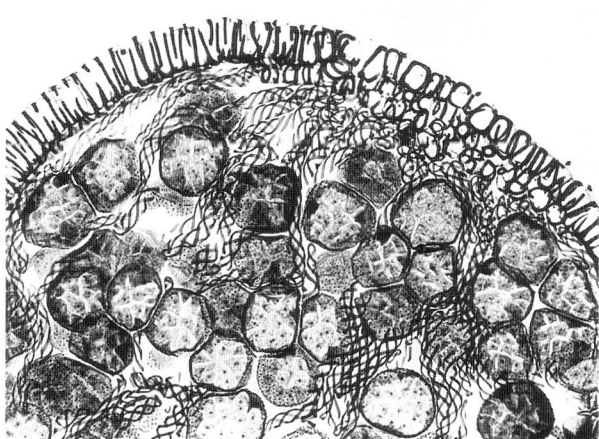

Fig. 6-9. Liverwort elaters, conspicuous in the sporangium of *Conocephalum*, are characterized by spiral wall thickenings and hygroscopic function. Changes in relative humidity cause elaters to undergo rapid shape changes that aid in spore dispersal. Note also the spiral thickenings in sporangial cell walls.

The relative systematic position of hornworts is also relevant to our discussion of gaps between charophytes and embryophytes, because some botanists have suggested that hornworts and *Coleochaete* may be closely related (Sluiman, 1985). This view is primarily based upon the presence of single large plastids with pyrenoids in cells of all species of *Coleochaete* and many species of hornworts. The sporophytes of hornworts, however, are generally regarded as being rather derived, on the basis of the presence of an unusual basal meristem, as well as a specialized spore dispersal mechanism involving dehiscence and twisting of the capsule along two or more suture lines. In addition, hornworts produce "pseudoelaters" that may function in spore disperal but are not considered to be homologous to the elaters of liverworts (Fig. 6-10). Among hornworts, the genus *Notothylas* exhibits several sporophytic characters that have been interpreted as plesiomorphic and, thus, possibly representative of an ancestral type.

Putatively primitive aspects of *Notothylas* sporophytes include its smaller size, as compared with other hornwort genera, determinate growth and synchronous spore production, a very small or absent columella (a central column of sterile

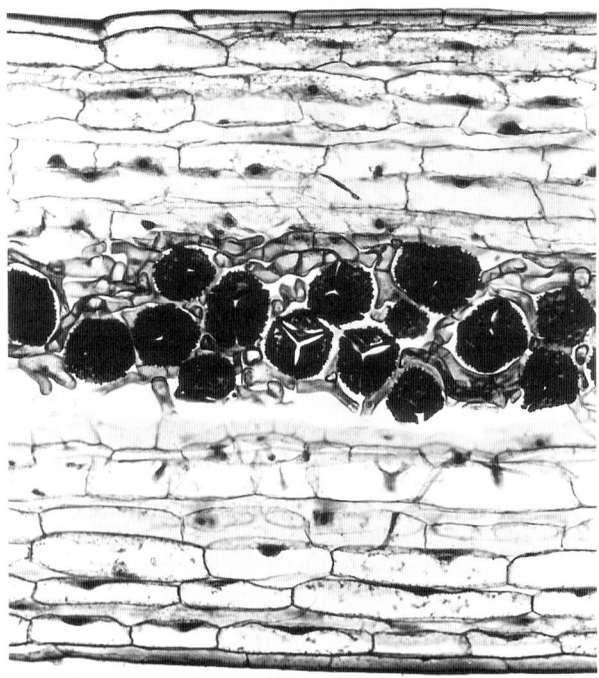

Fig. 6-10. Hornwort pseudoelaters (such as those of *Anthoceros*) are nonreproductive structures associated with spore dispersal that are not regarded to be homologous to liverwort elaters. Elater structure has been cited as an indicator of the direction of hornwort diversification.

tissue), and spherical elaters that are unlikely to function effectively in spore dispersal (Renzaglia, 1978; Mishler and Churchill, 1985). *Notothylas* also differs from other hornworts in development of the sporophyte foot (from the lower tier of embryonic cells, as compared with the lower two tiers of cells in other taxa).

Stomata similar to those of higher plants occur on sporophytes of some hornworts, but are absent from *Notothylas*. Finally, although most mature hornwort sporophytes split and twist along one or two lines, allowing slow, continuous spore dispersal, *Notothylas* sporophytes only occasionally split in this fashion; generally they are indehiscent (Renzaglia, 1978). These features of *Notothylas* have been interpreted by some authors as resulting from reduction and thus representing a highly derived condition. However, modern cladistic analyses suggest that many or all of these characters are plesiomorphic, and that *Notothylas* is the sister taxon to the remaining hornworts (Mishler and Churchill, 1985). This hypothesis needs to be tested by phylogenetic analysis of molecular data.

Seedless plant sporophytes are characterized by the occurrence of specialized cells at the junction between the parental gametophyte and the developing sporophyte generation, known as the placenta. Cells in this region typically exhibit extensive wall ingrowths or projections that may become so complex as to resemble a labyrinth (Wandlabrinthe) (Fig. 6-11). In a series of experiments on the placental region of the moss *Funaria hygrometrica,* Browning and Gunning (1979a, b) established that wall ingrowths substantially increase the surface area of cell membranes, facilitating transport of sugars produced by gametophytic photosynthesis into sporophytes. Renault et al. (1992) documented the transport of hexose sugars across the placental junction of the moss *Polytrichum formosum.* There is also evidence for transport of amino acids across this junction in the moss *Polytrichum* (Caussin, 1983). Hence, placental cells characterized by wall ingrowths are known as placental transfer cells, implying function in the translocation of photosynthates from parental gametophytes into developing sporophytes (embryos). Among mosses, placental transfer cells may occur on both sporophytic and gametophytic sides of the junction, or just on the sporophytic side. These cells occur on both sides in liverworts, but only on the gametophytic side of hornworts (Ligrone and Gambardella, 1988). This distribution pattern suggests that the plesiomorphic condition is exhibited by hornworts, that is, placental transfer cells occurred first in the gametophyte, only later developing in the sporophyte. Import of photosynthate from the parental gametophyte has been demonstrated to be necessary for sustained growth of hornwort sporophytes (Thomas et al., 1978), as well as that of mosses and liverworts (see Ligrone and Gambardella, 1988). The origin of placental function was probably a critical step in the evolution of the land plant embryo.

As has previously been mentioned in Chapter 4, there is evidence for the occurrence of placental transfer cells similar to those of land plants in the charophycean species *Coleochaete orbicularis* (Fig. 6-12) (Graham and Wilcox, 1983; Delwiche et al., 1989). In *C. orbicularis,* wall ingrowths develop in regions of enveloping cortical cells that are in contact with zygotes, suggesting the diffusion from zygotes of chemical effectors. Subsequently, zygotes accumulate large amounts

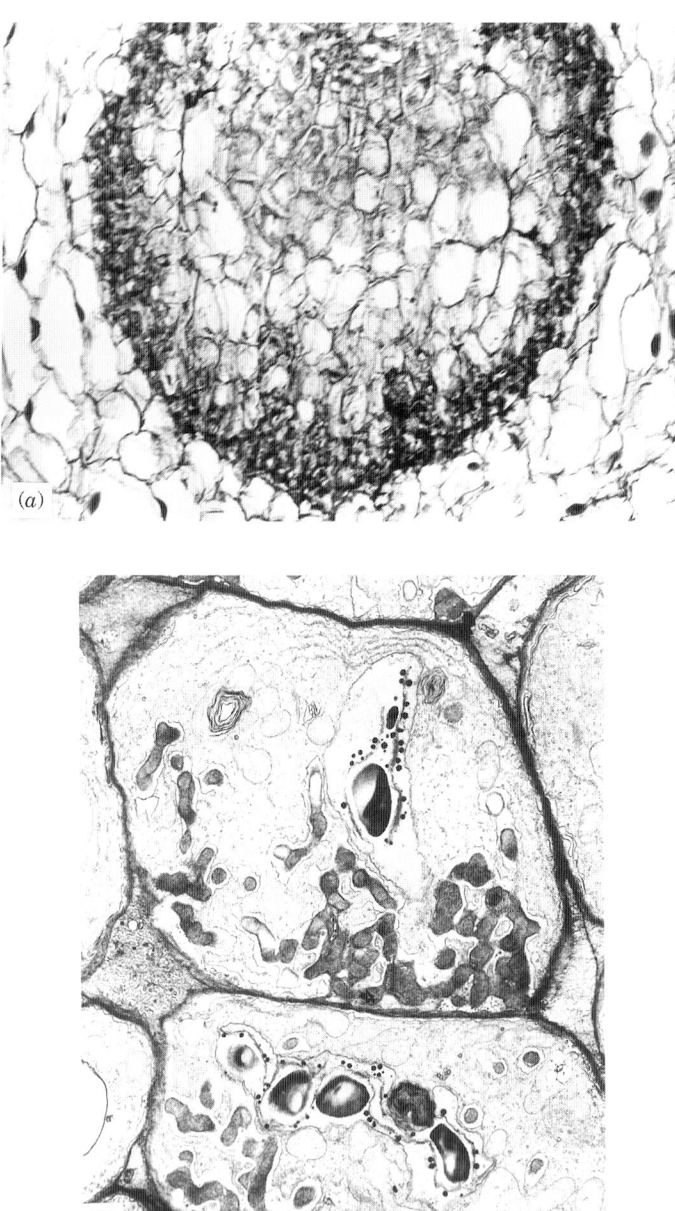

Fig. 6-11. The placental region of hornworts represents the interface between sporophyte and gametophyte tissues. *a.* Gametophytic cells of this region are typically modified in ways that imply developmentally controlled regulation of photosynthate flux from gametophyte to sporophyte. *b.* Transmission electron microscopy of embryophyte placental cells (those of the hornwort *Anthoceros* shown here), demonstrates occurrence of highly branched, densely staining wall ingrowths, or labyrinths (Photo from Graham and Kaneko, 1991). Some experimental evidence suggests that bryophyte placental cell wall ingrowths function to greatly increase cell membrane surface area involved in transport of photoassimilates such as glucose or amino acids. Plant "placental transfer cells" are thus functionally analogous to tissues occurring at the junction between female placental mammals and their embryos.

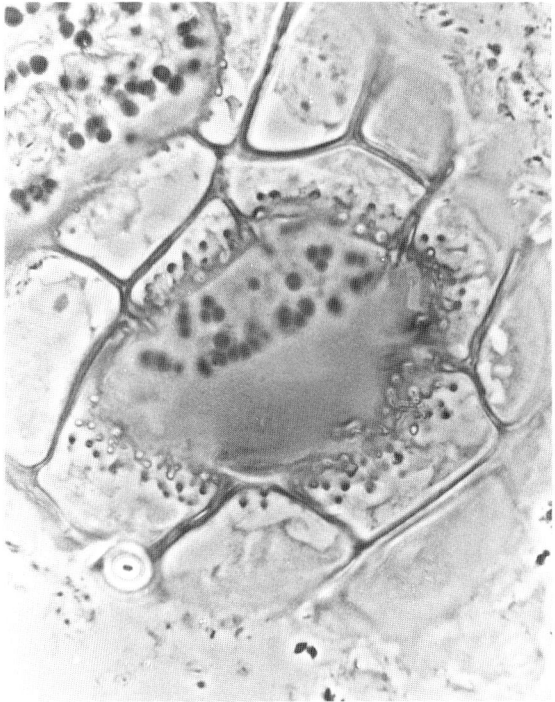

Fig. 6-12. Localized wall ingrowths similar to those of embryophyte placental transfer cells occur in cortical cells surrounding zygotes of the charophyte *Coleochaete orbicularis,* but are otherwise unknown in green algae. Wall ingrowths in *Coleochaete* cortical cells are likely to be functionally analogous to plant placental transfer cells, but are not necessarily homologous. Transfer cell morphology may have arisen more than once in response to similar selective pressure for enhancement of photosynthate flux across apoplastic junctions. Photosynthate flow from haploid to diploid cells in charophycean algae may, however, be homologous to movement of sugars across the placental junction of embryophytes and may even be based on ancestral transport mechanisms that also underlie phloem loading in plants.

of lipid and starch storage materials. This represents circumstantial evidence that photosynthates are transported from the parental thallus into zygotes. It should be noted that abundant photosynthate storage also occurs in zygotes of other species of *Coleochaete,* and in zygotes of zygnematalean and charalean algae, in which cells adjacent to zygotes lack wall ingrowths. It has been suggested that placental transfer cells may have evolved more than once (Duckett and Renzaglia, 1988). It is quite possible that wall ingrowths (Wandlabrinthe) appeared independently in *Coleochaete orbicularis* and in various embryophyte lineages, and the presence of wall ingrowths in one species of *Coleochaete* is not good evidence that this genus was directly ancestral to the plant kingdom. However, evidence for occurrence of similar acetolysis-resistant, autofluorescent compounds (probably polyphenolics) in the placental junctions of *Coleochaete* and *Anthoceros* (Delwiche et al., 1989) argues for the homology of charophyte and embryophyte placental junctions.

Resistant, polyphenolic materials occur in all species of *Coleochaete* investigated, specifically localized to walls of cells in the vicinity of zygotes, whether wall ingrowths are present or not. The phenolic compounds most likely have an ecological function, the inhibition of microbial degradation so that algal thalli and attached zygotes remain attached to substrates in the well-illuminated littoral zone until zygote germination can occur. This facilitates attachment of meiospores in a favorable environment for growth of germlings (Delwiche et al., 1989). It is also possible that the phenolic compounds, which appear in associated cortical cells rather late in zygote development, may function as a valve that shuts off the flow of photosynthate to maturing zygotes. The presence of polyphenolic compounds in the placental junctions of *Coleochaete* and bryophytes is strong evidence for the occurrence of homologous interactions between zygotes/embryos and parental tissues. The chemical resistance of the charophycean and bryophytan placentas also suggests that fossil evidence related to the origin of placental transfer cells and land plant embryos may yet be found (Delwiche et al., 1989). Because the origin of the plant embryo was such an important event, further study should be focused on the possible occurrence in charophycean algae of preadaptations that might have contributed to embryo evolution, and this subject is considered in more detail in Chapter 8.

Sporopollenin-walled spores resulting from meiosis are characteristic of all modern embryophyte groups, but not of charophycean algae. However, walls of charophycean zygotes are known to contain sporopollenin, which most likely represents a preadptation necessary for the origin of land plant spores (Graham, 1990). The only charophycean meiospores that have been studied ultrastructurally are those of *Coleochaete,* and these are surrounded by a layer of scales, rather than a coherent wall (Fig. 6-13). The biochemical composition of these scales is unknown.

Brown and Lemmon (1990c) summarized the evidence that in monoplastidic sporocytes (spore mother cells) of land plants, plastids divide twice, then migrate into tetrahedrally arranged polar regions prior to nuclear division. MTOCs associated with plastid tips generate four cones of microtubules, known as the quadripolar microtubular system (QMS). This complex apparatus is thought to function as a scaffold for intracellular movements in meiosis. In mosses, the four plastids remain locked in the tetrahedral position while cones of QMS microtubules then merge in pairs to form a bipolar spindle in preparation for the first nuclear division (Fig. 6-14). However, in the pteridophyte *Isoetes,* the plastids aggregate in pairs at division poles, bringing associated microtubules along (Brown and Lemmon, 1989; 1990c). After telophase, these plastids migrate back to tetrad poles and reestablish a QMS. In all cases involving monoplastidic sporocytes, preprophase plastid division and the quadripolarity that is established in prophase ensures that each product of meiosis (spore) receives both a plastid and a nucleus. The frequency of occurrence of monoplastidic sporocytes among seedless plants and the ubiquitous association of monoplastidic sporocytes with a QMS suggest that monoplastidic sporocytes and the QMS represent plesiomorphic characters

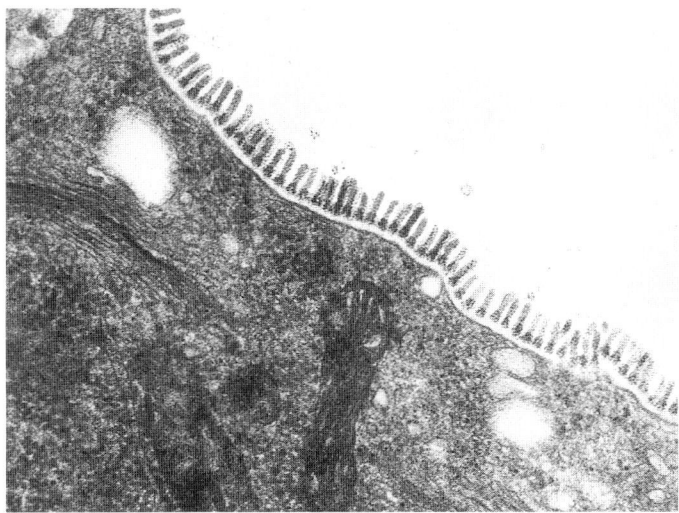

Fig. 6-13. Meiospores of *Coleochaete pulvinata* (and zoospores of *C. scutata*) are covered with unique electron-dense, pyramidal scales.

for embryophytes and were probably present in early land plants. Similar preprophase plastid division and migration occur in germinating zygotes of *Coleochaete* (Graham, 1990) and saccoderm desmids (Biebel, 1973), but whether a QMS is present is unknown. A callosic layer is formed around individual developing spores of *Coleochaete* (Graham and Taylor, 1986a, b), as in liverworts, where it appears to be involved in the patterning of the developing exine (the acetolysis-resistant outer wall of spores) (Brown and Lemmon, 1990b). However, it is not known whether callose involvement in sporogenesis is plesiomorphic for embryophytes. Callose typically encases developing microspores of seed plants, but has not been found to be associated with developing moss or hornwort spores (Brown and Lemmon, 1990b). It is possible that loss of callose is associated with derived aspects of sporophyte development in some bryophyte groups. Further comparative studies of sporogenesis in seedless plants and charophytes may illuminate these and other aspects of the evolution of walled spores.

Archegonia and antheridia, otherwise known as "jacketed" gametangia, are among the hallmarks of seedless plants. There are two major morphological types of jacketed gametangia, emergent gametangia typified by stalks and long necks (Figs. 6-6 and 6-7) and submergent gametangia for which stalks are generally not evident, and whose necks may barely protrude beyond the gametophyte surface. It is interesting, and probably phylogenetically significant, that the early developmental stages of archegonia and antheridia, whether submergent or emergent, are quite similar. In all cases, a single surface cell divides periclinally (in a direction parallel to the gametophyte surface). The upper cell becomes a jacket initial, and the lower cell undergoes divisions that ultimately lead to formation of gametes (egg or sperm) (Fig. 6-15). The jacket of emergent gametangia characteristic of

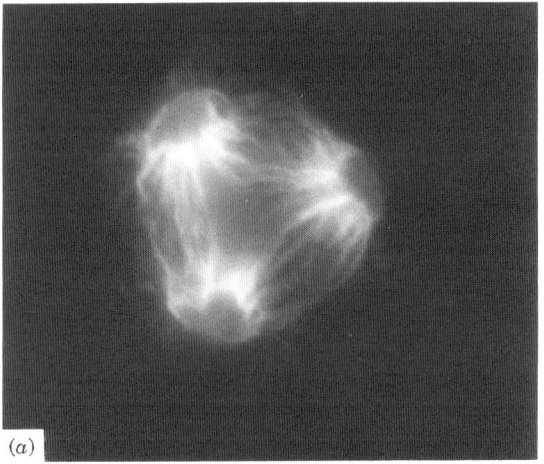

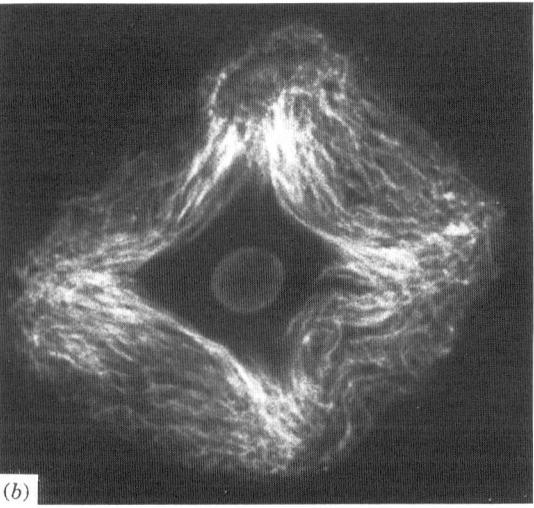

Fig. 6-14. Sporocytes of bryophytes (including *Sphagnum*, shown here in *a*), and pteridophytes (such as *Isoetes*, in *b*), produce a distinctive quadripolar microtubular system (QMS) that functions in positioning of four plastids prior to meiosis, so as to ensure equal distribution to the four spores resulting from meiosis. QMS microtubules later form the meiotic spindle. Similar plastid positioning and distribution occurs in *Coleochaete*, but whether a QMS occurs is not known. Immunofluorescence photos generously provided by R. Brown and B. Lemmon were previously published in the *American Journal of Botany* (1990), and are used with permission.

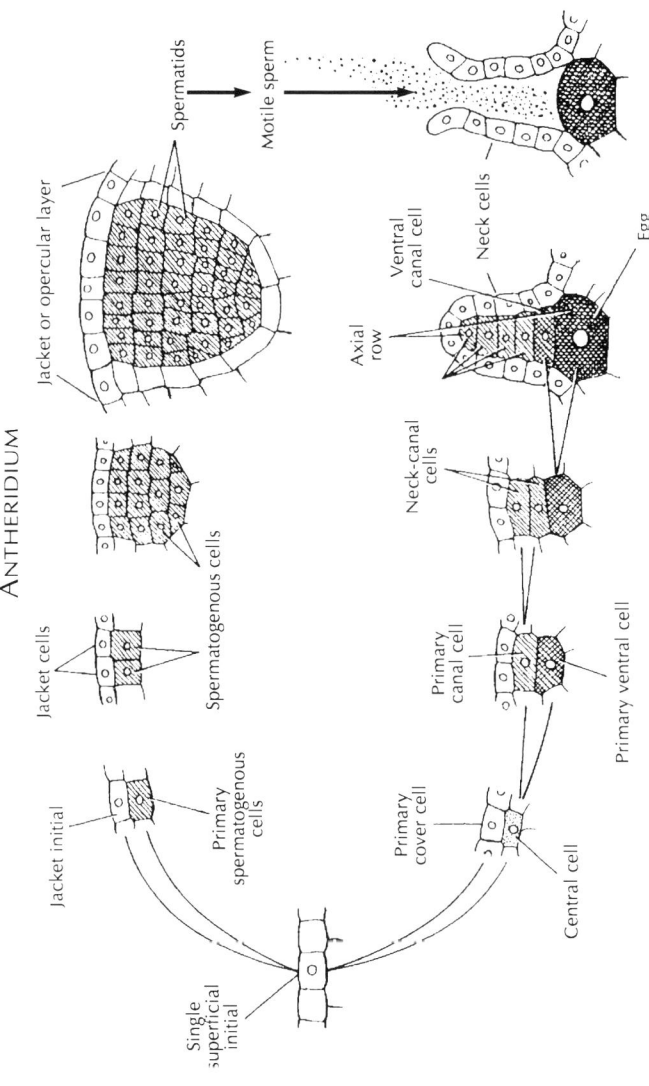

Fig. 6-15. Development of gametangia of pteridophytes (exclusive of leptosporangiate ferns) involves paradermal division (parallel to the thallus surface) of a single epidermal cell to form an exterior jacket initial and an interior cell, which ultimately generates spermatogenous tissue in antheridia, and the egg (and sister cells) of archegonia. The resulting gametangia may be partially or completely submerged within gametophytic tissues. From *Morphology and Evolution of Vascular Plants*, 3d ed., by Ernest M. Gifford and Adriance S. Foster. Copyright © 1989 by W. H. Freeman and Company. Reprinted by permission.

mosses, liverworts, and most ferns is typically one cell layer thick. The jacket cell layer of submergent gametangia is generally apparent only at the surface, other gametangial tissues generally being contiguous with those of the gametophyte (Gifford and Foster, 1989). Completely or partially submergent gametangia are produced by hornworts, *Lycopodium, Selaginella,* and primitive ferns such as *Marattia.* Many biologists think of emergent, stalked gametangia as characteristic of archegoniate plants. However, the repeated occurrence of embedded gametangia in taxa that are often placed in basal phylogenetic positions suggests that submergent gametangia may well be the primitive form of land plant gametangia, and the kind of gametangia likely to have been characteristic of early land plants. The sunken archegonia of the putatively primitive hornwort *Notothylas,* being smaller and more fragile than those of other hornworts, with fewer neck canal cells (Renzaglia, 1978), may provide an extant example of an ancestral archegonial type. However, the antheridia of hornworts, which typically occur as groups of stalked, jacketed structures within sunken chambers (Fig. 6-16), are clearly derived. Although similarities in early development suggest that the jacketed gametangia of land plants are homologous, some workers have suggested that jacketed gametangia may have had independent origins in various land plant lineages. Certainly gametangial structure among extant archegoniates is sufficiently

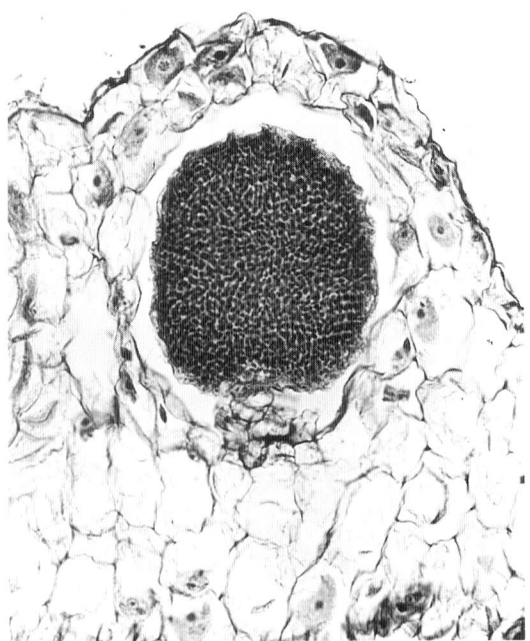

Fig. 6-16. Hornwort antheridia are typically submerged within gametophyte tissues. Submerged gametangia may represent the plesiomorphic type in archegoniates, but antheridia of *Anthoceros* shown here, also exhibit some presumably derived features and cannot be assumed to reflect the characteristics of early land plant gametangia.

variable as to suggest the effects of strong selection pressure on the evolution of plant gametangia, resulting in rapid diversification and development of derived morphologies. Thus, identification of plesiomorphic types has been difficult. Comparison of the early developmental stages of land plant gametangia with those of charophycean algae is likely to be more productive than comparison of mature structures.

The widespread view that stalked, jacketed gametangia are representative of early plants has led to some unlikely comparisons with algal gametangia. For instance, the eminent botanist Bradley Moore Davis proposed the origin of land plant gametangia from structures similar to plurilocular sporangia of brown algae such as *Ectocarpus* (see Lyon, 1904, for discussion). This hypothesis invokes the existence of hypothetical, morphologically complex green convergents to brown seaweeds, the Thallasiophyta, for which, as we have seen, there is no evidence. Emergent land plant gametangia have also been compared with the morphology of charalean gametangia, which are superficially similar in that gametes are protected by nonreproductive cells (Bold et al., 1987) (Figs. 6-17 and 6-18). The elegant ultrastructural studies of male and female gametangial development in *Chara* by Pickett-Heaps (1968; 1975) have demonstrated that these are the most morphologically complex gametangia produced by green algae (Fig. 6-19). *Chara* gametangial development begins with division of a nodal cell to form a basal or

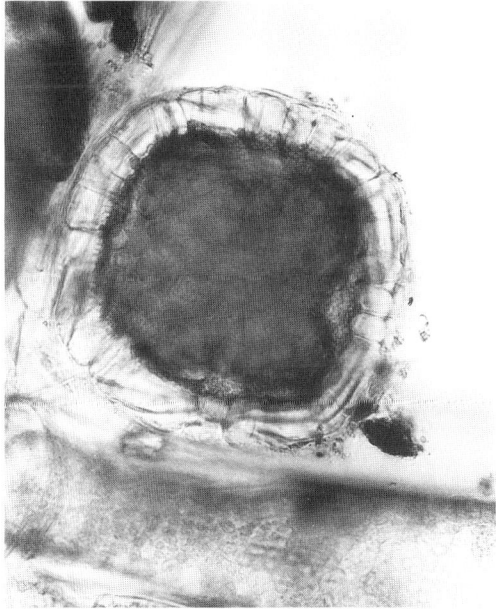

Fig. 6-17. Antheridium of *Nitella tenuissima*. A highly complex internal branching system generates filaments of spermatogenous cells, which are enclosed during development by shield cells arranged in petal-like arrays.

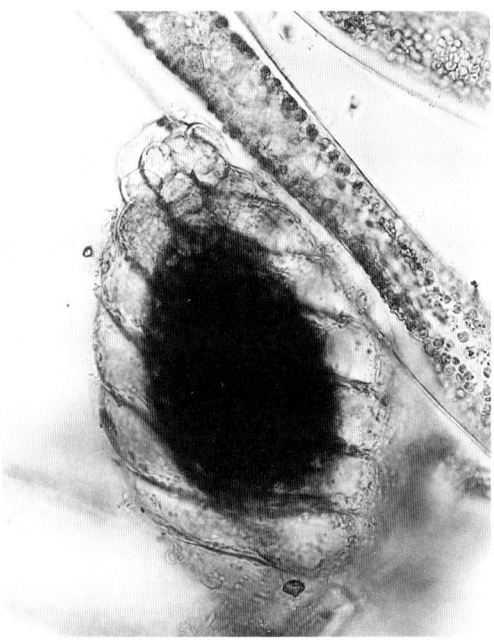

Fig. 6-18. Female gametangium of *Nitella tenuissima* illustrates enclosure of the oogonium (egg cell) by elongate, spirally twisted cells.

stalk cell, and an upper cell that will divide further. Some of the resulting cells become fertile (egg cells, spermatogenous filaments), and others generate sterile, protective layers. During antheridial formation, the upper cell is cut up by divisions in three planes, and each of the eight cells formed undergoes two additional periclinal divisions. The eight inner cells ultimately generate the spermatogenous filaments, whereas the outermost eight cells form a contiguous layer of shield cells. In female gametangial development, a cell distal to the stalk cell divides five times to produce a whorl of sterile cells that elongate, winding spirally around the developing oogonium. In common with land plant gametangia, the male and female gametangia of *Chara* are multicellular and consist of sterile and fertile cells, but the developmental processes involved are quite different. Actually, the development of the five cells that enclose the charalean egg more closely resembles cortication of zygotes in *Coleochaete* than development of any land plant archegonium. The mature antheridium of *Chara* is a highly complex and clearly derived structure that does not resemble any embryophyte antheridium. The complex and wonderful gametangia of charalean algae are certainly impressive, but difficult to relate to those of land plants, particularly if the hypothesis is accepted that embedded gametangia are the plesiomorphic type for land plants. Of the major differences that separate embryophytes from charophytes, transitions in gametangial structure are probably the least well understood.

Chara Oogonial Development

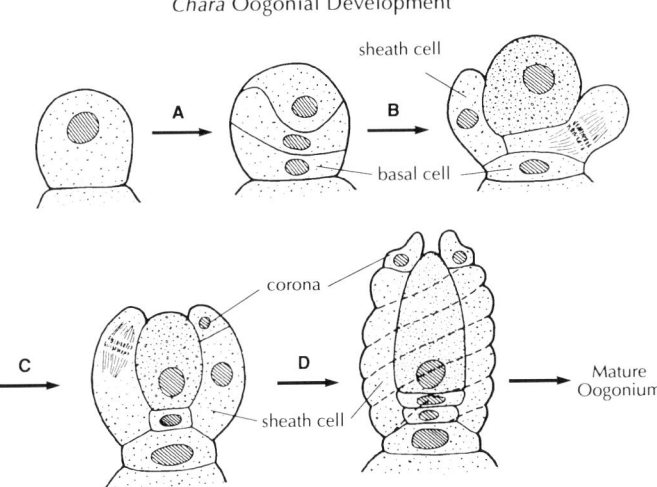

Chara Antheridial Development

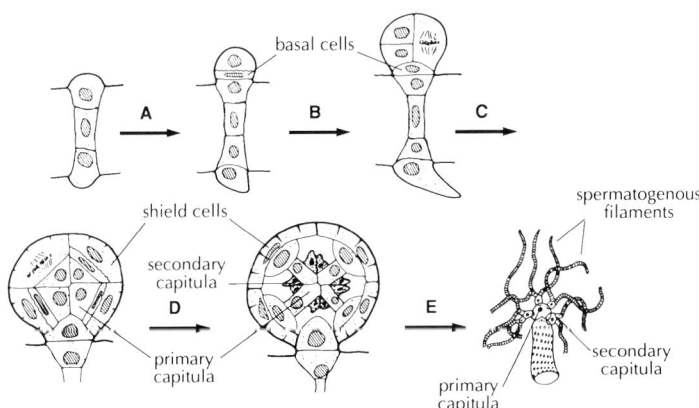

Fig. 6-19. Development of *Chara* gametangia. Illustration from Pickett-Heaps (1975), *Green Algae*, Sinauer Associates, Inc., Publishers, used with permission.

There are several differences in spermatozoid microanatomy that separate advanced charophytes from the least derived embryophytes. These are related to the mode of development of centrioles (flagellar basal bodies), the structure of the MLS, the arrangement of basal bodies with respect to the MLS during development, and the presence or absence of an extracellular matrix. The following characteristics have been proposed as representative of the ancestral land plant sperm type (Renzaglia and Duckett, 1988): cell ovoid, rather than elongate and helically coiled; nucleus rounded, relatively uncondensed; flagellar basal bodies lacking proximal microtubule extensions, inserted side by side above an MLS, and derived by separation of a bicentriolar centrosome; and MLS associated with an anterior mitochondrion. The spermatozoid of *Notothylas,* about 5 μm in length and 0.5 μm in diameter, is probably the smallest land plant spermatozoid known (Renzaglia and Carothers, 1986) and may illustrate some primitive spermatozoid features. The ovoid spermatozoids of *Lycopodium* (Robbins and Carothers, 1978) (Fig. 6-20) probably reflect plesiomorphy of shape in comparison with sperm of mosses, liverworts, and ferns, which are helically coiled. Among charophytes, the sperm of *Coleochaete* are ovoid, but those of charalean algae are helically coiled (Fig. 6-21). Because the ovoid shape is regarded as the plesiomorphic type for land plants, the helically coiled shape of charalean sperm is considered to be convergent to that of mosses, liverworts, and ferns.

Developing spermatozoids of *Coleochaete* are similar to those of land plants in the presence of a morphologically distinct anterior mitochondrion associated

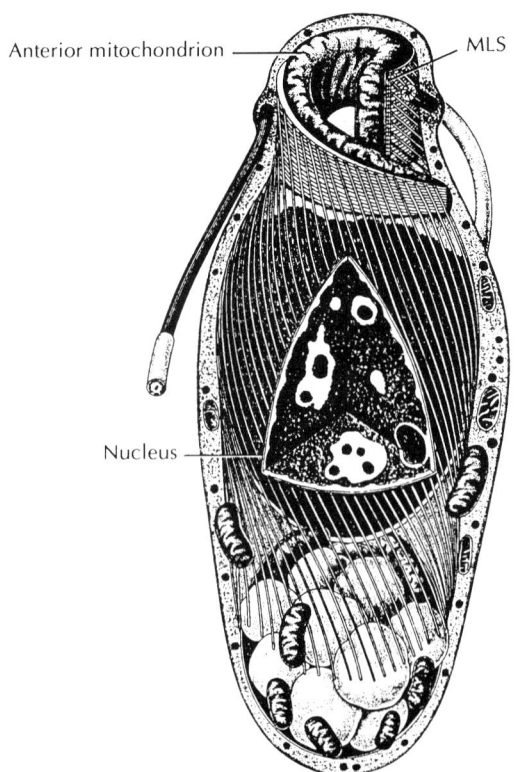

Fig. 6-20. Ovoid sperm of *Lycopodium* may illustrate the plesiomorphic shape for land plant spermatozoids. Illustration from Robbins and Carothers (1978), used with permission of the *American Journal of Botany.*

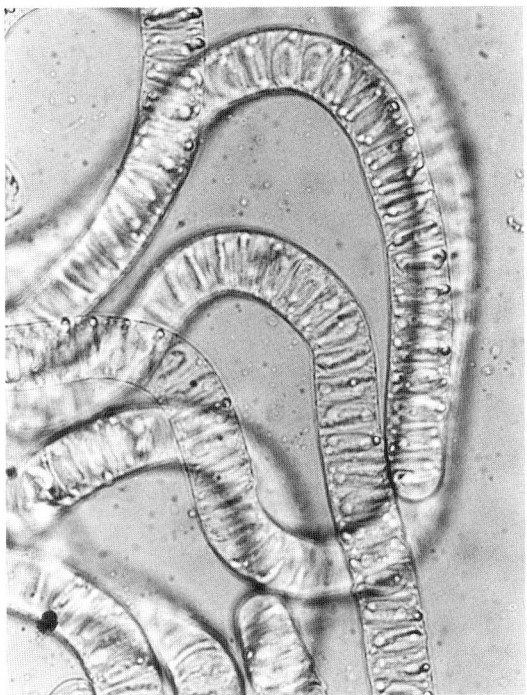

Fig. 6-21. Spermatogenous filaments of *Chara zeylanica,* obtained by compression of maturing antheridia. Each colorless cell produces a single coiled sperm.

with the MLS (Graham and Repavich, 1989) (Fig. 6-22), but later in sperm development, numerous mitochondria of more normal morphology occupy the cell anterior. Anterior mitochondria have not been identified in any other charophycean sperm (or flagellate spores), but this may reflect the general lack of detailed information on charophycean reproductive cell development more than anything else. Sperm of *Coleochaete* and Charales are similar to each other, but different from sperm of embryophytes in mode of centriolar development, details of MLS structure, and the occurrence of an extracellular layer of organic scales. Scales are not observed to occur on spermatozoids of land plants, but the diamond-shaped scales that cover *Coleochaete* and charalean sperm (as well as zoospores of other charophytes) resemble some scales produced by putatively primitive green flagellates. Scales are thus plesiomorphic for charophycean algae. The zoospores of *Klebsormidium* are unusual in lacking scales (Marchant et al., 1973); this suggests that scale loss occurred at least twice in the charophyte-embryophyte clade, first in divergence of the ancestors of *Klebsormidium,* and later in divergence of the direct ancestors of the embryophytes. The adaptive significance (if any) of scale retention in *Coleochaete* and Charales, and of scale loss in *Klebsormidium* and embryophytes, is not understood.

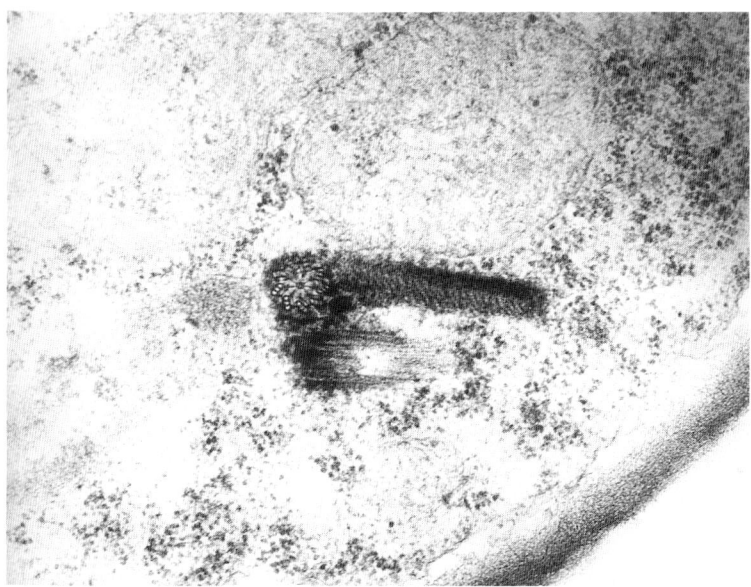

Fig. 6-22. Flagellar basal bodies of developing sperm of *Coleochaete pulvinata*. Note orthogonal relationship of basal bodies, location of MLS development, and association of MLS with an anterior mitochondrion. From Graham and Repavich (1989), *American Journal of Botany*.

Development of centrioles in land plants that produce biflagellate sperm involves "de novo" production of bicentriolar centrosomes in spermatid mother cells. Centrosomes first appear as dark-staining bodies on the outer nuclear surface. Later they consist of a coaxial pair of centrioles and are positioned at the spindle poles of spermatid mother cells. Upon division of these cells to form spermatids, a single bicentriole is distributed to each daughter cell (Robbins and Carothers, 1975; Robbins, 1984). During spermatid maturation into spermatozoids, the centrosomes separate at the midpoint, then rotate so that the two resulting centrioles are parallel (Moser et al., 1977) (Fig. 6-23). The occurrence of nearly identical centriolar development processes in various seedless plants having biflagellate sperm (summarized by Renzaglia and Duckett, 1991) has been cited as evidence in support of a hypothesis of monophyly of the embryophytes (Graham and Repavich, 1989). Centriole pairs of the hornwort *Phaeoceros* (Moser et al., 1977) and the liverwort *Blasia* (Renzaglia and Duckett, 1987) are connected by electron-dense material that may have been derived from the striated connectives characteristic of charophycean reproductive cells. In charophytes these connectives persist throughout spermatozoid development and are present in mature sperm, but centriolar connectives disappear during hornwort and liverwort sperm maturation.

Centrioles of charophycean algae, however, develop orthogonally, as do the centrioles of most protists. In this process, daughter centrioles originate from

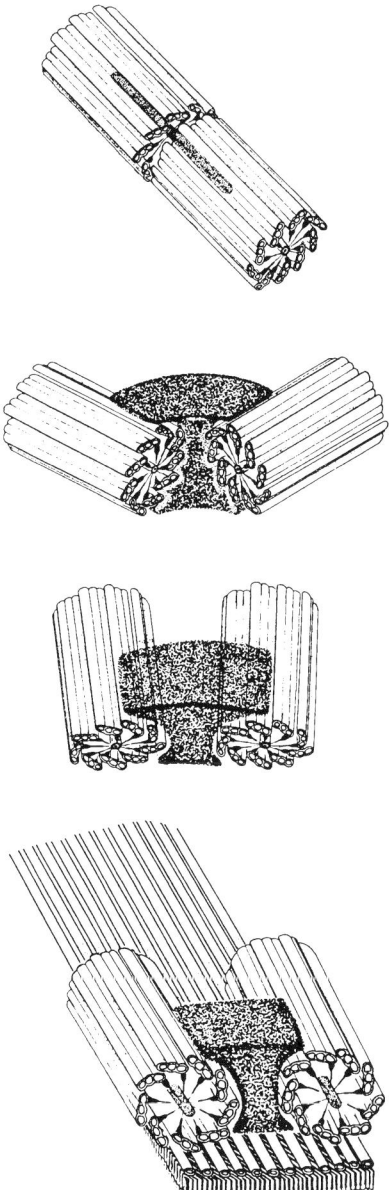

Fig. 6-23. Model for separation and reorientation of centrioles derived from a bicentriolar centrosome in the hornwort *Phaeoceros* (from Moser et al. [1977], with permission of the *American Journal of Botany*). Centrioles are connected by amorphous material possibly related to striated connectives of charophytes. Most land plant sperm blepharoplasts (the complex consisting of flagellar basal bodies, MLS, and other structures) lack such connecting material. Note development of the MLS beneath the two parallel basal bodies and the 45° angle between MLS lamellae and the long axis of the microtubular spline. These characteristics distinguish biflagellate sperm of embryophytes from those of charophytes.

the side and base of parental centrioles (Fig. 6-22). Interestingly, however, coaxial bicentrioles resembling those of land plants have been observed to occur in colchicine-treated sperm of *Nitella* (Turner, 1970) and in a few other protists (Robbins and Carothers, 1975). This suggests that the transition from orthogonal to coaxial centriolar development that apparently occurred during the divergence of embryophytes, may not have been difficult. It does however, represent a significant gap between charophytes and embryophytes that is poorly understood. A possible mechanism for the colchicine-induced change from orthogonal generation of centrioles to bicentriolar centrosome production in *Nitella* might have been colchicine binding to gamma-tubulin, a highly conserved component of eukaryotic centrosomes. Gamma-tubulin is thought to interact with alpha- and beta-tubulins (Sterns et al., 1991). This interaction may be important in patterning centrioles and/or directing the manner of their duplication. The possibility that evolutionary changes in gamma-tubulin may explain transition from orthogonal centriole replication to bicentriolar centrosome formation in plants requires investigation.

The development of a distinctive multilayered structure (MLS) in association with the flagellar apparatus of reproductive cells is a characteristic shared by charophytes and embryophytes. However, charophytes and embryophyte MLSs differ in that the angle between lamellae and the long axis of the microtubular layer is invariably 90° in charophytes, but 45° (or slightly less) in embryophytes (Figs. 6-22 and 6-23). It has been suggested that conformational change in the tiny links (keels) that join these two layers may explain the transition (Graham and Repavich, 1989). However, the molecular and genetic basis for this change has not been elucidated.

Yet another difference in sperm development of charophytes and embryophytes concerns the positioning of the developing MLS with respect to centrioles. In spermatids of land plants that produce biflagellate sperm, an MLS develops beneath the parallel centrioles (Fig. 6-23). In contrast, charophyte sperm MLSs (or homologs) invariably develop beside the posterior centriole, which lies at a right angle to the anterior centriole (Graham and Repavich, 1989) (Fig. 6-22). The basis for this phylogenetic transition in the developmental process is not understood, but may be related to morphological variation in positioning of the flagellar base and cytoskeletal components in mature sperm.

As previously mentioned (Fig. 6-21), charalean algae produce long, thin, helically coiled spermatozoids that superficially resemble sperm of mosses and liverworts in external morphology and orientation of flagellar basal bodies with respect to the MLS. If bryophyte and charalean sperm are viewed from a similar perspective, from their posterior ends, and with the major cytoskeletal band of microtubules (spline) facing upward, the flagella appear to extend backward toward the viewer. In contrast, viewed from the same angle, flagella of most charophycean reproductive cells (including *Coleochaete* sperm, but not those of Charales), extend toward the side. The derived flagellar arrangements exhibited by mature sperm of bryophytes and Charales are considered to represent an example of homoplasy (parallel or convergent evolution), on the basis of developmental observations. During charalean spermatid development, the position of flagellar basal bodies with respect to the MLS initially resembles that of other charophycean algae, but undergoes rearrangement to the position characteristic of bryophyte sperm. Such developmental rearrangement does not occur in sperm of *Coleochaete* or bryophytes. A massive striated connective may constrain potential developmental rearrangements of blepharoplast components in *Coleochaete*. In *Nitella,* the fairly delicate striated connective is deformed by changes in blepharoplast organization (Turner, 1968), but in *Chara,* the connective apparently disappears prior to centriolar rearrangement (Pickett-Heaps, 1968). Because various bryophyte sperm apparently possess the evolutionary remnant of a connective (see Fig. 6-23, for example), it may be supposed that the ancestral embryophyte sperm type also had one. It has been suggested that if backward-directed flagella were of adaptive advantage, the ancestors of embryophytes may have found a better way of achieving favorable morphology than rearranging or disassembling the flagellar connective (as charalean algae do), by changing the mode of centriolar development (Graham and Repavich, 1989). Parallel positioning of centrioles resulting from separation of bicentriolar centrosomes allows the MLS to develop beneath both centrioles and results in backward positioning of flagella. Rearrangement of blepharoplast components would thus become unnecessary, as would deformation or developmental disassembly of the connective. The basal body connective so characteristic of charophytes has almost disappeared entirely from land plant sperm, and the reason for its loss is not understood.

It is apparent that the phylogenetic chasm between algae and land plants that loomed large in Bower's time has diminished considerably in the light of modern comparative studies. In this chapter the structural differences in vegetative and reproductive structure that separate charophytes and embryophytes have been defined. With this background, it is now possible to explore the origin of various fundamental features of land plants from the perspective of evolutionary modification of the ancestral, preadaptive characters exhibited by charophytes. Such a discussion is bound to be relatively speculative, but is valuable in terms of testable hypotheses that emerge from the exercise. In the next chapters, fundamental processes that have influenced morphology, development, and reproduction in both charophytes and embryophytes are examined.

7

THE EVOLUTION OF PLANT MORPHOLOGY: CELL WALLS, CYTOSKELETON, CYTOKINESIS, INTERCELLULAR COMMUNICATION, AND HISTOGENESIS

Although plants share many conservative cellular, biochemical, and molecular features with other eukaryotes, including higher animals, plant development contrasts dramatically with that of animals. Animal development is influenced by morphogenetic movement of cells, but plant cell movement is constrained by the presence of a more or less rigid cell wall and the fact that plant cell lineages are cemented together. Plant morphology is therefore generated by different processes, including control of division planes, variation in division rates, and differential cell expansion (Goldberg, 1988). Consequently, evolution of plant structure and morphogenesis has involved phyletic changes in fundamental aspects of cell wall chemistry, the cytoskeleton, cytokinesis, intercellular communication, and meristematic structure. In this chapter, we examine features of charophycean algae and seedless plants which are important in understanding the evolutionary origin of plant structural and developmental characteristics, beginning with a consideration of the evolution of plant cell walls. Currently, plant scientists are engaged in a debate regarding the nature of plant cell walls (Stafford, 1991b; Staehelin, 1991; Sack, 1991; Robinson, 1991). Are plant cell walls analogous to the extracellular matrices of animal cells, or are they integral components of the plant cell itself? An evolutionary perspective provides some insight into these questions.

CELL WALLS

In terms of cell wall biochemical composition and the organization of putative cellulose-synthesizing complexes, the charophycean algae are more like land plants than are other green algae that have been studied. For example, the major hemicellulosic neutral polysaccharide of *Mougeotia* cell walls is a 4-linked xylan with 2,4 xylosyl branchpoints, similar to major hemicelluloses of land plants. In contrast, the xylans of marine green algae examined are 3-linked and unbranched. In addition, an arabinoglactan similar to one of the pectic carbohydrates of plant primary walls is present in *Mougeotia,* and sulfate esters are not formed as in green seaweeds (Hotchkiss et al., 1989). The cell walls of *Mougeotia* and *Klebsormidium* are similar in that both are characterized by low cellulose content, low levels of water-soluble rhamnose, and abundant water and alkali-soluble carbohydrates. Cell walls of *Chara* and *Nitella* contain 20 to 26 percent cellulose, similar to the 20 to 30 percent cellulose content of higher plant cell walls, and relatively less pectic and hemicellulosic polysaccharide than *Klebsormidium* and *Mougeotia.* Thus, it has been suggested that cell walls of *Klebsormidium* and *Mougeotia* represent a primitive state in the evolution of the plant cell wall, with charalean walls exhibiting a more derived condition, particularly with respect to cellulose content (Domozych et al., 1980; Hotchkiss et al., 1989). Although celluloses of *Coleochaete scutata* have been partially characterized (Okuda and Brown (1992), the biochemical composition of *Coleochaete* cell walls has not otherwise been investigated.

In higher plants, cellulose microfibrils are generated at the surface of the plasma membrane by a moving enzyme system consisting of particles in organized arrays (Giddings and Staehelin, 1991). As visualized by freeze-fracture preparations and transmission electron microscopy, the particle arrays appear as solitary globules on the surface of the outer plasma membrane leaflet and are opposed to a rosette composed of six particle subunits in the inner leaflet (Brown, 1985) (Fig. 7-1). The diameter of both globules and rosettes is 24 nm; rosette subunits are 8 nm in diameter (Hotchkiss and Brown, 1987). Although the evidence so far is circumstantial, these particle groupings are thought to be the synthetic enzyme system that generates cellulose from UDP-glucose and, therefore, are known as terminal synthesizing complexes. No one, however, has yet succeeded in demonstrating in vitro activity of cellulose synthase (formal name—UDP-glucose: (1-4)-β-D-glucan glucosyl transferase) (Delmer, 1991). The arrangement of subunits within the enzyme complex appears to determine the size, shape, crystallinity, and aggregation of cellulose microfibrils. Among

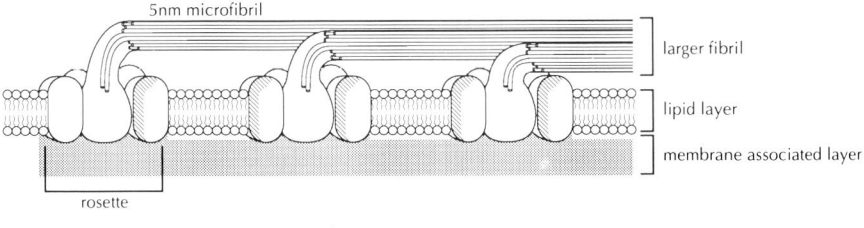

Fig. 7-1. Cellulose microfibrils are spun out at the cell surface from cellulose-synthesizing complexes, which in land plants and most charophytes consist of particles arranged in rosettes. This model represents cellulose deposition in the desmid *Micrasterias*. Diagram adapted from Giddings and Staehelin (1991), used by permission of Academic Press Ltd.

algae investigated, there are three main types of organization of the terminal complexes: linear formations characteristic of noncharophycean algae; rosette/globules, resembling those of the moss *Funaria,* pteridophytes, and flowering plants, exhibited by some charophycean algae (Brown, 1985; Hotchkiss and Brown, 1987); and a unique type of terminal complex reported from *Coleochaete scutata* (Okuda and Brown, 1992). The cellulose synthesizing systems of charophytes further support the hypothesis of relationship to embryophytes, but reveal some diversity consistent with ancient divergence from embryophyte ancestors.

Although solitary rosette/globules similar to those of land plants are associated with primary cell wall development in Zygnematales, hexagonal arrays of rosettes that occur during secondary wall development in some zygnematalean algae are unique to this group (Brown, 1985). The spacing of the aggregates within the hexagonal array correlate with the spacing of cellulose fibrils in secondary walls of *Micrasterias* (Giddings and Staehelin, 1991). Terminal complexes of *Coleochaete scutata* occur in groups of 8 to 50 particles on the exoplasmic plasmalemma face, and complementary aggregates showing 8 particles surrounding a central particle are present on the plasmatic face (Okuda and Brown, 1992). It has been suggested that *Coleochaete*'s terminal complexes (TCs) may have originated independently from rosette/globule TCs, or that they may have been derived from rosette/globule TCs (Okuda and Brown, 1992). The rosette/globules of *Nitella* are indistinguishable from those of land plants (Hotchkiss and Brown, 1987).

In charophytes, as in land plants, terminal cellulose-synthesizing complexes are closely associated with underlying cytoplasmic microtubules and microfilaments. However, Schmid and Meindl (1992) found that microtubule disorganizing chemicals (amiprophos methyl and colchicine) had no effect on the pattern of microfibril deposition in the primary or secondary cell walls of *Micrasterias,* indicating that microtubules are not directly involved in cellulose pattern formation. Instead, their freeze-etch images revealed a close spatial relationship between microfilaments (presumably actin) and plasma membrane rosettes. These workers suggested that the role of microtubules in cell wall deposition may be one of structural support, rather than patterning. Secondary wall formation in the desmid *Micrasterias* is characterized by production of dictyosome vesicles that carry membrane studded with hexagonally patterned particles to the plasma membrane (Kiermayer and Meindl, 1989). This finding suggests that the pattern of terminal complex subunits is determined within the Golgi apparatus.

The bacterium *Acetobacter* can produce cellulose, but because synthetic complexes are not mobile within the cell membrane, cellulose fibrils do not encase cells as they do in plants and algae. Mobility of eucaryotic terminal complexes is thought to be somehow associated with microtubules and microfilaments that lie beneath the cell membrane (Brown, 1985), suggesting that phyletic changes in cytoskeletal elements may have been involved in the evolutionary origin of algal and plant cell walls. The gene that encodes cellulose synthase in *Acetobacter xylinum* has been cloned and seems to specify a catalytic subunit of about 83 kDa (see discussion in Delmer, 1991). It would be interesting to compare this sequence with the cellulose synthase gene sequences of charophytes and embryophytes when these are available.

In plants, callose (β,1-3 glucan) is also generated by plasma membrane-bound synthase complexes (Roberts, 1990) which are activated by Ca^{2+} (Kauss, 1985). Calcium levels may become elevated when cells are perturbed, perhaps explaining the accumulation of callose at wound sites in plants (Delmer, 1991). Callose deposition can be localized by formation of a fluorescent complex with decolorized aniline blue (Kauss, 1985). This technique has been used to identify callose in the hair cells of *Coleochaete* (Marchant, 1977), where it has a wound plugging function, and in the walls of zygnematalean algae (Dubois-Tylski, 1981). Callose is also associated with reproductive cell development in *Coleochaete* and the desmid *Closterium*, as in various land plants (see Chapter 8 for further discussion). Callose-containing wall appositions form in mature internodal cells of charalean algae under the influence of certain ionophores. Local influx of calcium ions is thus implicated in the exocytosis of polysaccharide-containing vesicles. This phenomenon (which also occurs in the liverwort *Riella*) is surprising because callose deposition in plants is usually associated with activity of a plasma membrane-bound callose synthase (Foissner, 1990). Additional study is needed to determine whether charophyte algae normally generate callose at the cell membrane or via the endomembrane system.

In plants, polysaccharides other than cellulose and callose are synthesized in the lumen of the endomembrane system, packaged in dictyosome vesicles and transported, probably under the direction of microtubules, to the cell membrane, where calcium-dependent fusion occurs (Northcote, 1989). Among these polysaccharides are pectins, which in the past were regarded as a relatively inert, sticky matrix in the middle lamella. However, recent work suggests that pectins play a more dynamic role in plant cell walls than previously thought (Roberts, 1990; McCann and Roberts, 1991). New findings suggest that the ordering of pectins in plant cell walls may be as important as cellulose organization, and both may be associated with cytoskeletal dynamics. Pectin organization may influence wall porosity and thus regulate aspects of intercellular movement of substances. Three pectin-containing layers have been detected in electron micrographs of onion cell walls: an outer zone containing methyl-esterified pectins, an inner zone with unesterified pectins, and a middle lamella, containing densely staining rodlike polymers that label with antibodies to pectin. Pectins of higher plants contain phenolic side groups believed to provide the opportunity for the coupling of pectins to each other and to other cell wall materials. It has been suggested that middle lamella pectins may cross-link with primary cell wall pectins on either side, thus coupling adjacent cells and possibly mediating synchronized elongation (McCann and Roberts, 1991). The presence of phenolic groups might explain the electron density of the rod-shaped pectin aggregates in the middle lamella. Similar electron-dense, rodlike structures occur in charophyte walls (Figs. 7-2, 7-3, 7-4, and 7-5), but their biochemical composition has not yet been ascertained. In view of its potential morphogenetic importance, it would be of interest to determine whether or not phenolic coupling might also occur in charophycean pectins.

150   THE EVOLUTION OF PLANT MORPHOLOGY

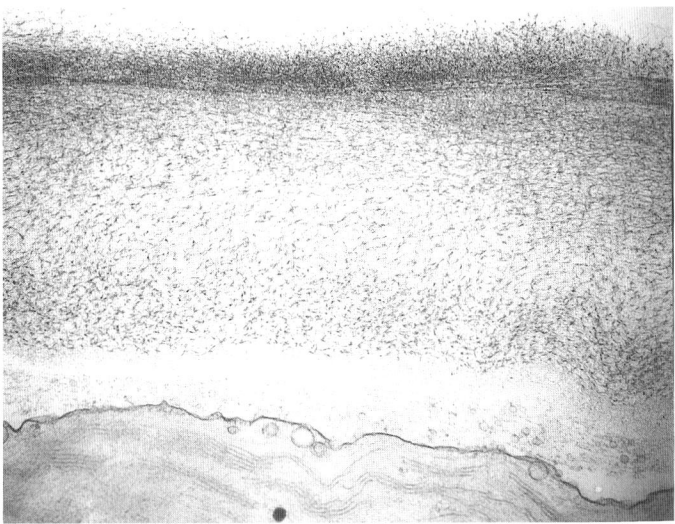

Fig. 7-2. Transmission electron micrograph of *Coleochaete orbicularis* outer cell wall sectioned perpendicular to the thallus surface, showing occurrence of several distinct layers. An amorphous layer adjacent to the cell membrane is separated by a translucent space from a wider layer containing electron-dense rodlike structures. The uppermost "cuticular" layer occurs only on the dorsal surface.

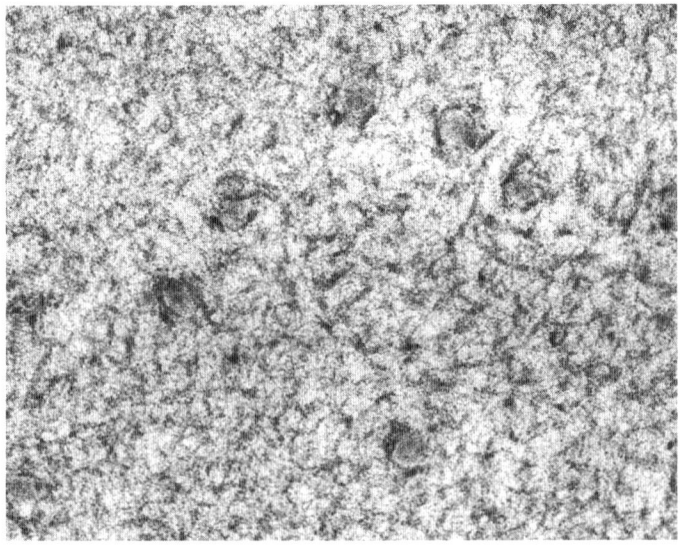

Fig. 7-3. Transmission electron microscopic view of the middle lamella of *Coleochaete orbicularis*, showing cross-sections of plasmodesmata, and densely stained rodlike structures that somewhat resemble pectinaceous aggregates in the middle lamellae of plant tissues.

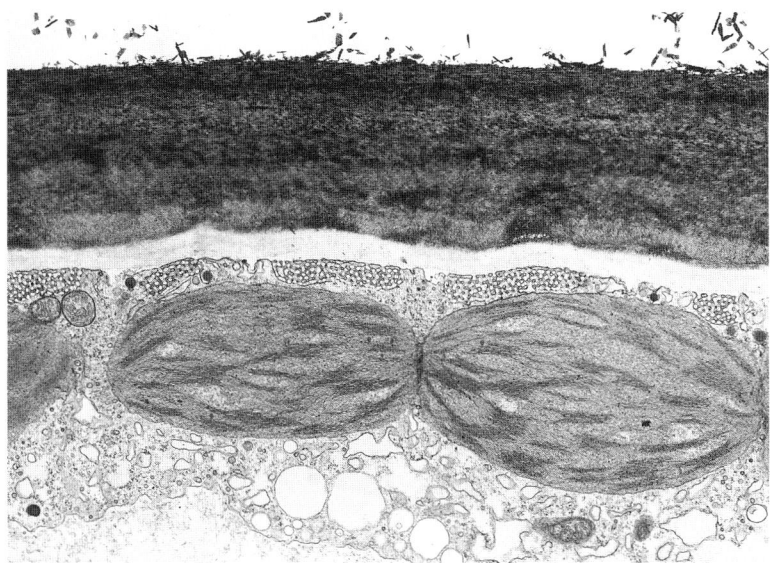

Fig. 7-4. Transmission electron micrograph of the densely stained outermost wall of *Chara zeylanica*. A thinner, less dense inner wall layer lies adjacent to the cell membrane and charasomes.

Golgi-derived vesicles of various kinds that appear to contribute to cell wall formation also occur in charophytes. Many desmids exhibit extraordinary complexity of form in single cells (Fig. 7-6) and thus have been used as model systems for the study of plant cell morphogenesis (Giddings and Staehelin, 1991). Experiments using chlorotetracycline have shown that vesicles related to primary wall formation in *Micrasterias* preferentially fuse at plasma membrane sites of Ca^{2+} accumulation (Kiermayer and Meindl, 1989). These observations may be relevant to the evolutionary origin of cell walls of charophytes (and hence, those of land plants). As previously noted in Chapter 4, the putative flagellate ancestors of charophytes are thought to have been related to prasinophytes, which are characterized by scaly extracellular matrices rather than coherent walls. Scales of prasinophytes, and scales produced on reproductive cells of charophytes, are synthesized in Golgi vesicles and transported to the cell membrane, where they are attached by matrix material also derived from dictyosome vesicles (Graham and Taylor, 1986a, b). Becker et al. (1991) reported that 2-keto-sugar acids, found in higher plant cell walls as a constituent of the rhamnogalacturonan II in pectic polysaccharides, represent chemical markers for prasinophycean scales and are absent from walls of other green flagellates, including *Chlamydomonas* and relatives. These authors suggested that higher plants inherited the capacity to produce 2-keto-sugar acids from charophycean ancestors, which in turn derived them from prasinophytes. Thus, the scale-producing endomembrane system of primitive scaly flagellates may have formed

152 THE EVOLUTION OF PLANT MORPHOLOGY

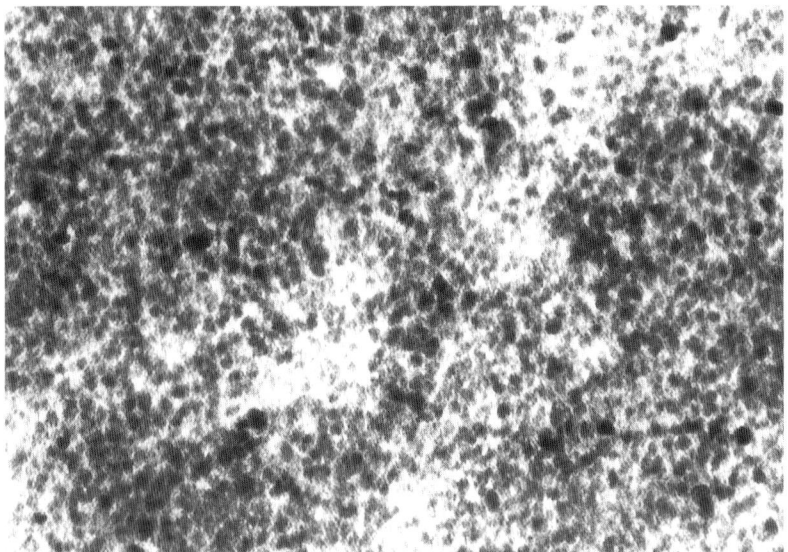

Fig. 7-5. Transmission electron micrograph of *Chara zeylanica* middle lamella. Electron-dense globules or rods resemble those observed in *Coleochaete* walls.

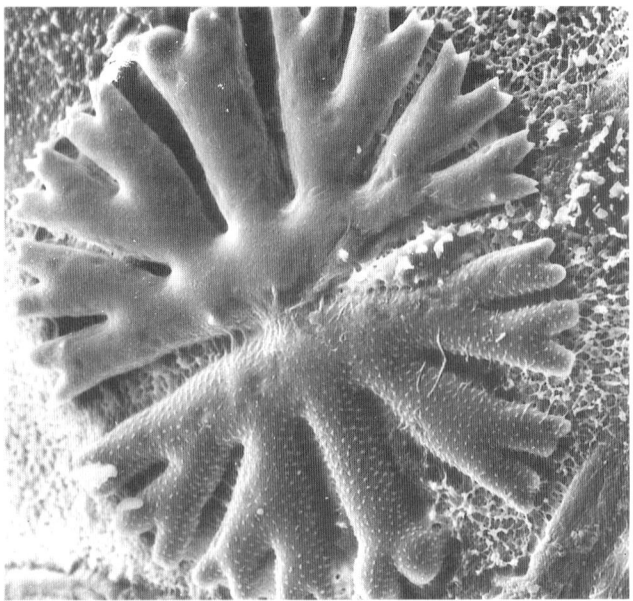

Fig. 7-6. Desmids such as *Micrasterias*, shown here in SEM view by Sonia Cook (University of California-Davis), exhibit highly complex morphology and are thus useful model systems for study of cellular differentiation in charophytes and plants.

the basis for the evolutionary origin of charophycean and land plant cell walls. Future comparative studies of cell wall biogenesis in prasinophytes related to the charophycean-embryophyte lineage may shed light on the steps involved in the biochemical evolution of plant cell walls.

The primary walls of growing plant cells contain about 10 percent protein in the form of glycoproteins (Darville et al., 1985; Roberts et al., 1985). A growing body of data suggests that four families of higher plant glycoproteins—hydroxyproline-rich glycoproteins (HRGPs), arabinoglactan proteins, repetitive proline rich proteins (RPRP), and the solanaceous lectins—may be related by virtue of divergence following gene duplication and homologous recombination. Glycoproteins may play a role in plant morphogenesis by forming highly cross-linked cages that affect cell wall extensibility, and in targeting deposition of lignin to specific wall locations (Roberts, 1990). Hydroxyproline-rich glycoproteins (HRGP) were initially thought to play a role in extensibility of the primary wall and became known as "extensin." This term now seems to be a misnomer, because a role in cell extension has not been demonstrated, but the term "extensin" is widely used in the literature. Extensin is interesting in that it is difficult to extract without the use of degradative treatments. When polysaccharides are removed from plant cell walls, an insoluble fraction (about 10 percent) remains, composed of roughly equal amounts of protein and an unknown, possibly phenolic, compound. Monomers of extensin can be cross-linked in vitro by peroxidase, but the extent to which this occurs in cells is uncertain and no linkages to cell wall polysaccharides are known (McCann and Roberts, 1991). Although little is known about the evolutionary origin of plant glycoproteins, it is significant that the extracellular matrix of prasinophytes contains 5 percent protein. Cell wall glycoproteins of charophytes have not been studied, but perhaps future biochemical work will illuminate aspects of glycoprotein evolution in plants.

Stebbins (1992) suggested that adaptive diversity in plants has been promoted by mutation of genes controlling cell wall glycoproteins. For example, inhibitors of the synthesis of hydroxyproline-rich proteins caused morphogenetic change in the number of rows of phyllomes (leaf analogs) per branch in several genera of "leafy" liverworts. The inhibitors also caused alterations in the level of branching. These structural modifications mimicked morphological/phylogenetic trends within the group (Basile, 1990; Basile and Basile, 1990). The primary cell wall can thus be viewed as a highly complex system, which is continuously modified by enzyme activity (Stebbins, 1992) and most likely has a much more important role in the determination of plant morphology than previously appreciated. In the discussion that follows, the role of the cytoskeleton in the formation and siting of primary plant cell walls is considered.

THE PLANT CYTOSKELETON

Various structural and dynamic features of the cytoskeleton are remarkably consistent throughout land plants (Gunning and Wick, 1985), suggesting common origin.

For example, actin has been localized in a wide variety of plants, and the same four microtubule systems characteristic of dividing higher plant cells—the cortical array, preprophase band (PPB), mitotic spindle, and phragmoplast system (Fig. 7-7)—are also found in bryophytes (Brown and Lemmon, 1990c). As previously discussed (Chapter 6), charophycean algae exhibit variations in the occurrence of these four microtubule systems that may illustrate stages in the evolutionary origin of the plant cytoskeleton (Graham, 1982b). In addition, we have previously noted that the cytoskeletal features of dividing liverwort cells suggest how some aspects of transition from charophycean to embryophyte cytoskeletal systems may have occurred. Here we examine in more detail the evidence provided by charophycean algae that relates to evolution of cortical arrays, PPBs, microtubule organizing centers (MTOCs), phragmoplasts, actin involvement in cytoskeletal dynamics, and plasmodesmata. This will set the stage for subsequent development of hypotheses regarding the evolutionary origin of some fundamental aspects of land plant morphology.

Ultrastructural and immunofluorescence studies have demonstrated the occurrence of cortical microtubule arrays in the very large internodal cells of charalean algae (Pickett-Heaps, 1967a), where their orientation is related to cell wall deposition patterns as in higher plants (Williamson, 1991). Evidence for this relationship includes observations that colchicine treatment results in a change in cell shape, from cylindrical to spherical (Wasteneys, 1992). However, cortical microtubules in Charales may behave differently from those of higher plants (Pickett-Heaps, 1967a). For example, the cortical array of charalean internodal cells appears to be organized and nucleated independently of nuclear influence. Furthermore, the emerging view regarding the charalean cortical array is that it is more fragmented than the highly organized array of higher plants (Wasteneys, 1992). Actin has been localized in internodal cells of *Chara* with both antiactin antibodies and fluorescent rhodamine-labeled phallacidin. However, the actin appears to occur only in prominent cables that are associated with cytoplasmic streaming in the endoplasm (the cytoplasmic region adjacent to the large central vacuole and internal to the gel-like cortical cytoplasm). Association of actin with the cortical microtubular array does not seem to occur (Wasteneys, 1992), in contrast to other plant cells in which actin is reported to be associated with the cortical array (Lancelle and Hepler, 1991).

Transverse, hooplike arrays of microtubules also occur in interphase cells of various zygnematalean algae. It is intriguing that cortical microtubules persist throughout the cell cycle in *Spirogyra* (Grolig, 1992), whereas they occur at interphase in *Mougeotia* (Galway and Hardham, 1991), as they do in land plants. In contrast, cortical microtubule arrays are not present in *Coleochaete* (Brown and Lemmon, 1990c). Because presence of cortical arrays in charophytes seems to be highly correlated with absence of centrioles (which, in turn, appears to be associated with lack of zoospore production), it has been suggested that MTOC material is localized with centrioles when these are present (as in *Coleochaete*), but is otherwise distributed when centrioles are not present (Graham and Kaneko, 1991). That is, in the absence of centrioles, MTOCs "may change in form and position during the cell cycle" (Brown and Lemmon, 1992). This hypothesis

THE PLANT CYTOSKELETON 155

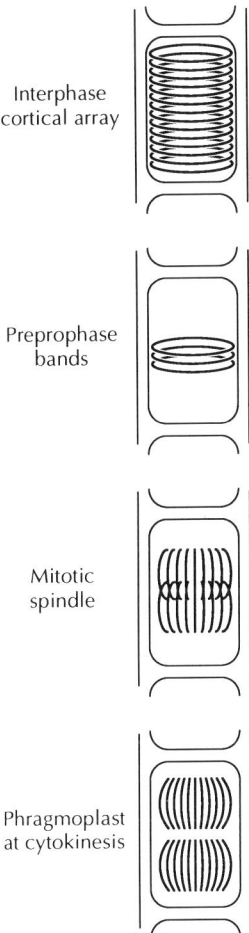

Fig. 7-7. Four distinct microtubule systems have been defined in higher plant cells, these occurring in different stages of the cell cycle. Interphase cells are characterized by cortical arrays of hooplike microtubules that are involved in patterning of the cell wall. Preprophase cells contain localized rings of microtubules whose location predicts the site where cytokinesis will later occur. Mitotic cells exhibit microtubular spindle arrays that are somehow involved in chromosomal separation and movement of sister chromatids to opposite cell poles. Cells in the process of cytokinesis possess a phragmoplast composed of microtubules and other elements. Phragmoplast microtubules control the fusion of vesicles to form the nascent cell plate. All of these arrays except preprophase bands have been documented to occur in charophycean algae. The evolutionary origin of PPBs and their precise function in plant cells has yet to be elucidated.

could be tested by determining whether antibodies to MTOC material localize with the point of origin of cortical or centriole-based microtubules. However, until recently, the nature of MTOCs has been obscure.

An emerging body of data suggests that gamma-tubulin is a highly conserved component of eukaryotic centrosomes (Zheng et al., 1991; Sterns et al., 1991; Kimble and Kuriyama, 1992) that may represent the long-sought MTOC material. The *mip* A gene of *Aspergillus nidulans* encodes gamma-tubulin that is associated with the spindle pole body, the fungal MTOC (Zheng et al., 1991). Gamma-tubulin is a minor protein, present at less than 1 percent of the level of alpha- and beta-tubulins. Homologous gamma-tubulin sequences have been cloned from yeast, a diatom, maize, and animals. Immunogold labeling localizes gamma-tubulin to pericentriolar material. It is estimated that there might be 10^4 to 10^5 gamma-tubulin molecules per centrosome, and that gamma-tubulin probably interacts with alpha- and beta-tubulins in some way (Sterns et al., 1991). The distribution of a gamma-tubulin-related polypeptide in two cell types, and in three flowering plant species, was consistent with a role in the organization of microtubules (Liu et al., 1993). Application of gamma-tubulin antibodies to cells of charophytes and seedless plants could prove to be a powerful technique for following evolutionary changes in MTOC localization and function postulated to have occurred during plant evolution.

When most embryophyte cells enter mitosis, cortical microtubules disappear, whereupon a localized band of microtubules appears during preprophase. Actin microfilaments have been localized in the band (Ding et al., 1991). The preprophase band (PPB), first described by Pickett-Heaps and Northcote (1966), is of very great morphogenetic interest because it predicts the plane of cytokinesis (Mineyuki and Gunning, 1990). Although PPBs appear to be absent from most charophyte cells, it has been suggested that a region of persistent cortical microtubules that sometimes occurs in *Mougeotia* might represent a precursor of the plant PPB (Galway and Hardham, 1991). Grolig (1992), however, suggests that these microtubules represent an intermediate stage in the depolymerization of the cortical array and are not homologous to higher plant PPBs. Further immunofluorescence studies of microtubular dynamics in charophycean cells are needed to determine whether or not these organisms produce cytoskeletal arrays that might be ancestral to PPBs.

Charophycean algae also provide some evidence for the capacity of plastids to nucleate microtubules (Pickett-Heaps and Wetherbee, 1987). This may represent the ancestral character state from which was derived the plastid-based MTOC commonly observed in monoplastidic dividing cells of land plants by Brown and Lemmon (1990c). Antibodies to MTOC material would be helpful in following this transition by comparative examination of monoplastidic mitosis in charophytes, bryophytes, and pteridophytes. Such comparative studies might also shed light on the apparent transfer of mitotic spindle MTOC function from plastid envelope to nuclear envelope that may have been associated with the origin of polyplastidic meristematic cells within each of these groups.

Goto and Ueda (1988) used fluorescence microscopy to follow changes in the distribution of microfilament bundles (F-actin) during the cell cycle of *Spirogyra*.

Their technically excellent work revealed the occurrence of four groups of bundles: those dispersed in the cytoplasm near the cell surface, parallel arrays associated with the cell membrane, bundles at the edges of chloroplasts, and microfilaments surrounding the nucleus. The parallel arrays and chloroplast bundles persisted through the cell cycle, but the nuclear microfilaments were not observed at anaphase. The dispersed microfilaments aggregated at prophase to form a ring at the cell periphery that marked the division site. Subsequently, a cleavage furrow developed at this region, then progressed centrifugally at cytokinesis. When cytokinesis was completed, microfilaments associated with the furrow disappeared (Goto and Ueda, 1988). This work is of importance in elucidating the evolution of division site determination in plants, a topic that is considered further in the section of this chapter dealing with phragmoplasts and cytokinesis. However, progress in this area requires a comparative survey of actin dynamics through the cell cycle of other charophycean algae. Related work includes documentation of the occurrence of actin in *Coleochaete* and *Mougeotia* (Marchant, 1976) and demonstration by Hashimoto (1992) that actin filaments, but not microtubules, are involved in division of chloroplasts of the charophyte *Closterium ehrenbergii*.

In addition to tubulin and actin, other evolutionarily conserved proteins that probably occur in plant cells should be sought in charophycean algae. For example, plant cells may possess an integrin-like protein that might generate connections between extracellular material and the intracellular cytoskeleton (see discussion in Delmer, 1991). MARCKS (myristolated alanine-rich C kinase substrate) is a specific substrate of protein kinase C that also binds actin and calmodulin. This molecule may be a regulated crossbridge between actin and the cell membrane that links the calcium-calmodulin and protein kinase C transduction pathways (Hartwig et al., 1992) (see Chapter 9 for further discussion of signal transduction). Plant cells also contain proteins antigenically related to intermediate filaments (= vimentin). Antibodies to intermediate filaments label cortical microtubules, PPBs, chromosomal kinetochores, and phragmoplasts. Intermediate filaments also share similar sequences and topological domains with nuclear lamins (Shaw et al., 1991). During cell division of animal cells lamins become hyperphosphorylated, causing subunits to become soluble, which results in the breakdown of the nuclear envelope. Evidence for the occurrence of lamins in plants is inconclusive (Shaw et al., 1991). It is, however, significant that charophycean algae and land plants share the feature of open mitosis, i.e., early breakdown of the nuclear envelope. This suggests the involvement of nuclear proteins similar to lamins and a similarity in their regulation. Unfortunately, charophycean algae have not been investigated for the presence of integrin, MARCKS, intermediate filaments, or nuclear lamins.

PHRAGMOPLASTS AND CYTOKINESIS

In higher plants there appears to be a dynamic relationship between spindle and phragmoplast microtubule systems, and the phragmoplast is believed to control the formation of new cell walls and sites of intercellular communication (plasmodesmata). The siting of cell walls and plasmodesmata is integral to plant morphogenesis, but the mechanisms for determining planes of cytokinesis are not well understood. It has been suggested that studies of "lower" plants may provide clues to the evolution of division site control mechanisms (Wick, 1991). For example, the transition from tip growth in development of a filamentous protonema to a two-dimensional prothallus is a system that has proven useful in following correlative changes in cytoskeletal components of ferns (Wada and Murata, 1991). Examples of the value of bryophyte experimental systems include immunofluorescence analysis of cytoskeletal involvement in control of morphogenesis in *Funaria* and *Physcomitrella patens*. The developmental switch from tip growth in the caulonema stage of moss protonemata to meristematic growth in bud development is correlated with the appearance of PPBs at the latter stage (Doonan, 1991). However, PPBs were noted to occur in *Funaria* protonemata when specialized tmema cells are formed. Sawidis et al. (1991) interpreted this finding as evidence that PPBs are present in cells in which the mother cell wall expands or is ruptured close to the cell plate connection site. The siting of divisions leading to branch formation in the moss protonema is presaged by the development of a ring of actin (Quader and Schnepf, 1989).

The cytoskeletal work on bryophyte and pteridophyte gametophytes suggests that comparative analysis of the cytoskeletal control of development in unbranched versus branched filamentous charophytes is likely to illuminate the evolutionary origin of plant cytoskeletal control of cytokinetic location and, hence, morphology. A series of ultrastructural studies of cell division in various charophycean algae have suggested a hypothetical transformation series for the evolutionary origin of the plant phragmoplast and cell plate (Graham and Kaneko, 1991). Charophycean algae therefore represent a particularly useful system in which to analyze patterns of evolutionary change in cytoskeletal dynamics as related to cytokinesis.

Whereas cytokinesis is effected by a cleavage furrow (a centripetal invagination of the cell membrane) in the putatively primitive taxa *Chlorokybus* and Klebsormidiales (Lokhorst and Star, 1985; Lokhorst et al., 1988; Floyd et al., 1972) (Fig. 7-8), and furrowing also occurs in Zygnematales (Goto and Ueda, 1988, for example), a small phragmoplast has also been observed in some filamentous zygnematalean algae. The latter cytokinetic system is proposed to represent an intermediate stage in the evolutionary development of the plant phragmoplast (Fowke and Pickett-Heaps, 1969; Pickett-Heaps and Wetherbee, 1987; Galway and Hardham, 1991). In *Zygnema*, persistent spindle microtubules associate with the leading edge of the cleavage furrow and are apparently involved in guiding Golgi vesicles to this site. However, a phragmoplast array does not form in the *Zygnema* species examined (Grolig, 1992). Although some species of *Mougeotia* apparently do not produce phragmoplasts, Galway and Hardham (1991) elegantly demonstrated the occurrence of a small phragmoplast in a species of *Mougeotia*. The closely related *Spirogyra* also produces small phragmoplasts consisting of microtubules, Golgi vesicles, and unidentified electron-dense material. Phragmoplasts of *Spirogyra* differ from those of land plants in being hollow rather than disk shaped. It has been suggested that *Zygnema, Mougeotia*, and *Spirogyra* form

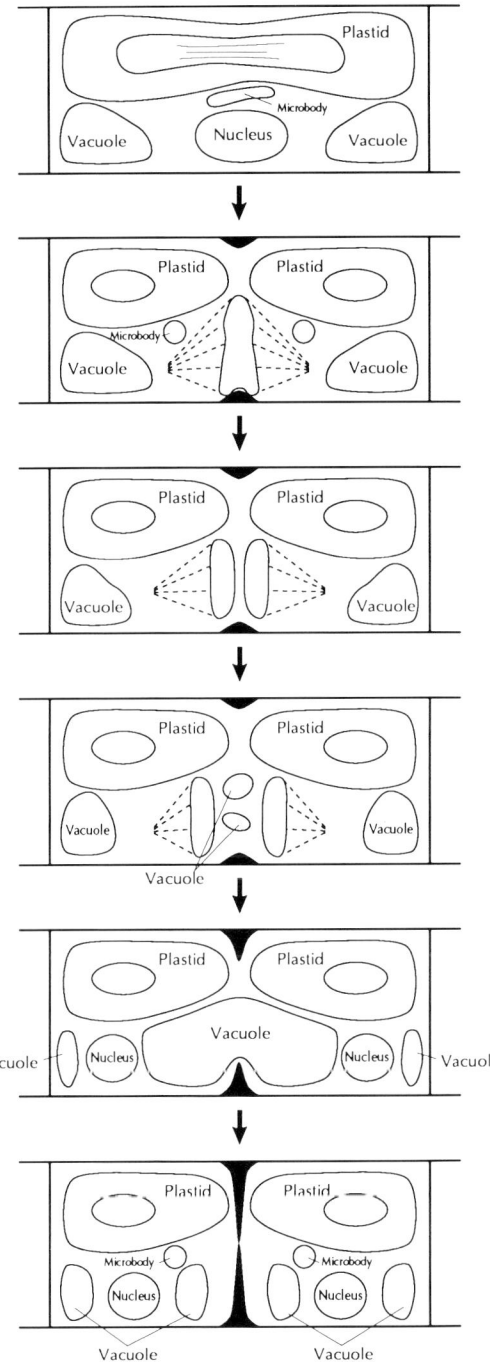

Fig. 7-8. Illustrations of mitosis and cytokinesis in the charophyte *Klebsormidium* (from Floyd et al., [1972], with permission of the *Journal of Phycology*). Early events include division of the single plastid and pyrenoid, coordinated division of the single peroxisome, and development of a precocious cleavage furrow. Cytokinesis involves extension of the cleavage furrow through a vacuolate region.

a series illustrating increasingly more complex phragmoplast development within Zygnematales (Grolig, 1992), but the adaptive advantage of phragmoplast elaboration in this group of algae is not understood. It may be conjectured that even a small phragmoplast contributes to increased coherence of filaments, inasmuch as other unbranched, filamentous charophytes (*Klebsormidium* and relatives), which lack phragmoplasts, are also more easily dissociated into short filaments or single cells. This hypothesis might be tested by comparing the coherence of zygnemataceaen species (within *Mougeotia,* for example) that lack or possess phragmoplasts. Increased adherence of cells into filaments might confer an advantage in the form of resistance to damage caused by drag or herbivory, or increase in photosynthetic surface area useful in the periphytic life-style, but these ideas require testing.

In *Spirogyra,* an actin-based cleavage furrow extends inward, impinging upon the central phragmoplast in formation of cross walls (Goto and Ueda, 1988; Grolig, 1992). Because actin filaments are recruited into cleavage furrows of dividing animal cells, the involvement of actin in cleavage is most likely very ancient and probably occurs in other charophytes as well as in ancestral forms. Interestingly, *Spirogyra* also exhibits a phragmosome-like structure, wherein the nucleus is suspended in the center of a vacuolar region by a scaffold of cytoplasmic strands (Grolig, 1992) (Fig. 7-9). Phragmosomes occur in highly vacuolate higher plant cells; they are complex, actin-based arrays of organelles that, like

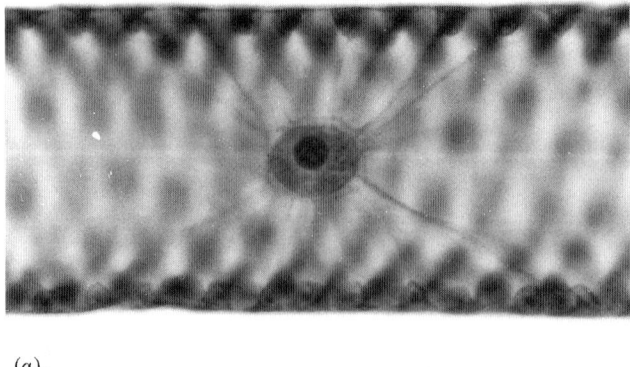

(a)

Fig. 7-9. *a.* Interphase nuclei of *Spirogyra* are suspended by conspicuous cytoplasmic threads (shown here in a Fast Green stained preparation). It has been suggested that this represents a structure similar to actin-containing organellar bands termed phragmosomes, which are observed in vacuolate cells of higher plants *b.* In such cells, phragmosomal actin filaments mark the division plane, and the phragmoplast develops within the phragmosome. Reproduced from Lloyd, (1991). Cytoskeletal elements of the phragmosome establish the division plane in vacuolated higher plant cells. In: Lloyd (ed.), *The Cytoskeletal Basis of Plant Growth and Form,* Academic Press, pp. 245–257, used with permission of Academic Press.

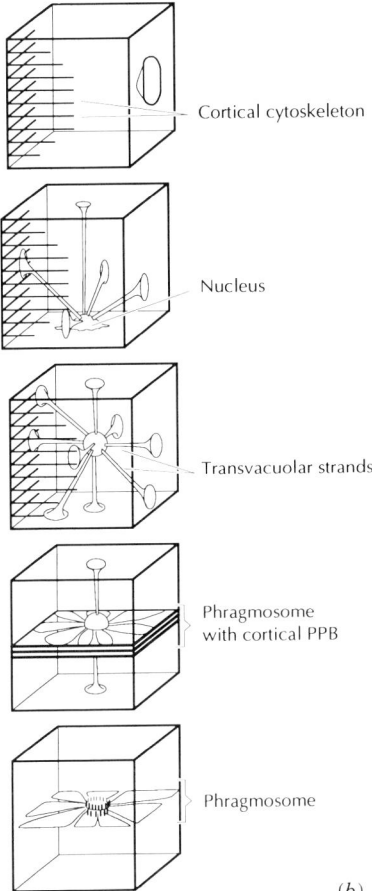

Fig. 7-9. *(Continued)*

PPBs, predict the polarity of cell divisions (Venverloo and Libbenja, 1987; Lloyd, 1991). The possibility that the origin of phragmosomes extends back to the charophycean algae is exciting, because it suggests that phragmosomes may be a more ancient correlate of wall site localization than PPBs. Wall site determination in land plants might thus be based on an ancient, fundamental mechanism also present in charophycean algae (Graham and Kaneko, 1991).

Marchant and Pickett-Heaps (1973) observed that phragmoplasts similar to those of land plants occurred in *Coleochaete scutata*, but that phragmoplast form differed, depending on division plane direction. In cells having the division plane oriented parallel to the thallus edge (circumferential division), wall formation partially occurred by means of fusion of vesicles from the wall area inward (centripetal development). This ingrowing wall intercepted a microtubular array interpreted as a small, central phragmoplast, similar to that of zygnematalean algae.

In contrast, cells undergoing radial division (division in a plane perpendicular to the thallus circumference) formed a more typical phragmoplast that was involved in cell plate development from the center of the cell toward the parental wall (centrifugal cell plate development), as in land plants (Fig. 7-10). A small vestige of a peripheral furrowing system was noted at the periphery of radially dividing cells (Marchant and Pickett-Heaps, 1973).

A similar ingrowth of wall material is present in metaphase cells of *Coleochaete orbicularis* (Fig. 7-11). This wall protrusion predicts the site of cell plate fusion with the existing wall, but the docking mechanism has not been elucidated. Interestingly, Brown et al. (1993) used tubulin immunolocalization to establish that *C. orbicularis* produces a typical land plant-like phragmoplast prior to both radial and circumferential cytokinesis. These workers noted that the structure of mature *Coleochaete* phragmoplasts, fusiform bundles of microtubules bisected by a dark line, was indistinguishable from that of higher plant phragmoplasts. In radially-dividing cells, as expected by analogy with land plant cytokinesis, expansion of phragmoplasts was associated with centrifugal formation of a new cell plate. Surprisingly, though, wall deposition in circumferentially divisions was almost completely centripetal, and did not follow the midline of the phragmoplasts. This was interpreted as reflecting the evolutionary imposition of a cytoskeletal innovation (phragmoplast) onto a more primitive cytokinetic mechanism (furrowing) inherited from ancestral unbranched, filamentous charophytes.

Cytokinesis in *Chara* also involves a phragmoplast in which the cell plate is formed by fusion of vesicular elements derived from the Golgi apparatus. Among

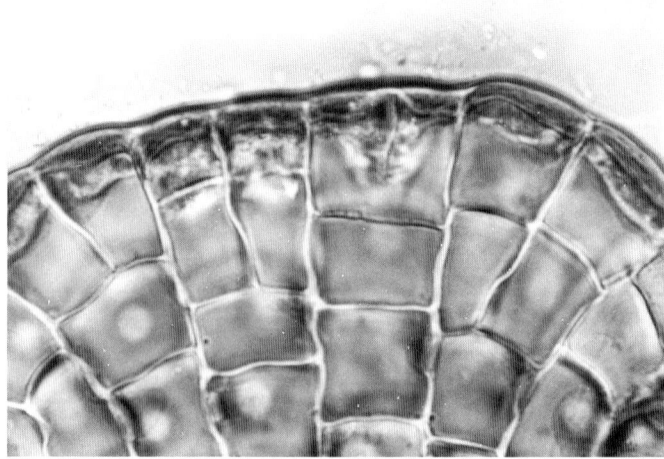

Fig. 7-10. Although dividing cells of *Coleochaete orbicularis* produce phragmoplasts similar to those of land plants, plantlike cell plates develop only during radial divisions (right). Circumferential cytokinesis is apparently accomplished primarily by ingrowth of a cleavage furrow (left).

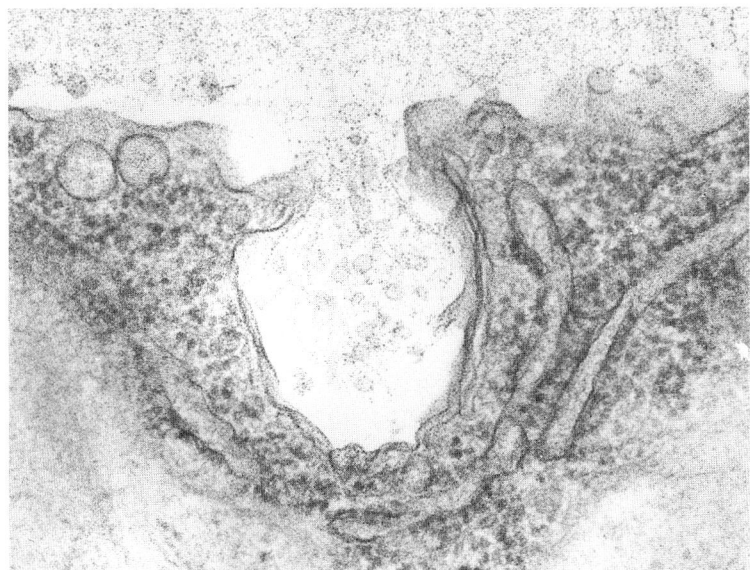

Fig. 7-11. Radial divisions in *Coleochaete orbicularis* involve production of a small precocious furrow that defines the site at which cell plate fusion with the parental cell wall will later occur.

charophytes, cytokinesis in *Chara* most closely approximates that of higher plants. However, microtubules are reportedly less abundant in *Chara* phragmoplasts than in those of land plants, and fusion of *Chara* cell plate vesicles is reported to occur along the whole plate simultaneously, rather than starting in the center and growing outward as in higher plants (Pickett-Heaps, 1967b). The basis for these differences is not apparent and deserves additional study.

It has been suggested that the land plant cytokinetic system evolved by a progressive decrease in the activity of the peripheral wall-forming system coupled with increased development of the central phragmoplastic system (Graham and Kaneko, 1991). This would seem to require phyletic change in the mechanism used to direct Golgi vesicles transporting polysaccharides and membrane-bearing cellulose-synthesizing complexes to the area where the cell plate is to form. In the ancestral system exhibited by primitive charophytes, Golgi vesicles are guided to and fuse with peripheral cell membranes, whereas in derived phragmoplasts, Golgi products are channeled (by microtubules and microfilaments) into the cell center, where vesicles fuse with each other. Studies of wall development in desmids suggest that changes in localization of calcium (via targeted deposition of calcium-binding proteins in membranes) could be related to evolutionary modification in the site of vesicle fusion during cytokinesis, but further study is indicated. Although considerable progress has been made toward molecular dissection of the secretory pathway in animal and fungal cells (Rothman and Orci, 1992), equivalent understanding of Golgi function and control of vesicle fusion in plants has not been achieved. Perhaps developmental studies of the "primitive" phragmoplasts of charophycean algae will illuminate fundamental aspects of the more highly derived cytokinetic systems of plants.

PLASMODESMATA AND DESMOTUBULES

In land plants, plasmodesmata are regarded to be important channels of cell-to-cell communication that may influence plant development (Gunning and Overall, 1983), but the evolutionary history of embryophyte plasmodesmata is incompletely known. Although plasmodesmata appear to be absent from *Chlorokybus*, Klebsormidiales, and Zygnematales, plasmodesmata are formed during cell plate

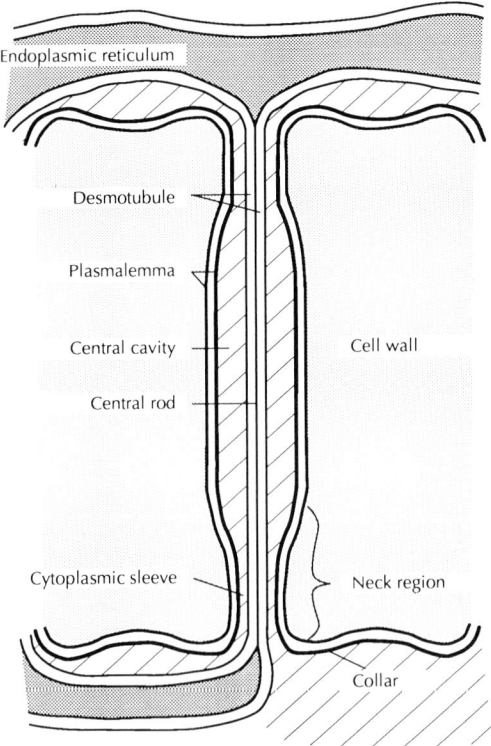

Fig. 7-12. Higher plant plasmodesmata are characterized by the occurrence of a central membranous tubule (desmotubule) originating from endoplasmic reticulum trapped in wall development during cytokinesis. Reproduced from Oleson and Robards (1990), the neck region of plasmodesmata. In: Robards et al. (eds.), *Parallels in Cell to Cell Junctions in Plants and Animals,* Springer-Verlag, New York, p. 250. Used with permission of Springer-Verlag. Branched charophytes produce plasmodesmata, but whether these are structurally equivalent to those of higher plants has not been clearly established. Evolutionary origin, development, and function of desmotubules in higher plants are not well understood.

development in *Chara* and *Coleochaete*. However, it has been suggested that charophycean plasmodesmata lack a structure characteristic of higher plant intercellular connections, the desmotubule (Robards and Lucas, 1990) (Fig. 7-12). Desmotubules are thought to originate by the entrapment of endoplasmic reticulum (ER) in the growing plant cell plate, but their role in intercellular communication is not understood. Plasmodesmata appear in several other green algal lineages (Mattox and Stewart, 1984). With the possible exception of *Aphanochaete elegans,* however, for which ER is reported to enter plasmodesmata (Stewart et al., 1973), desmotubules have not been demonstrated to occur in these algal groups. Some ultrastructural evidence suggests that central elements resembling desmotubules may occur in charophycean plasmodesmata (references in Marchant, 1976) (see photos of *Chara* plasmodesmata in Pickett-Heaps, 1975, and Kwiatkowska and Maszewski, 1986, for examples). However, additional work utilizing techniques for improved cellular preservation needs to be done to determine whether desmotubules equivalent to those of land plants occur in charophycean plasmodesmata. If desmotubules are indeed absent from plasmodesmata of charophytes, but generally present in those of embryophytes, this character should be added to the list of autapomorphies of the embryophyte clade.

Recent work has demonstrated that antibodies to connexin proteins of animal gap junctions recognize molecules associated with higher plant plasmodesmata (Meiners et al., 1991). This evidence suggests that there is biochemical homology of some elements of plant plasmodesmata and animal gap junctions. In view of the apparent evolutionary conservation of connexins, various algal plasmodesmata should be examined using connexin antibodies. Such studies may illuminate the process of parallel evolution of plasmodesmata in algae and plants and clarify the evolutionary origin of land plant plasmodesmata.

PARENCHYMA, MERISTEMS, AND MORPHOGENESIS

Kaplan and Hagemann (1991) have suggested that higher plant structure and development are described more adequately by the "organismal theory," which interprets morphology as a property of whole organisms, rather than the result of aggregation of individual units (cells), as described by the "cell theory." They note that plant cell division and growth involve successive partitioning of protoplasts by insertion of walls penetrated by plasmodesmata, which perpetuate cytoplasmic continuity. The plant body is thus considered to be a protoplasmic tube that has been secondarily partitioned into a multicellular structure, analogous to a house whose external architecture is unchanged by the erection of new internal walls. Plasmodesmata then function as doorways whereby intercellular communication may be maintained or selectively blocked. The organismal "theory" suggests that plant form is an emergent property of the organism as a whole. As evidence for this concept, Kaplan and Hagemann (1991) cite Poethig's (1987) conclusion that patterns of cellular differentiation in plants, and gross plant morphology, appear to be independent of cell lineages. Comparative studies of charophyte development may illuminate the process by which fundamental aspects of land plant morphology evolved. Such information may be useful in evaluation of alternative concepts of plant morphology and development, i.e. "organismal versus cell theories."

As we have earlier observed (Chapter 4), charophycean algae exhibit a continuum of morphological complexity from unicells to unbranched filaments to branched filaments, culminating in forms exhibiting even higher levels of organization. Examples of the latter include the complex thalli of charalean algae and thalloid species of *Coleochaete*, which superficially resemble parenchymatous gametophytes of the moss *Sphagnum* and various ferns. Transitions in level of charophyte morphological complexity are correlated with changes in internal cellular organization. For example, well-developed phragmoplasts and plasmodesmata do not occur in charophytes having morphologies below the level of branched filaments. It has been proposed that the ancestral mechanism for accomplishing cytokinesis, an inward growth of the cell membrane known as furrowing, may have been incompatible with the degree of cytoplasmic reorganization required for branch formation by monoplastidic algae (Graham and Kaneko, 1991). During cytokinesis in morphologically simple charophytes such as *Chlorokybus* or *Klebsormidium*, the new wall may begin emergence from the cell periphery as early as preprophase, actively extending inward during metaphase, with cleavage well-advanced at anaphase (Stewart et al., 1973; Lokhorst et al., 1988) (Fig. 7-8). Even in zygnematalean algae that exhibit small central phragmoplasts, the peripheral wall-forming system may begin development precociously (Fowke and Pickett-Heaps, 1969). Because the majority of charophytes are monoplastidic, it is possible that precocious inward growth of cross-walls presents a physical barrier to plastid reorientation or other cellular processes that must occur in preparation for division at right angles to a filament axis in branch formation. Change in plastid orientation may, however, be physically compatible with centrifugal phragmoplast formation at telophase. The need for changes in plastid orientation during branch formation thus may have operated as a selective force resulting in a phylogenetic trend toward loss of the peripheral wall-forming system, coupled with increased reliance on the phragmoplast mode of cytokinesis. This hypothesis requires further evaluation, however, and there may be alternative explanations for the evolutionary elaboration of the phragmoplast in charophycean ancestors of plants.

Interestingly, transition from furrowing to a cell plate mode of cytokinesis also seems to have occurred in other green algal lineages. For example, the chlorophycean genus *Fritschiella* is a highly branched, morphologically complex terrestrial form that evolved a cell plate (McBride, 1967), although not a phragmoplast, in parallel with cell plates of higher charophytes. Since *Fritschiella* cells are monoplastidic, perhaps changes in chloroplast orientation necessary for branching operated as a selective force in the evolution of a centrifugally developing cell plate. Parallel evolutionary development of phragmoplast, cell plate, and plasmodesmata also occurred in chroolepidacean algae (Chapman and Henk, 1986) (Figs. 7-13 and 7-14). Chroolepidacean algae (including *Cephaleuros* and *Trentepohlia*) resemble charophytes in having persistent spindles, but unlike charophytes and plants, have closed

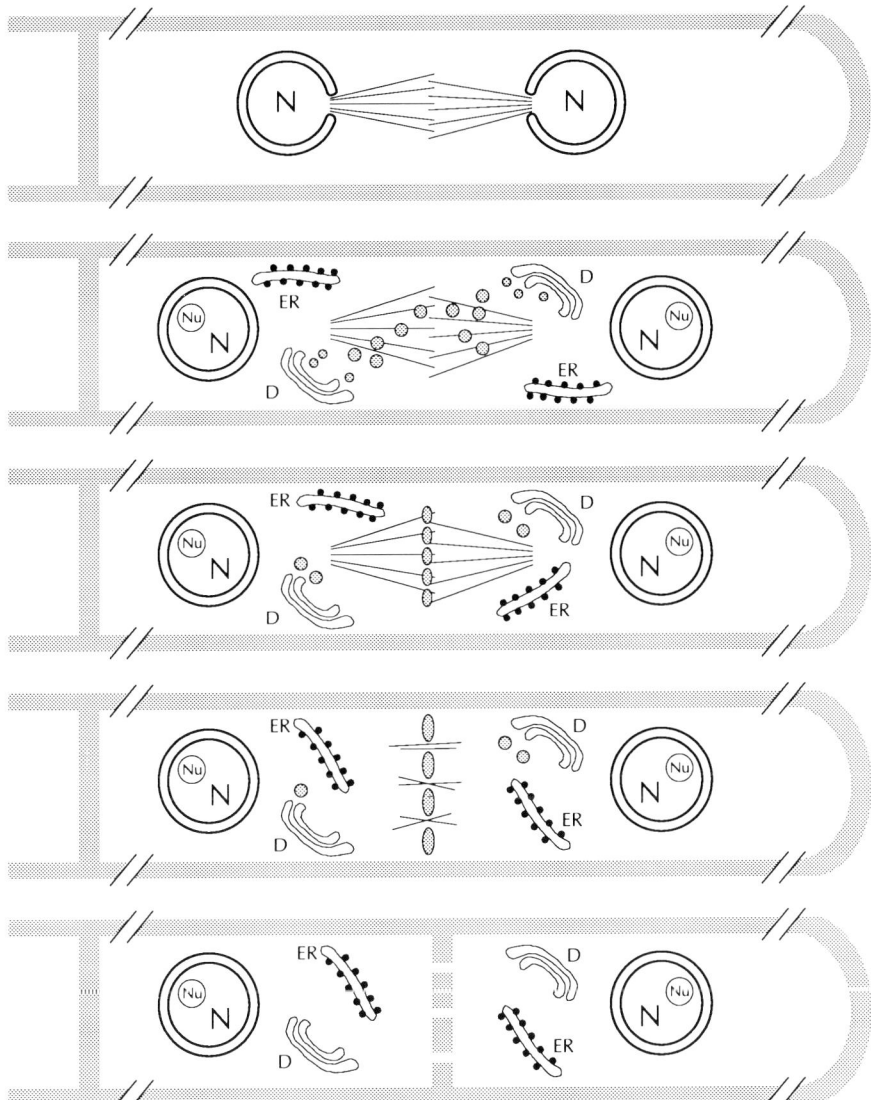

Fig. 7-13. Diagram of cytokinesis in *Cephaleuros parasiticus,* a member of the terrestrial green algal group Chroolepidaceae, from Chapman and Henk (1986), with permission of the *Journal of Phycology.* The nuclear envelope remains virtually intact during mitosis, and cytokinesis involves a microtubular array somewhat similar to plant and charophyte phragmoplasts, production of a cell plate, and generation of plasmodesmata. The phylogenetic affinities of *Cephaleuros* and the Chroolepidaceae are indistinct.

168 THE EVOLUTION OF PLANT MORPHOLOGY

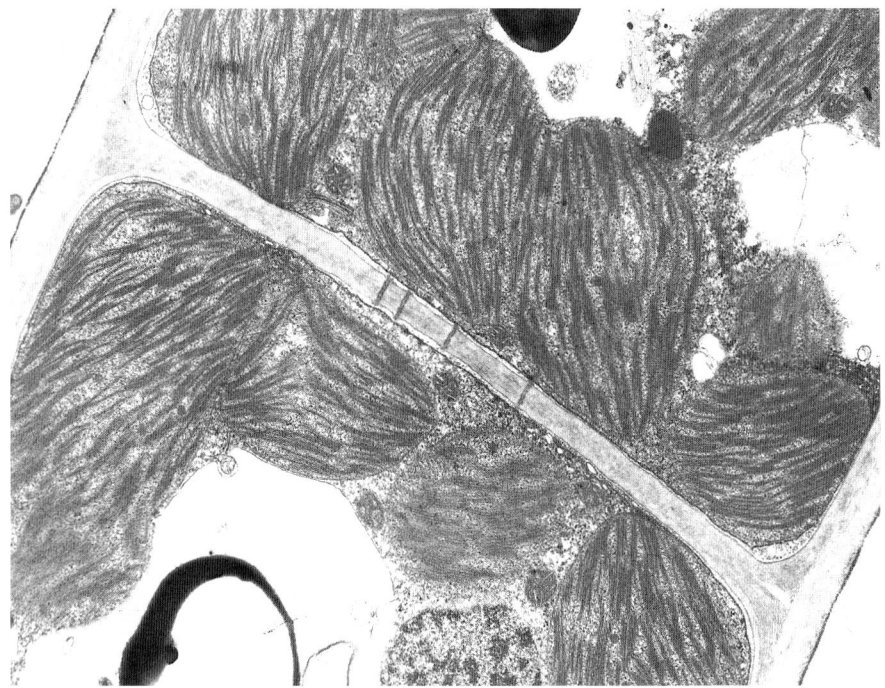

Fig. 7-14. *Trentepohlia aurea,* another member of the terrestrial Chroolepidaceae, illustrates occurrence of plasmodesmata and polyplastidy in this group of green algae.

mitosis. These terrestrial algae are highly branched, but polyplastidic (Fig. 7-14), arguing against the hypothesis that evolution of cell plates is related to branching in monoplastidic forms. It is possible that the ancestors of chroolepidacean algae were monoplastidic and that evolution of cell plates preceded polyplastidy, but there is no evidence for this conjecture. The selective advantages provided by evolutionarily advanced cytokinetic mechanisms in *Fritschiella* and chroolepidacean algae are not apparent, unless, as in charophytes, cell coherence and ability to branch are enhanced. Further study of cytokinesis in these algae may shed light on analogous selective pressure/cytokinetic adaptations in charophytes.

The persistent spindle, which is characteristic of all charophytes, including forms that are unicellular or composed of unbranched filaments, may have been the evolutionary precursor of the phragmoplast. An evolutionary trend toward increasing morphological complexity in extant charophytes then appears to have been coupled with phyletic change in the mode of cytoplasmic partitioning. Transition from the furrowing mode of cytokinesis, typical of unicellular or colonial protists, to small, then increasingly more embryophyte-like phragmoplasts in higher charophytes marks the origin of intracellular wall development in plants. At cytokinesis unicellular charophytes make copies of individuals, whereas branched, filamentous forms essentially subdivide the individual into compartments whose contents are interconnected by plasmodesmata. The pattern of evolution of cytokinesis in charophycean algae is thus compatible with the organismal concept of plant morphology and development as presented by Kaplan and Hagemann (1991).

The complexity of cytoskeletal and endomembrane systems controlling intracellular wall development in higher charophytes and plants argues strongly for a concept of cell walls as integral components of plant cells. As we have earlier noted in this chapter, however, evolutionary origin of these systems is likely to be related ultimately to scale biogenesis in ancestral green flagellates. Conceptually, prasinophyte cell coverings can be related to extracellular matrices, because the scales do not generally form a coherent cell enclosure. Charophycean algae, as phylogenetic intermediates between unicellular flagellates and land plants, thus provide evidence in support of divergent views of the nature of plant cell walls. An evolutionary perspective suggests that both viewpoints in the debate regarding the nature of plant cell walls are valid.

The morphological variation exhibited by modern charophytes may illustrate changes in fundamental developmental processes necessary for the evolution of plant morphogenesis, tissue formation, and meristems. The evolution of branching in monoplastidic charophytes probably generated a mechanism for accomplishing changes in plastid orientation, which is a necessary component of morphogenesis in monoplastidic land plant tissues. Roy Brown and Betty Lemmon (1984) pointed out that if plastids in root meristems of *Selaginella* and *Isoetes* did not migrate prior to division, and the cell division axes were positioned to ensure that each daughter cell received a plastid, the resulting tissue would resemble a cube of cells rather than an axis. These workers also suggested that the preprophase band serves to stabilize or anchor the plastid isthmus in preparation for plastid division at the future site of cytokinesis. This hypothesis is supported by observations on the association of PPBs with the plastid isthmus in hornwort cells. As previously mentioned in Chapter 6, dividing cells of hornworts are invariably monoplastidic. In such cells, preprophase plastid migration, the earliest sign of impending division, occurs before any change in the arrangement of cortical microtubules. The plastid then develops a constriction at the midpoint, the isthmus, which intersects the plane of subsequent division. A PPB appears after division polarity has been thus established and is unusual in its asymmetry, having tightly associated microtubules on the side of the cell over the plastid isthmus, but more widely spaced tubules elsewhere. The function of this putatively primitive PPB was suggested to be orientation of the spindle axis relative to the future plane of cell division (Brown and Lemmon, 1988). The fact that PPBs do not occur in morphologically complex charophycean algae, but are found nearly universally in mitotic cells of land plants, suggests that they originated prior to the divergence of the common ancestor of modern embryophyte groups. The PPB may have first appeared in transitional plants that had inherited monoplastidy and plastid migration from charophycean ancestors, in response to selective pressure for plastid stability with respect to the developing spindle. PPBs might then have been retained by descendants having derived polyplastidic cells in which PBB function is less obvious, such as division site preparation in cells whose walls undergo expansion, as suggested by Sawidis et al. (1991).

Two species of *Coleochaete* produce thalli characterized by coherent cell lineages derived by circumferential and radial divisions that bisect marginal cells (Fig. 7-15). These divisions are at least superficially similar to those occurring in land plant meristems, and the resulting morphology has been described as thalloid or parenchymatous. As previously noted (Chapter 5), it has been suggested that various other *Coleochaete* species provide putative examples of ancestral morphologies, and that a transformation series leading to the origin of tissue organization analogous to that of land plants can be deduced (Hagemann, 1978; Graham, 1982b). *Coleochaete soluta* has been cited as an illustration of a morphological intermediate, i.e., a branched, filamentous form in which branching (radial) divisions occasionally bisect cells (Fig. 7-16). This does not necessarily indicate that tissuelike morphology in *Coleochaete* is homologous to land plant parenchyma. However, available evidence indicates that embryophytes, modern *Coleochaete*, and Charales share a common ancestor that was likely to have been a branched filament similar to certain extant *Coleochaete* species (Chapter 4). If, as proposed in Chapter 5, the algal ancestors of land plants were inhabitants of the freshwater littoral zone, they were likely to have been subjected to physical and biotic selection pressures similar to those experienced by modern *Coleochaete*. Thus, morphological variation among extant *Coleochaete* species provides a useful model for exploring the evolutionary origin of parenchyma.

The meristem of thalloid *Coleochaete* species, like that of the *Charales*, is terminal, or "apical" (Fig. 7-17). The *Coleochaete* meristematic region, however, extends around the thallus periphery and thus is not so localized as meristematic regions of Charales or land plants. In other words, every peripheral cell of thalloid *Coleochaete* species is capable of division. Transitional stages between such broadly

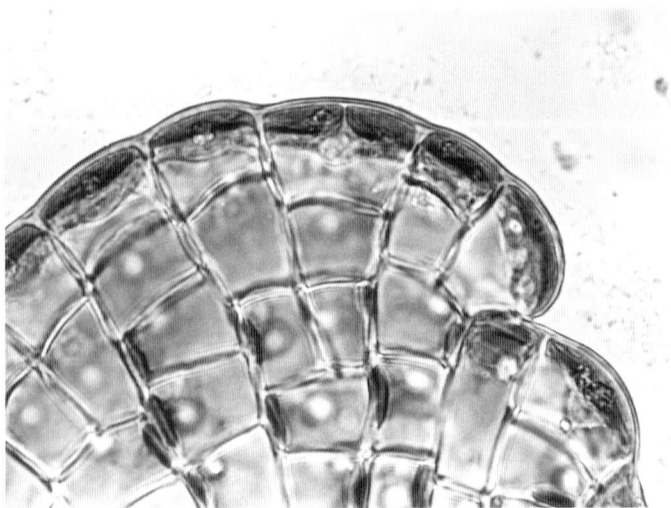

Fig. 7-15. At the peripheral meristem of thalloid *Coleochaete* species (such as *C. orbicularis*, shown here), radial divisions are homologous to branching divisions of filamentous species, but in contrast to branching divisions and similarly to plant meristematic divisions, bisect parental cells.

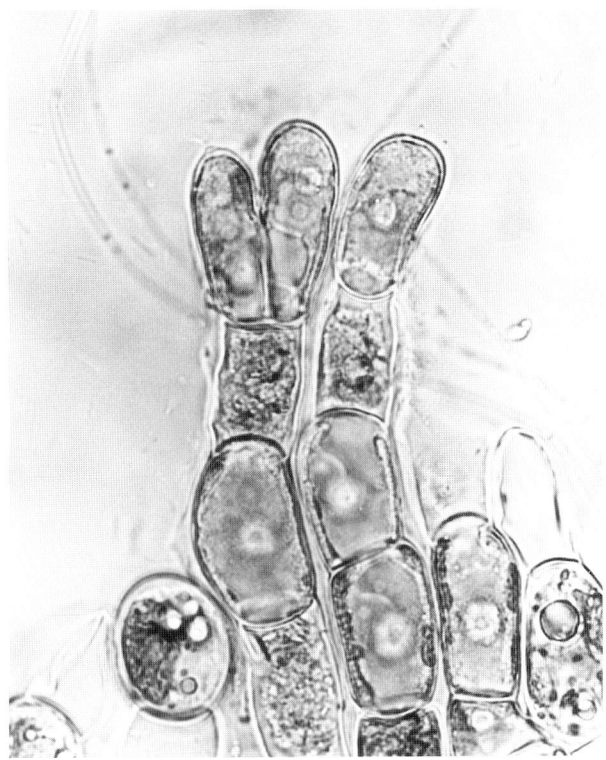

Fig. 7-16. The peripheral meristem of the filamentous species *C. soluta* is characterized by Y-shaped cells. Usually, one arm of the Y is cut off to form the first cell of a branch, but occasionally, as shown here, radial divisions bisect parental cells as in thalloid species of *Coleochaete*. It is possible that similar morphological intermediates also occurred during the transition from branched filamentous charophytes to the first parenchymatous representatives of the land plant lineage.

peripheral meristems and highly localized apical meristems, such as those of seedless plants, are not apparent. However, such transitions occur in some analogous brown algal systems whose study might illuminate the mechanisms underlying alteration of meristematic form. The phaeophytan group composed of *Padina, Zonaria,* and *Dictyota* can be arranged to form a putative transformation series illustrating transition from broadly peripheral meristematic activity (as in *Coleochaete*), through small groups of apical meristematic cells, to single apical cells (analogous to meristems of various seedless plants). When apical cells of *Dictyota* divide, dichotomous branching of the parenchymatous thallus occurs, yielding morphology reminiscent of liverwort and hornwort morphology (Fig. 7-18). Of course, it is possible that the brown algal transformation should be read in the opposite direction, i.e., from localized to broad meristematic zones.

An impressive correlative immunofluorescence/transmission electron microscopic study during the cell cycle of *Dictyota dichotoma* meristematic cells

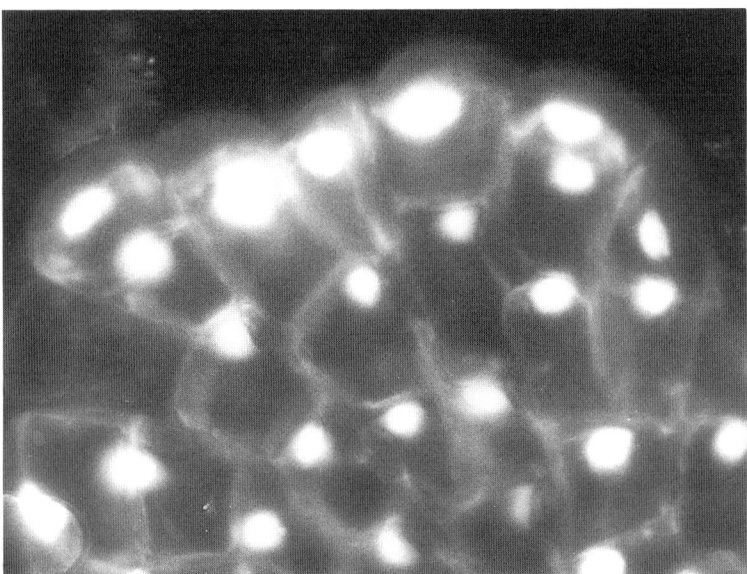

Fig. 7-17. The apical meristem of thalloid species of *Coleochaete* consists of a continuous row of peripheral cells, shown here in fluorescence view of DAPI-treated thalli. DAPI binds specifically to DNA, thus illuminating nuclei. Note that nuclei of peripheral meristematic cells appear to be significantly larger than those in nondividing interior cells.

demonstrated that microtubule organization is a function of centriolar sites, and that cortical microtubule arrays are absent. In these respects, the cytoskeleton of *Dictyota* is similar to that of *Coleochaete,* and differs from that of land plants. This suggests that, in general, parenchymatous organization and localized meristematic function are not dependent on the occurrence of cortical microtubule arrays or preprophase microtubule bands. By analogy, early land plants composed of tissues generated by a localized meristem might have retained ancestral cytoskeletal features such as centrioles, and lacked cortical and PPB microtubules. In other words, it is possible that the land plant cortical array and PPB arose after the appearance of parenchyma and localized meristems. The occurrence of polar spindle organizers with astral microtubules in certain liverworts (Brown and Lemmon, 1990a; Brown and Lemmon 1992), supports this conjecture. However, the Dictyotales do not produce phragmoplasts at cytokinesis; rather, cytoplasmic partitioning occurs when a membrane is laid down between microtubule arrays that enclose nuclei (Katsaros and Galatis, 1992 and references cited therein). It is possible that origin of land plant issues and meristems was somehow constrained by phragmoplasts or other factors such that early loss of centrosomes and gain of cortical microtubules became necessary. Further analysis of meristematic function in seedless plants may illuminate this issue.

Whereas individual apical meristematic cells of *Coleochaete* usually divide in one of two directions (radially or circumferentially), the single apical cell of

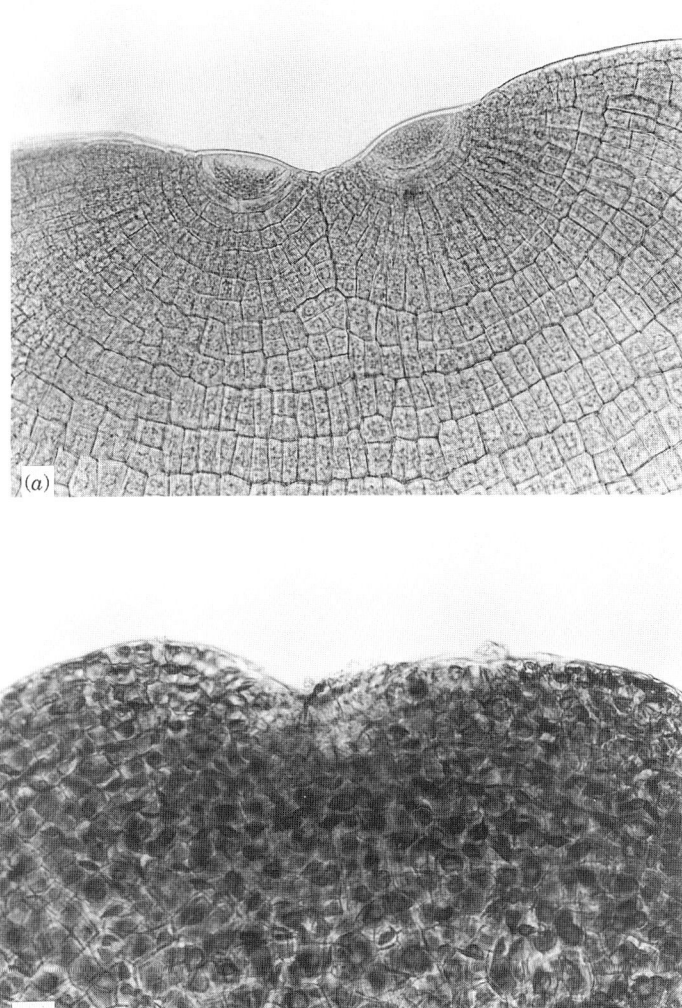

Fig. 7-18. *a.* Transition from a meristem composed of a continuous row of peripheral cells capable of division to a meristem consisting of a single apical cell with multiple cutting faces appears to have occurred in the parenchymatous brown algal group Dictyotales (*Padina, Zonaria, Dictyota*). Division of the single apical cell of *Dictyota* results in dichotomous branching resembling that of thalloid liverworts and hornworts, such as *Notothylas*. *b.* Presumably, division and separation of apical cells generates dichotomous lobes.

charalean algae and its immediate derivative do not divide in more than one direction. However, individual nodal cells generated by the charalean apical cell typically produce a whorl of branches (Fig. 7-19), presumably by altering the direction of successive cell divisions. Nodal cells thus appear to be derived regions of meristematic activity. The pattern of meristematic activity of charalean nodal cells might be explained by production of successive spindles at increasingly larger angles to the first, within a plane parallel to division of the apical cell. If so, this mechanism predicts that the young branches produced by a nodal cell will be of increasing length, correlating with elapsed time since each branch initial was produced, and, indeed, such variation in the length of young branches can be observed (Pringsheim 1863; Shen, 1966). Interestingly, similar production of whorled branches occurs during spermatid formation in *Coleochaete pulvinata* (Graham and Wedemayer, 1984) (Fig. 7-20). This pattern could also be explained by angular movement of successive spindles, but this explanation requires verification.

Fig. 7-19. Production of whorls of branches by nodal cells of *Chara zeylanica* is probably based on angular change in the position of successive mitotic spindles. This radial developmental pattern may suggest evolutionary mechanisms for accomplishing positional change in successive mitotic spindles that occurs when land plant meristematic cells use alternate cutting faces.

PARENCHYMA, MERISTEMS, AND MORPHOGENESIS 175

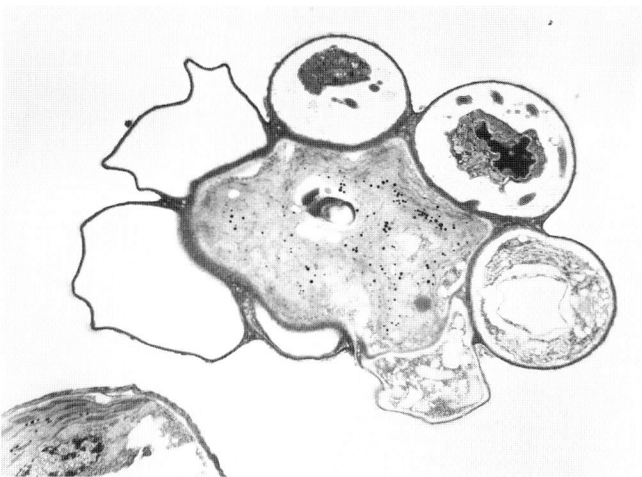

Fig. 7-20. Radial patterning in the production of antheridial branches in the putatively primitive *Coleochaete* species *C. pulvinata* suggests that angular change in position of successive mitotic spindles may also have been characteristic of the common charophycean ancestor of Coleochaetales, Charales, and embryophytes. From Graham and Repavich (1989), *American Journal of Botany.*

In contrast to charophycean meristems, the plesiomorphic apical meristematic cells of bryophyte gametophytes have four cutting faces and thus can generate thalli that are more than a single cell thick. An example is provided by the apical cells of several hornworts (including the putatively basal genus *Notothylas*) (Fig. 7-21). These are wedge-shaped cells from which derivatives are produced

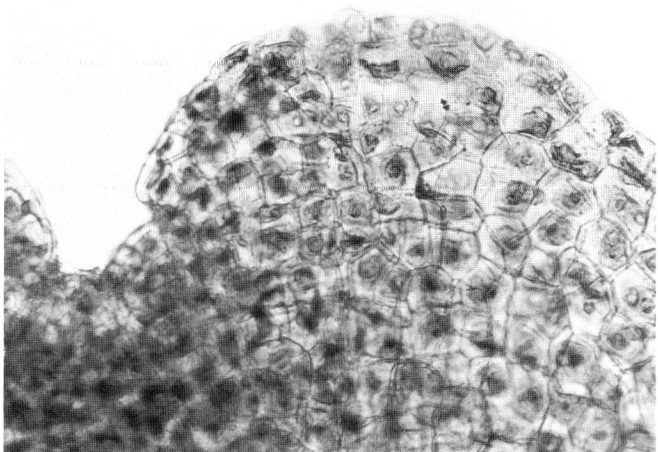

Fig. 7-21. The single apical cell of *Notothylas orbicularis* is difficult to distinguish from similar-appearing derivatives.

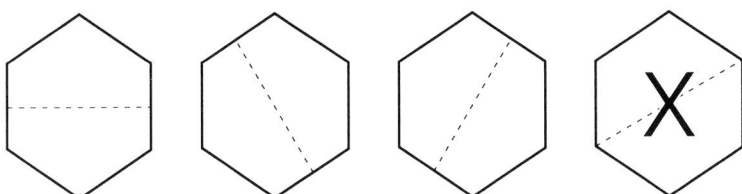

Fig. 7-22. Diagram illustrating the avoidance in tissues of higher plants of four-way junctions. In other words, placement of new cell walls directly opposite existing three-way cell junctions is avoided. Adapted from an illustration in Lloyd (1991), Cytoskeletal elements of the phragmosome establish the division plane in vacuolated higher plant cells. In: Lloyd (ed.), *The Cytoskeletal Basis of Plant Growth and Form,* Academic Press, pp. 245–257. Used with permission of Academic Press.

on four sides, in a spiral segmentation pattern: lateral, ventral, lateral, dorsal (Renzaglia, 1978). The pattern of successive cell divisions suggests highly controlled shifts in the angular positioning of successive spindles. The mechanism for accomplishing changes in plane of cell divisions in embryophyte meristems may thus be based on mechanisms similar to those in charophycean algae, but transitional stages in the evolution of localized apical meristematic cells of embryophytes have not yet been documented. Perhaps dynamic studies of spindle orientation changes in the meristematic cells of land plant gametophytes, nodal regions of Charales, and meristematic regions of *Coleochaete* may shed light on the mechanism by which shifts in cutting face (alterations in plane of division) are made. Such information might then allow development of hypotheses regarding transition to the land plant meristem.

The tissues of land plants are characterized by positioning of new walls so that only three walls in a plane meet at a point. In other words, production of four-way cell junctions is avoided (Fig. 7-22) (Cooke and Lu, 1992). Lloyd (1991) regards this phenomenon, first described by Sinnott and Bloch (1939), as central to plant tissue geometry. Three hypothetical mechanistic explanations have been proposed: transvacuolar strands contributing to phragmosomes contact the cell periphery where the path is shortest, thereby avoiding vertices; cell-cell alignment is somehow blocked where three cells meet; new walls are preferentially placed where there are high concentrations of plasmodesmata, perhaps reflecting chemical influences from adjoining cells. Comparative observations of wall placement in thalloid charophytes and parenchymatous land plant gametophytes (Cooke and Lu, 1992) may prove useful in testing these hypotheses. In summary, charophycean algae provide useful model systems for elucidation of interactions between the plant cytoskeleton and morphogenesis that are complementary to "lower" plant systems. Continued comparative study of algal and plant cell division should shed considerable light on unsolved problems in the field of plant morphogenesis.

8

Evolution of Plant Sexual Reproduction

Success in breeding and genetic engineering efforts in modern higher plants is dependent on an understanding of the fundamental bases of plant sexual reproduction and the life cycle of plants (Yang and Zhou, 1992). Yet many critical questions remain unanswered regarding reproduction, not only of agriculturally important seed plants, but of all embryophytes. For example, the molecular events leading to gamete formation are not understood, and the molecular basis of sperm-egg recognition in fertilization is unexplored. Only recently has in vitro fertilization of isolated plant sperm and egg cells been accomplished, and electroporation (blasting holes in cell membranes by electrical discharge) was required (Kranz et al., 1990). The relative roles of female gametophyte influence versus egg gene expression on embryogenesis are also unknown. Unfortunately, the enclosure of egg cells and embryos within successive layers of integument, ovary, and other tissues of higher plants renders access difficult (Goldberg, 1988). Because gametes and zygotes of seedless plants and charophycean algae are relatively more accessible, these organisms may provide useful model systems in which to explore fundamental questions related to plant gametogenesis, fertilization, and embryogenesis. It is therefore useful to consider the evolutionary origin of plant reproduction, with a focus on charophycean algae.

The evidence we have examined earlier indicates that land plants inherited sexual reproduction from charophycean ancestors. We have further observed that aspects of the relationship between the two alternating generations of land plants are likely to have been derived from innovations first appearing in charophycean algae. In addition, the phylogenetic connection between the occurrence of sporopollenin in charophycean zygotes and the origin of sporopollenin-walled land plant spores has been explored. In this chapter, a speculative approach is taken in further consideration of the evolution of land plant sexual reproduction, the plant life cycle, and walled spores. The goal is to define important questions that remain to be answered and to highlight taxa likely to serve as appropriate model systems for the study of fundamental aspects of land plant reproduction.

THE ORIGIN OF SEXUAL REPRODUCTION IN CHAROPHYTES

The origin of sexual reproduction in eukaryotes is poorly understood, and there is no consensus as to whether sex arose once or many times in different groups of protists. It is widely accepted that more primitive forms of sexual reproduction involve the fusion of morphologically identical flagellate gametes (isogametes), that anisogamy (slight dissimilarity in the size of flagellate gametes) is derived from the isogamous condition, and that oogamy (fusion of a larger, nonflagellate egg with a smaller, flagellate sperm) is even more highly derived. The algal evidence for this hypothetical transformation series (Fig. 8-1) comes from comparative studies of chlorophytes such as Chlamydomonas and other members of the

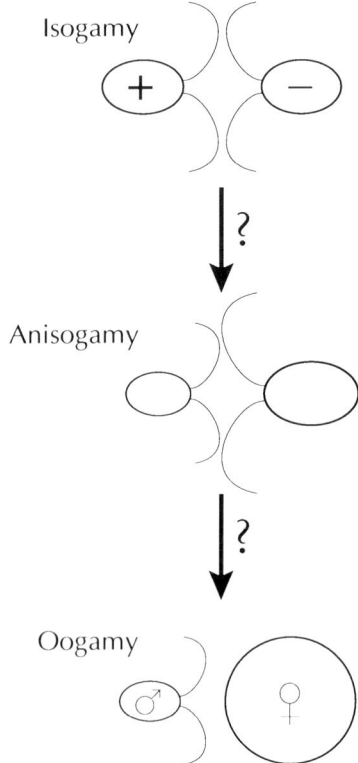

Fig. 8-1. Diagram illustrating hypothetical transformation series from isogamy to anisogamy to oogamy.

Volvocales, and of chromophytes such as brown algae and diatoms. However, we cannot assume that charophycean sexual reproduction is homologous to that of other protists, including other chlorophytes, inasmuch as the various green algal lineages are perhaps independently derived from flagellate ancestors that would be classified as prasinophytes. Sexual reproduction in the modern prasinophyte group is unknown, so it is possible that the various green algal lineages arose from the same or different prasinophytes and that sex arose after their divergence. That those charophycean algae appearing to have diverged earliest from the lineage leading to plants (Chlorokybales and Klebsormidiales) lack sexual reproduction suggests that sex did not arise until later in this line. It is, of course, possible that the various green algal lineages shared a common ancestor that had sexual reproduction, but sexual reproduction was subsequently lost from the ancestors of modern *Chlorokybus, Klebsormidium, Stichococcus,* and *Raphidonema.* Because these forms occur in terrestrial as well as aquatic habitats, it might be argued that sexual reproduction was lost in response to selective pressures such as those proposed to influence loss of zoospores (see Chapter 5 for discussion of zoospore loss). However, soil algae such as the ubiquitous *Chlorococcum* maintain a sexual cycle involving flagellate gametes, so it cannot be assumed that terrestrial conditions influenced loss of sex in charophytes. In the absence of evidence for this scenario, origin of sexual reproduction in charophytes after divergence of Klebsormidiales is the more parsimonious explanation for the pattern of occurrence of sexual reproduction among charophytes.

Chloroplast t-RNA intron data (Chapter 3) and comparative cell wall biochemistry data (Chapter 7), previously discussed, suggest that zygnematalean algae diverged from the charophycean lineage prior to ancestors of modern Coleochaetales and Charales. Therefore, zygnematalean sexual reproduction could be viewed as the plesiomorphic type for charophytes, in which, consistent with theory regarding the origin of sex, there is fusion of isomorphic gametes. However, the method by which fertilization is achieved, conjugation, is unusual for green algae. Conjugation in Zygnematales involves aggregation of filaments or unicells within a gelatinous matrix and subsequent fusion of nonflagellate, naked gametes (Figs. 8-2, 8-3, and 8-4). Syngamy occurs either within a conjugation canal (in various desmids and zygnematacean algae such as *Spirogyra*) or after liberation into the gelatinous matrix (in the case of some placoderm desmids such as *Cosmarium*) (Bold and Wynne, 1985). Production of the conjugation tube involves the outgrowth of a branchlike protrusion from the sides of conjugating cells (Fig. 8-2). This represents a growth response related to sexual reproduction that probably involves changes in patterns of cell wall synthesis and cytoskeletal dynamics, but these have not been investigated. There is some evidence that conjugation in zygnematalean algae is mediated by sex pheromones (Brandham, 1969; Hoshaw et al., 1990). Similar chemical regulation of growth related to sexual reproduction occurs in other freshwater protists. Examples include development of dwarf male filaments in some species of the chlorophyte *Oedogonium* (Rawitscher-Kunkel and Machlis, 1962), growth of antheridial branches toward oogonia of the water mold *Achlya* (elucidated by the elegant

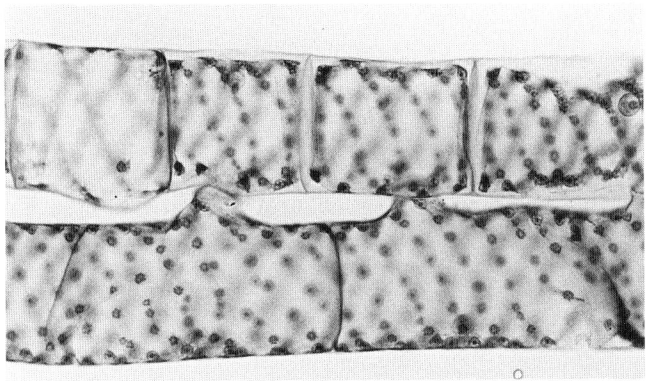

Fig. 8-2. Conjugation in *Spirogyra* (and related algae) begins with parallel alignment of compatible filaments and production of cellular protrusions in the vicinity of pairing.

experiments of Barksdale and Raper, as described by van den Ende, 1976), and most likely also growth of antheridial branches toward oogonia of the chromophyte *Vaucheria* (Fig. 8-5). However, the chemical basis for zygnematalean conjugation and possible relationships to protist or land plant sexual attractants are unknown. As previously mentioned, growth responses associated with sexual reproduction in *Coleochaete* (cortication of zygotes) and Charales (cortication of egg cells) may also be mediated by chemical influences and could be related to conjugation in Zygnematales. This hypothesis might be testable when the biochemical basis for sexual growth responses in charophytes has been determined.

The occurrence of flagellate reproductive cells (zoospores) in charophycean taxa (*Chlorokybus* and *Klebsormidium*) that are regarded as having diverged

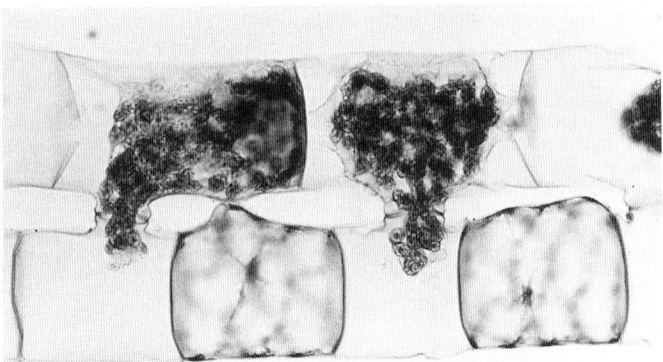

Fig. 8-3. Conjugation in *Spirogyra* continues with dissolution of intervening wall material, movement of cytoplasmic units across the tubular junction, and fusion with cytoplasm of the adjoined cell.

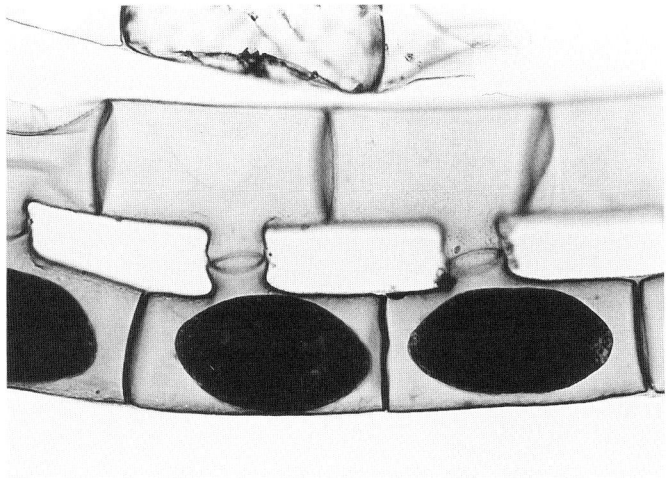

Fig. 8-4. Conjugation in *Spirogyra* culminates in formation of resistant zygotes that are retained within parental walls for a time prior to dispersal.

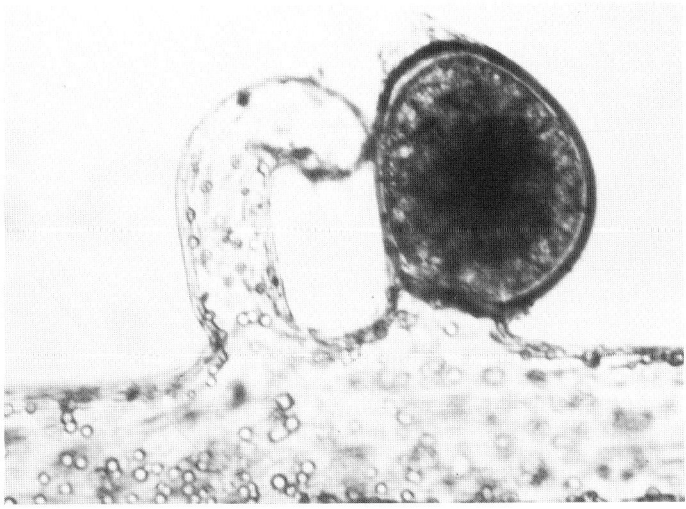

Fig. 8-5. Sexual reproduction in *Vaucheria* occurs when antheridial branches grow toward oogonia, depositing sperm very close to fertilization pores. This behavior suggests production of chemical growth stimulants for which there are receptors in target cells. The sexual growth responses observed in charophycean algae similarly suggest diffusible growth substance production and reception.

earlier than zygnematales suggests the possibility that the ancestors of zygnematalean algae could produce flagella, but that this capacity was lost. Indeed, it has been suggested that zygnematalean conjugation (and gametangiogamy of pennate diatoms) may have been derived from oogamy, in response to selection pressures of shallow water habitats (Nakahara and Ichimura, 1992). A possible means of testing this hypothesis might be to probe the genome of zygnematalean algae for genes related to flagellar production. Dutcher (1989) has found that the genes necessary for flagellar assembly and function in *Chlamydomonas* are clustered in linkage group XIX and are most likely nuclear-encoded. When flagellar genes have been more thoroughly characterized, it may become possible to use heterologous sequences as hybridization probes to search for remnants of a flagellar assembly system in protists such as zygnematalean algaê (or rhodophytes) that do not currently produce flagella.

It is also interesting to consider the impetus for evolutionary origin of sexual reproduction in the charophycean-embryophyte lineage. Theorists such as Kondrashov (1988) have suggested that selection against mutation has been a major factor in the evolution of sexual reproduction. This hypothesis is based on the assumption that most nonneutral mutations are deleterious and that selection will favor a decrease in the mutation rate. Although sexual populations are vulnerable to invasion of mutations (thought to be a possible reason for loss of sex), under conditions where the genomic deleterious mutation rate is substantial, the mutation load under diploid selection may be less than that under haploid selection. Change from haploidy to diploidy confers selective advantage by reducing the effect of partially dominant and recessive mutations, and the redundancy of diploidy protects against loss of function if an essential sequence is altered (complementation). Diploidy is regarded as favored when the degree of dominance of a single deleterious mutation is less than about 0.25 (Kondrashov and Crow, 1991), or less than about 0.5 (Valero et al., 1992). These genetic factors may have been important in the origin of sex in green algae and charophytes, particularly in shallow waters or moist terrestrial environments where mutagenic effects of ultraviolet radiation may have been substantial. However, the selective advantages of the perennation role played by freshwater algal zygotes should also be considered. Production of resistant, nutrient-filled zygotes that can withstand cold temperatures, desiccation, and microbial degradation and that are cued to germinate under appropriate conditions may have helped to overcome the "cost of meiosis" associated with sexual reproduction (as defined by Williams, 1975).

As we have noted (Chapter 5), resistant zygotes are often produced by algae inhabiting typical freshwater environments that fluctuate seasonally or successionally. Sexual reproduction in higher charophytes, whether by conjugation or fusion of egg and sperm cells, generates a resistant zygote (Fig. 8-6). The zygote walls of charophytes characteristically exhibit sporopollenin, the most chemically resistant biopolymer known. Resistant compounds said to be similar to lignin may occur in the walls of vegetative cells of desmids (Gunnison and Alexander, 1975), but sporopollenin is not known to be produced by charophytes except as the result of sexual reproduction (Graham, 1990). This suggests that

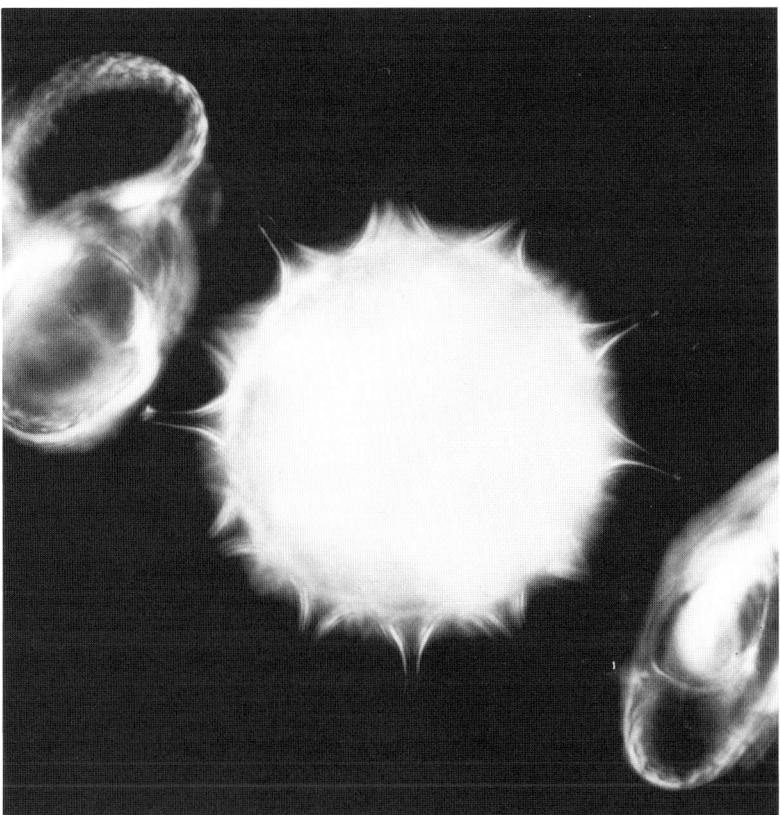

Fig. 8-6. Zygotes of the desmid *Cosmarium* are formed from fusion of the cellular contents of two parental cells which function as gametes. The abandoned parental cell walls are often visible near highly ornamented zygotes. Photo courtesy of Lee Wilcox (University of Wisconsin-Madison).

fertilization may be a prerequisite for sporopollenin production in charophytes, and that genetic regulation of the biosynthetic pathway leading to this polymer is directly associated with the fertilization process. Further research on enzymatic pathways leading to sporopollenin production in zygnematalean algae (the most primitive charophytes known to exhibit sex) may provide the data necessary to test this hypothesis.

As a group, charophycean algae provide a potentially useful system for evaluation of hypotheses that attempt to explain the origin of sexual reproduction in eucaryotes. Because the relationship of charophycean algae to the ancestry of plants is now clearly established, the study of evolutionary transitions in sexual reproduction that have occurred in this group is directly applicable to understanding the origin of sexual reproduction in land plants. It is possible that sexual reproduction originated independently in the ancestors of various groups of modern charophytes and land plants, i.e., that sexual reproduction is a homoplasious feature of the charophyte-embryophyte clade. However, the correlation of sporopollenin production with sexual reproduction in charophytes and similarities in the microanatomy of sperm of charophytes and embryophytes favor a hypothesis of homology. Further comparative study of sexual reproduction in charophytes and lower embryophytes at the biochemical and molecular levels may clarify this issue.

It is interesting to consider the possibility that the origin of sexual reproduction and resistant, sporopollenin-walled zygotes in charophytes might be related to the highly seasonal environment and arid continental conditions postulated to be associated with glaciation of the earth during the late Ordovician period (discussed in Chapter 2). In modern lacustrine environments, resistant materials in zygote walls may function to reduce microbial degradation processes that could occur during periods that are not favorable for germination and growth of propagules. Resistant biopolymers are also thought to contribute to the ability of charophycean zygotes to resist desiccation. The relationship between production of resistant zygotes by charophytes, and environmental factors such as onset of cold temperatures or low-water conditions arising from aridity, needs further examination. However, the circumstantial association of dramatic climatic shifts in the time period associated with the earliest evidence related to the origin of land plants, along with putative reproductive adaptations to cold and desiccation exhibited by higher charophytes, is intriguing. The hypothesis of association would be strengthened by the occurrence of fossils attributable to zygotes of modern charophytes in sediments older than those bearing trilete spores and cuticle-like sheets associated with the remains of early land plants. However, such fossil evidence is not currently available.

So far as is presently known, meiosis in charophycean algae occurs during zygote germination and yields haploid daughter cells. It has been suggested that the evolution of a chromosome disjunction system followed rapidly upon the origin of sex, and that similarities in the process of meiosis among eukaryotes argue for the origin of meiosis (and presumably sex) prior to divergence of the main eukaryotic lineages (Maguire, 1992). Steps proposed to have occurred during the evolution of meiosis include the occurrence of mutation(s) enhancing sister centromere cohesiveness, formation of a mechanism for binding homologous chromosome pairs (rudimentary synaptonemal complex), development of processes related to chromatid recombination, and innovations for maintenance of sister chromatids through anaphase of meiosis I (Maguire, 1992). Recombination (crossing over) is widely regarded as having been derived from preexisting DNA-repair systems. The formation of Halliday-type structures, transitory complementary base-pairing of matching regions on homologous chromatids (heteroduplexes), may have brought homologues together initially; mutation is presumed to have generated a molecular mechanism that would further promote synapsis. Binding (during meiosis I) and subsequent release (during meiosis II) of sister chromatids is postulated to be controlled by strategically timed production of topoisomerase II inhibitors. Topoisomerase II activity resolves catenated DNA at replication fork junctions. Inhibition of enzymatic activity would promote sister chromatid cohesiveness, whereas promotion would release chromatids from association. Meiosis appears to have been interpolated into the preexisting mitotic process, with meiosis II representing completion of the disrupted mitotic sequence (Maguire, 1992).

If the evolutionary bases that Maguire (1992) has proposed as those underlying unique meiotic events (adherence of sister chromatids and homologous chromosomes, and recombination) are indeed processes related to DNA replication and repair, they may be quite ancient and thus widely present in early eukaryotes. These processes could have served as preadaptations allowing evolution of meiosis in various ancestral protist groups. Similarity in meiosis as it occurs in diverse eukaryotes is not necessarily evidence for a single ancient origin of reduction division. Meiosis may have been polyphyletic, arising after divergence of major eukaryotic lineages, in association with independent evolution of sexual reproduction. Progress in understanding molecular aspects of DNA organization in protists may provide the means for testing alternative hypotheses related to the origin of sex and meiosis. Molecular dissection of meiosis may also suggest mechanisms for delay of meiosis, important in the development of hypotheses explaining the origin of alternation of generations in protists.

THE ORIGIN OF LAND PLANT GAMETANGIA

As previously noted in Chapter 6, the phylogenetic gap between charophycean algae and land plants appears greatest when gametangia of the two groups are compared. There is considerable controversy among plant morphologists as to whether archegonia and antheridia of embryophytes are of monophyletic or polyphyletic origin. So far, results of studies of gametangial structure and development in charophytes have contributed little to the resolution of this question.

As discussed in Chapter 6, the gametangia of the Charales are the most complex found among charophycean algae, but they are generally regarded as highly specialized along independent lines in relationship to the phylogeny of "jacketed" gametangia of land plants. Charalean gametangia, however, provide useful models for investigation of cell-cycle phenomena associated with spermatogenesis (Maszewski and Kolodziejczyk, 1991). Moreover, antheridial development in *Chara* appears to be influenced by endopolyploidization of nongenerative antheridial cells (Kwiatkowska et al., 1990). Endoreplication (production of multiple copies of the nuclear genome) is a common, but poorly understood, phenomenon in plants and animals. Perhaps further study of cellular increases in DNA C-values in Charales will illuminate the roles of endoduplication in land plants.

Although the antheridia of *Coleochaete* are smaller and considerably less complicated than those of charalean algae, they are among the most complex gametangia found in green algae (Wesley, 1930; Graham, 1984). *Coleochaete* species can be arranged to illustrate a transformation series in the evolutionary development of more complex male gametangia from simpler ones (Figs. 8-7, 8-8, 8-9, and 8-10). The species that exhibit thalloid organization produce more

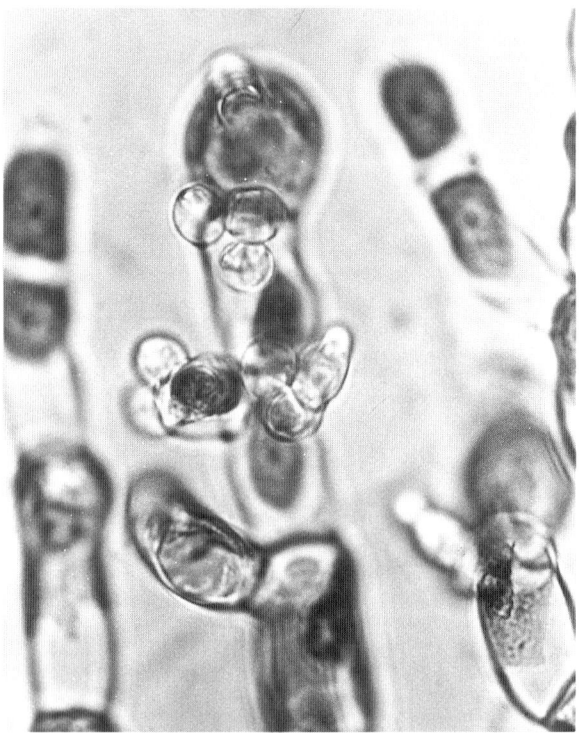

Fig. 8-7. Antheridia of *Coleochaete pulvinata* are small cells produced in whorls or clusters from underlying initials. Although parental cells contain well-developed green plastids, antheridial cells are colorless or pale green and possess amyloplasts. The terminal cell is an oogonium with protruding trichogyne. Hydrolysis of the wall at the trichogyne tip allows fertilization to occur.

complex antheridia than filamentous *Coleochaete* species, in which sperm are produced in unicellular branches. The complex gametangia of *Coleochaete* are derived by repeated subdivision of gametangial initials, generating both fertile and vegetative products. In a careful study of spermatogenesis in *Coleochaete scutata,* Ophelia Wesley (1930) ascertained that an initial cellular bisection, followed by two nearly simultaneous asymmetric divisions, gave rise to four small sperm-producing cells underlain by two vegetative cells. She noted that these multicelled antheridia were the most complex gametangia to be found in green algae outside Charales. Callose deposition is associated with gametogenesis in *Coleochaete* (Figs. 8-11 and 8-12) (unpublished observations), as in seedless plants, and highly controlled asymmetric, diagonal cell division is associated with production of spermatogenous cells of *Coleochaete orbicularis* (Fig. 8-13), as also occurs in various groups of seedless plants (Graham et al., 1991). It is not difficult to visualize the derivation of simple land plant antheridia, especially of the submergent type, from an evolutionary precursor similar to antheridia of

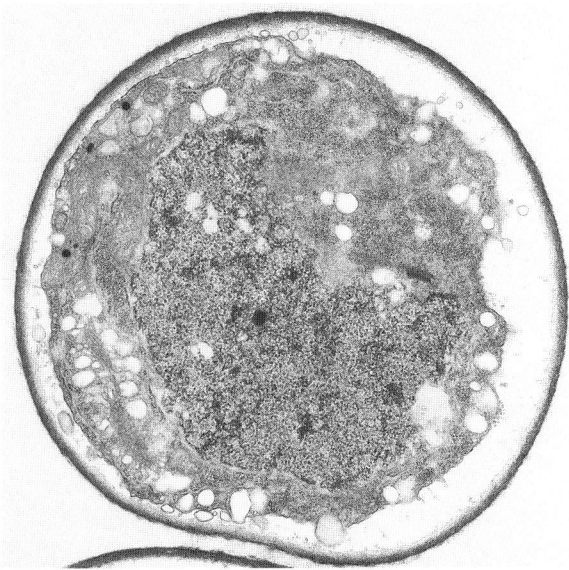

Fig. 8-8. Sperm of *Coleochaete pulvinata* develop within single-celled antheridia by reduction in cell volume, coupled with increase in cell density. An MLS has appeared near the nucleus, and dictyosome vesicles migrate to the cell membrane, fusing with it and releasing contents. These changes resemble stages in development of embryophyte sperm. From Graham and Wedemayer (1984), *Journal of Phycology.*

Coleochaete. However, innovations that might have served as the basis for evolutionary origin of land plant archegonia are not evident in *Coleochaete* or Charales.

It has been suggested that land plant gametangia arose independently in various groups of embryophytes that evolved separately from charophycean ancestors (see Sluiman, 1985, and references cited therein). In contrast, Graham et al. (1991) suggested that the fundamental basis of the embryophytic relationship (nutritional and developmental interaction between zygote and parental gametophyte cells) had evolved prior to the origin of archegonia. Delay in zygotic meiosis could have generated small, multicellular, diploid embryos (sporophytes) in the absence of recognizable archegonia. Graham et al. (1991) further proposed that the origin of sporophytes resulted in selection for archegonia, which could have evolved in parallel within different embryophytic lineages. This hypothesis could explain the diversity of gametangial form and development observed to occur among archegoniates (for example, submergent versus emergent, stalked gametangia), but requires further evaluation. A third hypothesis, that land plant gametangia are monophyletic, is supported by the relative uniformity of early gametangial development in various groups of archegoniates. This consistency includes initiation of gametangial formation by periclinal division of a surface cell, with the upper derivative serving as a jacket initial and the lower cell leading to gamete production (see Chapter 6 for diagrams). Progress in understanding the

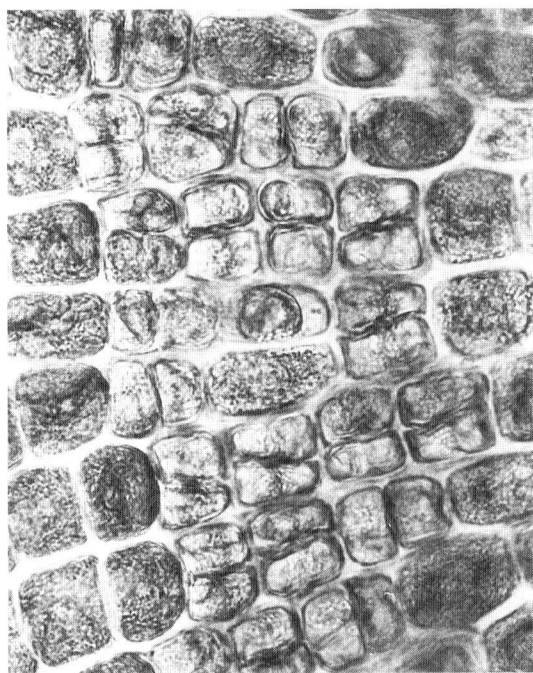

Fig. 8-9. As in *C. pulvinata,* sperm of *C. scutata* develop within small cells cut off from the surface of an underlying cell, but development differs in that the first division of the antheridial initial bisects the cell. Each daughter cell then undergoes two unequal divisions, yielding an antheridial complex consisting of four spermatogenous cells lying dorsal to two vegetative cells. From Graham (1984), *American Journal of Botany.*

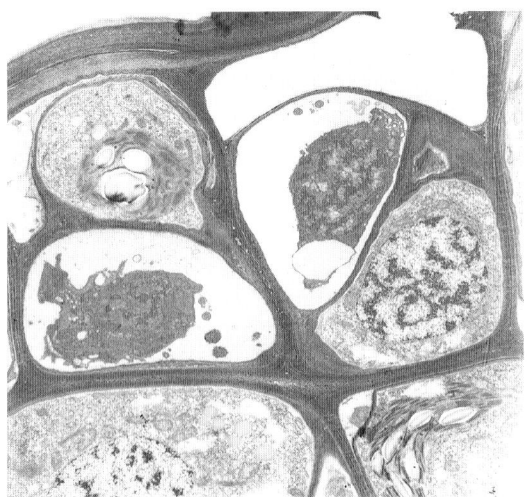

Fig. 8-10. Transmission electron microscopy shows the multicellular antheridia of *C. scutata* containing mature sperm, empty cell walls from which two sperm have been released, and portions of the two vegetative cells of the six-celled complex. From Graham (1984), *American Journal of Botany.*

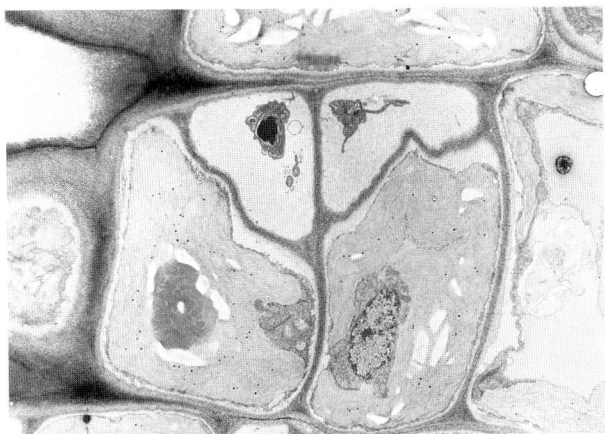

Fig. 8-11. Sperm production in *Coleochaete orbicularis* is similar to that of *C. scutata*, in that the first division bisects the initial cell. Subsequently, one or more spermatogenous cells are cut off by diagonal, unequal divisions. Two vegetative cells remain as part of the antheridial complex. The opaque material surrounding sperm correlates with aniline blue fluorescence and resorcinol blue staining, indicating the presence of callose. The callosic material may isolate developing sperm from adjacent cells of the same lineage by blocking plasmodesmatal continuity.

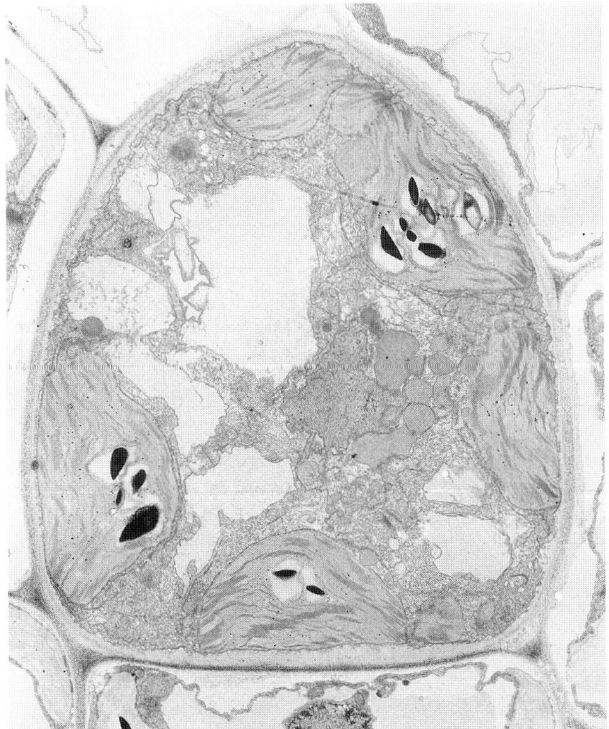

Fig. 8-12. Slightly electron-dense material identified by correlative staining reactions as callose also surrounds developing egg cells of *Coleochaete orbicularis*. Callose presumably blocks plasmodesmatal continuity with the sister cell below.

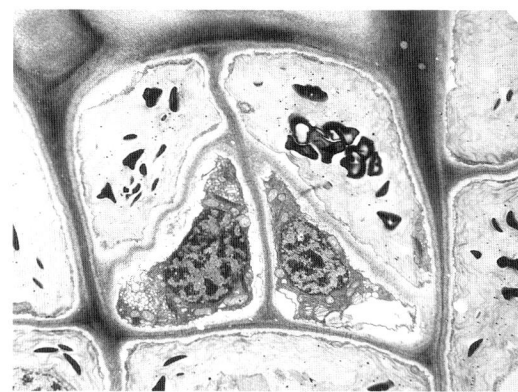

Fig. 8-13. Antheridia of *Coleochaete orbicularis* are composed of at least four cells derived from subdivision of single antheridial initials. These initials are internal cells that frequently occur near developing eggs. The first division of the antheridial initial divides the cell into two equal portions. Each daughter cell then undergoes at least one unequal, diagonal division to form a smaller spermatogenous cell and a larger vegetative remnant. Spermatogenous cells derived from the same initial undergo coordinated development to form sperm. The cells shown here are approximately midway through the development process.

evolutionary origin of land plant gametangia is most likely to come from comparative study of early gametangial development in charophytes and archegoniates. It may also be useful to examine possible differences between cells of the vegetative (or jacket) cell lineage and cells of the germline lineage of archegoniates, with respect to patterns of gene expression. Determination of the plesiomorphic (least derived) gametangial type among archegoniates would be useful in understanding the transition. As mentioned in Chapter 6, putatively primitive archegonia of the hornwort *Notothylas* are possible candidates for comparative study. However, hornwort antheridia appear to be rather derived, and it is not presently clear which archegoniate taxa exhibit the plesiomorphic antheridial type.

ORIGIN OF THE LAND PLANT EMBRYO, SPOROPHYTE, AND LIFE CYCLE (ALTERNATION OF GENERATIONS)

If the hypothesis for charophycean ancestry of embryophytes is accepted, the universal occurrence of zygotic meiosis in modern charophytes and the presence in these algae of transitional features related to the origin of the embryo and sporophyte indicate that the land plant life cycle originated by delay in meiosis. In other words, a significant body of molecular and cellular data that supports alliance of modern charophyte algae with the ancestry of embryophytes also supports the "antithetic" or "interpolation" hypothesis for the origin of the land plant life cycle. Invocation of the alternative "homologous" or "transformation" hypothesis requires postulation of extinct charophytes having alternation of physically independent, multicellular generations (Stebbins and Hill, 1980). Although the latter scenario is theoretically possible, parsimonious use of data presently available favors derivation of the sporophyte generation from the zygote, by delay in meiosis, as proposed by Bower (1908).

Some authors have accepted the meiotic delay mechanism, but consider the possibility that the embryophytic habit arose more than once among land plant lineages. However, as has been discussed previously (Chapter 4), considerable evidence for monophyly of the modern embryophytes includes the highly conserved architecture of the land plant chloroplast genome, the universal occurrence of preprophase microtubule bands in mitotic cells of embryophytes coupled with absence of PPBs from charophytes, the occurrence of bicentriolar centrosomes in various archegoniate taxa, the aspects of MLS structure that differ between charophytes and embryophytes, and the production of walled spores by land plants but not by charophytes. The probability seems low that this suite of character transitions occurred simultaneously during independent derivation of various embryophyte lineages from separate ancestors. In the absence of equally compelling evidence for polyphyly of the embryophytes, at present a hypothesis of monophyly is more strongly supported.

Sporic meiosis (production of a multicellular sporophyte generation) may, therefore, have been inherited by the ancestors of modern bryophytes and vascular plants from a common early embryophytic ancestor. The evidence provided by *Coleochaete* suggests that at least some nutritional and developmental interactions characteristic of the embryophytic habit could have arisen prior to the origin of a multicellular sporophyte. If so, the origin of chemical interactions between parental haploid cells and diploid cells derived from fertilization should be viewed as a series of evolutionary events distinct from the origin of processes related to delay of meiosis and production of a multicellular sporophyte. In phylogenetic (cladistic) analyses, the character "embryo," which implies chemically mediated nutritional and developmental interactions between generations, should probably be evaluated separately from the characters "multicellular sporophyte" or "sporic meiosis." Molecular and biochemical studies should be focused separately upon (1) cell-to-cell interactions related to evolution of the embryophytic habit and (2) processes that would be required to delay the onset of meiosis until repeated mitoses had generated a multicellular diploid generation. This is not to say that the two groups of character state changes are entirely unrelated. Future work may reveal that both meiotic delay and subsequent mitotic proliferation of diploid nuclei are controlled by chemical influences emanating from parental gametophytes. It is important, however, to consider the strong possibility that at least some nutritional and developmental relationships characteristic of plant embryos preceded the origin of a multicellular sporophyte generation.

The Origin of Cell-to-Cell Interactions in Plant Embryogenesis

A substantial body of experimental data obtained from study of nutritional relationships between sporophytes and gametophytes of various hornworts, liverworts, and mosses has been reviewed by Ligrone and Gambardella (1988). Bryophyte sporophytes are typically capable of photosynthesis, and photosynthate transport to gametophytes has been demonstrated to occur in hornworts and liverworts (Thomas et al., 1978; 1979). The flow of photosynthate from illuminated hornwort sporophytes has been shown to support nitrogen fixation by symbiotic cyanobacteria associated with gametophytes kept in darkness (Stewart and Rodgers, 1977). However, translocation studies reveal that bryophyte sporophytes depend on photosynthate provided by gametophytes, and that the net direction of nutrient flow is generally from gametophyte to sporophyte. Some evidence suggests that movement of photosynthates (sugars, amino acids, etc.) across the placental junction involves membrane-associated transporters (see Ligrone and Gambardella, 1988).

Renault et al. (1992) found that although sucrose was the main soluble sugar in sporophyte and gametophyte tissues of the moss *Polytrichum formosum,* mainly hexose sugars occur in the vaginula apoplast, the region of contact between two generations. These workers also noted high levels of soluble and cell wall-associated invertase activity in the same region. Their results indicate that at the placental junction, gametophyte sucrose is convert to hexose sugars, glucose is transported into the apoplast by carriers, sporophytes take up the hexose, then sucrose is reformed on the sporophytic side. This data is consistent with evidence for the occurrence of cell membrane hexose transport proteins in sugar-loading regions of plants (Sauer et al., 1990), described in Chapter 5.

Studies of phloem function indicate the presence of carrier-mediated proton/sugar cotransport at the sieve element-companion cell complex in the phloem of vascular plants (Turgeon and Beebe, 1991), and DeWitt et al. (1991) used a beta-glucuronidase (GUS) reporter gene to localize expression of a proton pump protein in phloem cells. When more is known about plant sugar transporters and proton pumps, it may be possible to use antibodies or other molecular methods in attempts to localize similar proteins in placental regions of land plants and charophytes. The success of such efforts would support a concept of homology of placental function across various embryophyte taxa, and between charophytes and embryophytes.

Nutritional interactions between plant generations may have arisen in a series of steps, beginning with facilitated movement of photosynthate from vegetative cells into nearby zygotes, in which macromolecular storage materials (starch and lipids) are typically accumulated. Short-range directional transport of osmotically active compounds such as sugars would thus reflect a source-sink relationship between parental cells and zygotes. The occurrence of membrane-bound transporter molecules at the junction would facilitate apoplastic movement of photosynthate across the interface, which is characteristically lacking in plasmodesmatal connections. Although there is as yet no direct evidence for facilitated apoplastic transport of photosynthate between cells of charophyte algae, as discussed in Chapter 5, there is substantial evidence that at least some charophytes have the cellular mechanisms required for uptake of exogenous sugars. Primitive osmotrophic capabilities may have preadapted charophycean algae for evolution of nutritional interactions between parent thalli and zygotes. It has been proposed that origin of the plant embryo was contingent on acquisition of the fungal genes' encoding capacity for extracellular digestion and absorption (Atsatt, 1991), implying that incorporation of fungal genes occurred just prior to divergence of embryophytes from charophytres. However, evidence that relatively simple charophytes, such as saccoderm desmids, can utilize exogenous sugars, suggests instead a more ancient capacity for sugar transport. The subsequent evolution of placental transfer cells whose wall labyrinths increase membrane surface area, and hence photosynthate flux, would further elaborate the nutritional interaction between generations. Presence of an efficient nutrient translocation system could be viewed as a preadaptation allowing subsequent elaboration of the diploid generation, the sporophyte. Future work focused on nutritional relationships between vegetative cells and zygotes of charophycean algae may indicate whether the above described evolutionary scenario provides a satisfactory explanation for the origin of nutritional relationship between alternating generations of seedless plants.

It is important to note that zygotes of zygnematalean and coleochaetalean algae contain green and, presumably, photosynthetically active chloroplasts, which are replicated and distributed to meiospores at zygote germination. This is consistent with the assumption that sporophytic photosynthetic capacity in embryophytes is derived from that of ancestral algal zygotes, which is implicit in the antithetic hypothesis for origin of alternation of generations in plants. Thus, the capacity of bryophyte sporophytes for photosynthesis does not necessarily represent support for the homologous hypothesis, as has been claimed (Thomas et al., 1979).

Developmental Interactions at the Placental Junction

Fertilization of *Coleochaete* egg cells is followed by significant enlargement of zygotes and directional growth toward zygotes by neighboring cells, which ultimately enclose zygotes (Figs. 8-14, 8-15, 8-16, 8-17, 8-18, 8-19, 8-20, 8-21, and 8-22). As previously mentioned (Chapter 4), in one species wall ingrowths develop on cortical cell walls that are in direct contact with zygotes. These localized developmental responses suggest that zygotes produce one or more diffusible growth substances for which vegetative cells have target receptors, and that reception leads to signaling and developmental cascades that result in directional growth and other changes in adjacent cells. Hypothetical diffusible growth substances are commonly invoked to explain many types of cell to cell interactions in higher plants, but in general little is known about the nature of such compounds.

One relevant example of chemically mediated cell-to-cell interaction in plants concerns the maize gene Knotted-1 (Kn1). This is a dominant mutation that causes abnormal leaf development. The gene is thought to encode a protein that influences the production in some cells of a signal that causes adjacent cells to proliferate. Interestingly, the Kn1 gene, which has been cloned and sequenced, contains a homeobox, the first found in plants (Vollbrecht et al., 1991). Homeoboxes are 180 base pair sequences that encode 60 amino acid homeodomains.

194 EVOLUTION OF PLANT SEXUAL REPRODUCTION

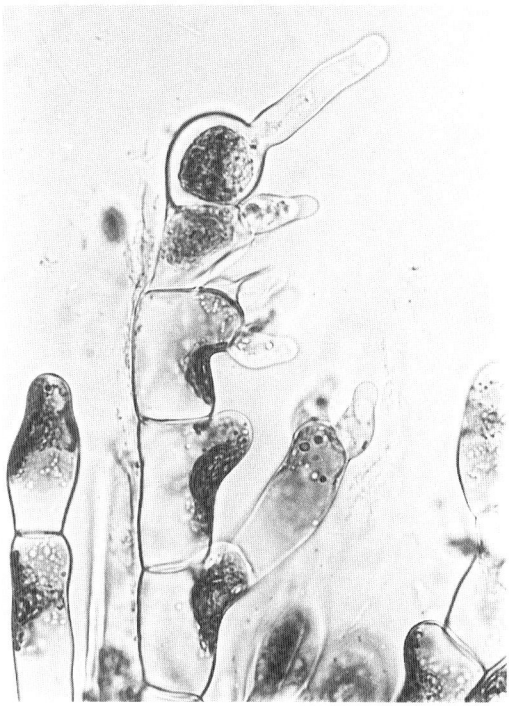

Fig. 8-14. Sexual reproduction and zygote development in the filamentous species *Coleochaete pulvinata*. Production of eggs, sperm, and zygotes in clonal cultures suggest that this species is both monoecious and homothallic (self-fertile).

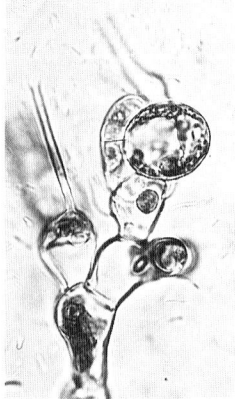

Fig. 8-15. Fertilization in *C. pulvinata* is signaled by enlargement of zygotes and concomitant growth of adjacent vegetative branches toward zygotes. Remains of the trichogyne can be observed; note that trichogynes are larger in diameter that the sheaths of setae (shown at left).

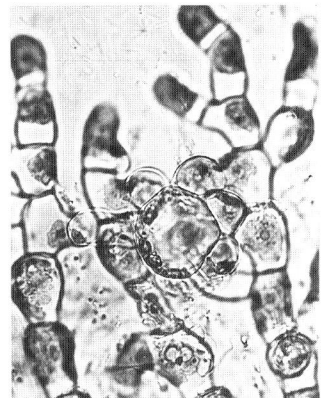

Fig. 8-16. Zygote development in *C. pulvinata* proceeds by further cell enlargement and further overgrowth of zygotes by neighboring cells. Growth of these cells appears to be influenced by proximity of zygotes, suggesting secretion by zygotes of diffusible substances for which vegetative cells are targets.

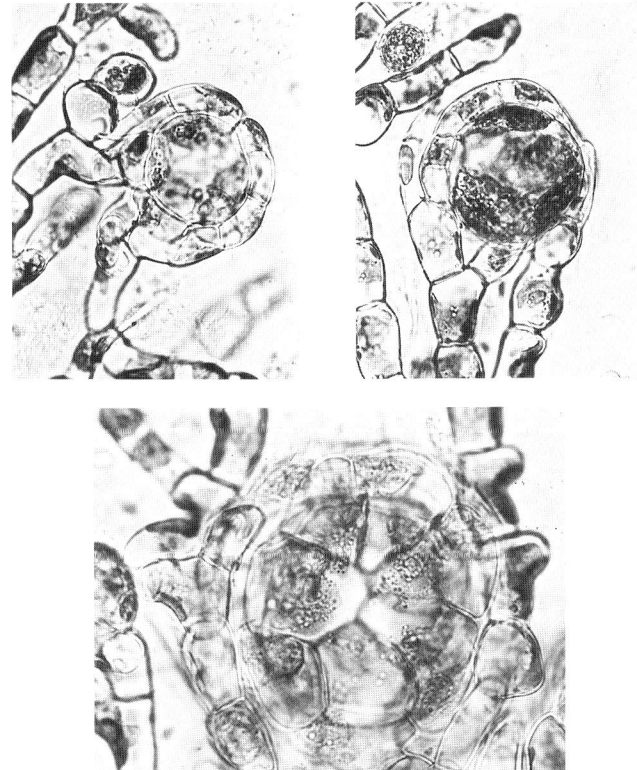

Figs. 8-17 (top left), 8-18 (top right), and 8-19 (bottom). Eventually, a corticating layer of cells is formed around large zygotes of *C. pulvinata*.

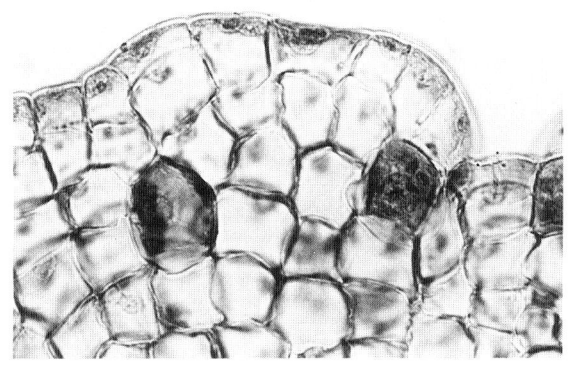

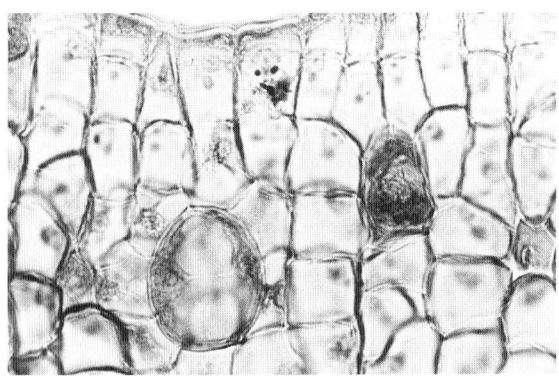

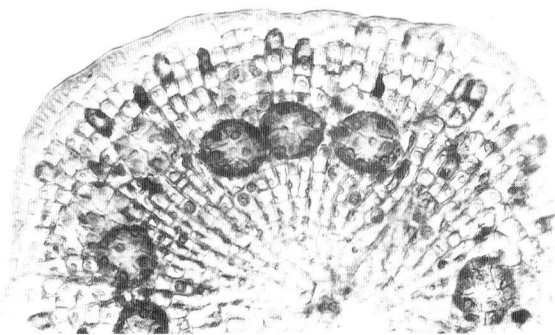

Figs. 8-20 (top), 8-21 (center), and 8-22 (bottom). Sexual reproduction in the thalloid species *Coleochaete orbicularis*. Eggs and young zygotes are surrounded in the thallus plane by continued growth of adjacent cell lineages. Fertilization results in zygote enlargement and overgrowth of neighboring cells to form a dorsal covering of cortical cells. Production of zygotes in a clonal culture suggests that this species is self-fertile. As thallus growth continues, successive rings of zygotes may be formed.

These are thought to be DNA-binding regions that influence developmental patterning and cellular specialization. Many developmental events in plants are regulated at the level of gene transcription, and thus potentially under the influence of homeobox genes. Because similar sequences occur in (and are better known from) animals and yeast, homeoboxes are regarded as ancient, conserved motifs (Chasen, 1992). This suggests that plants are likely to have inherited their homeoboxes from protist ancestors, and that charophycean development is probably influenced by the activity of homeobox sequences. Heterologous sequences could be used to probe charophyte, bryophyte, and pteridophyte genomes for the occurrence of homeobox domains. However, this strategy requires the use of highly conserved regions (Chasen, 1992).

The development of specialized gametophytic tissues that are associated with developing bryophyte sporophytes is hypothesized to reflect emission of growth regulators by sporophytes, but there are currently no empirical data related to hormonal regulation of sporophyte-gametophyte relationships in seedless plants (Ligrone and Gambardella, 1988). In general, cell-to-cell interactions in plants are very poorly understood in comparison with animal systems. Only recently have the genes encoding plant growth regulators and their receptors begun to be characterized. However, as discussed more fully in Chapter 9, there is substantial evidence that signal binding to surface receptors of target cells, initiation of intracellular second messengers (cyclic AMP, GMP, or calcium ion), involvement of calmodulin, and regulation of protein kinases—the signal-transmission elements characteristic of animal cells—also occur in plants. It is significant that a calcium-binding calmodulin and protein kinase have been isolated from zygnematalean algae (see review by Wagner and Grolig, 1992), suggesting that charophytes may also be equipped with the components required for cellular communication. When higher plant signaling systems are sufficiently well understood, it may be possible to use genetic probes or antibodies to dissect the chemical processes that control developmental relationships at the placental junction in charophytes and simple embryophytes. Such information will not only shed light on the evolutionary development of the plant embryo, but may also suggest useful strategies for the cultivation of plant embryos and establishment of plant in vitro fertilization systems.

CONTROL OF THE ONSET/DELAY OF MEIOSIS

In general, little is known about the genetic and biochemical controls that influence the timing of meiotic division in eukaryotes. It was recently found that meiotic initiation in the amphibian *Xenopus* is controlled by synthesis of the protein product of the *mos* proto-oncogene. This protein is required for activation of the maturation promoting factor (MPF) in meiosis I (Yew et al., 1992). In plants, meiosis-specific DNA transcripts reported to occur in meiocytes of *Lilium* (lily) were observed to be similar to those appearing in mouse spermatocytes (Hotta et al., 1985). This mRNA (termed zygRNA) was not detected in nonmeiotic tissues and was encoded by a group of single or low copy DNA sequences constituting about 0.1 to 0.2 percent of the nuclear genome. ZygDNA was unusual because it did not replicate during the synthesis (S) phase of the cell cycle along with the rest of the genome, but began replication at meiotic zygotene in conjunction with chromosome pairing. Replication of these sequences was not completed until after pachytene, when the synaptonemal complexes begin to disassemble, suggesting a role in regulating meiotic recombination. Hotta et al. (1985) speculated that zygRNA is translated into proteins that function in the formation of synaptonemal complexes and recombination nodules, but this possibility requires further study. Interestingly, meiocytes of lily also produce a protein that suppresses the synthesis of zygDNA. This protein thus represents a mechanism for regulating the timing of meiosis (Hotta et al., 1985). If homologous proteins function in regulation of meiosis in other plants, they (or antisense nucleic acids) could be used in attempts to experimentally modify the timing of meiosis in simple sporophytes (such as those of liverworts) or zygotes of charophycean algae. When the genetic regulation of meiosis in plants and protists has been further characterized, it seems not only possible but likely that evolutionary steps involved in the origin of the plant sporophyte and alternation of generations may someday be experimentally reproduced in the laboratory.

THE ORIGIN OF WALLED MEIOSPORES

Although there may be a number of valid evolutionary explanations for the origin of alternation of generations in embryophytes, the plant sporophyte has generally been regarded as a mechanism for amplification of the products of sexual reproduction and production of meiospores capable of resisting desiccation in the terrestrial habitat. It has been suggested that production of more than four meiospores per fertilization event may be a way of increasing progeny numbers and genetic diversity under conditions where the availability of liquid water for fertilization is limited (Bower, 1908; Searles, 1980). Comparison of the reproductive strategies of modern charophycean algae may be useful in understanding relationships between environmental constraints and reproductive responses that are of relevance to the transition to land.

As previously mentioned in this chapter, the Zygnematales appear to be the earliest diverging charophytes that exhibit sexual reproduction. Among these, members of the Mesotaeniaceae, the saccoderm desmids, may represent particularly useful model systems for elucidation of the evolutionary origin of plant sex and meiosis. *Mesotaenium, Spirotaenia, Netrium,* and *Cylindrocystis,* for example, can be induced to undergo conjugation and zygote germination in the laboratory, and in many cases all four meiotic products can be recovered (Figs. 8-23, 8-24, 8-25, 8-26, and 8-27). It is significant that zygote germination involves migration of the four plastids (derived in pairs from gametes) prior to segregation in meiosis (Biebel, 1973). This process is analogous (or perhaps ancestral) to premeiotic plastid migration in *Coleochaete* and land plants.

In contrast to some saccoderm desmids, zygotes of other zygnematalean algae typically produce fewer than four meiotic products at germination, as a result of

THE ORIGIN OF WALLED MEIOSPORES

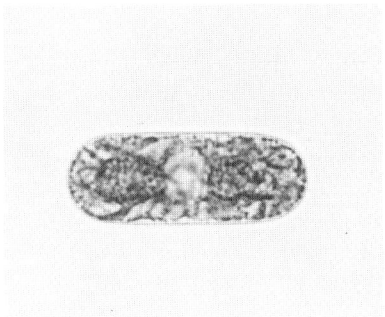

Fig. 8-23. Cultured cells of a heterothallic strain of *Cylindrocystis brebissonii*, isolated by Biebel (1973) from a small pond near Carlisle, Pennsylvania, and grown on Bold's Basal Medium agar. Photo from Biebel (1973), Morphology and life cycles of saccoderm desmids in culture, *Beih. Nova Hedwigia* 42:39–57, courteously supplied by P. Biebel.

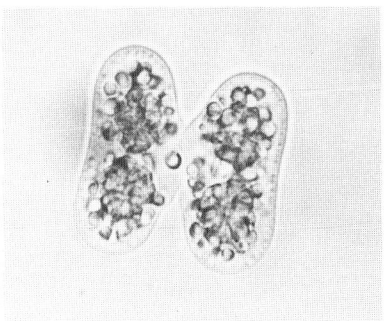

Fig. 8-24. Conjugation of *C. brebissonii* begins when cells pair lengthwise and develop wall thickenings at the contact region. These wall projections then develop into a conjugation tube. Photo from Biebel (1973), with permission.

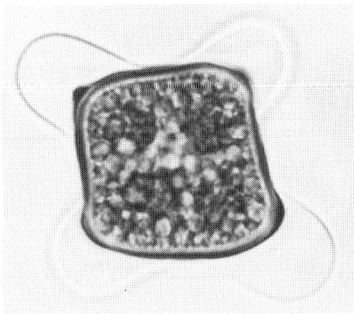

Fig. 8-25. Protoplast fusion results in zygote formation and development of a zygospore wall in *C. brebissonii*. Remnants of gamete cell walls are associated with zygospores. Photo from Biebel (1973), with permission.

200 EVOLUTION OF PLANT SEXUAL REPRODUCTION

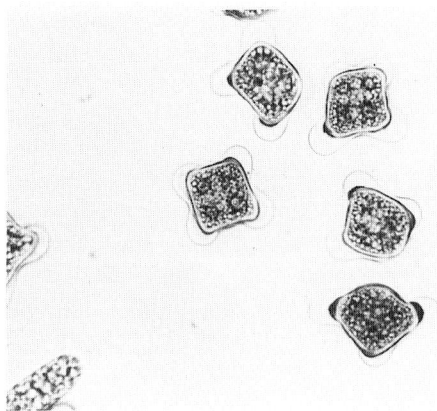

Fig. 8-26. A high yield (about 85 percent) of zygospores is produced within two weeks after *C. brebissonii* cells are suspended and diluted to an absorbance of 0.3 at 665 nm with distilled water, mixed with an equal amount of double strength BBM lacking a nitrogen source, and incubated at 20°C in a 16:8 light-dark cycle. Photo from Biebel (1973), with permission.

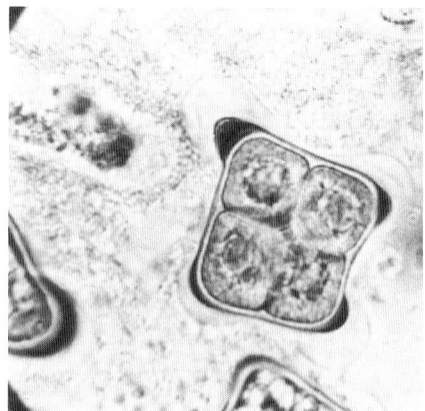

Fig. 8-27. Germination of *C. brebissonii* zygotes occurs after a dormancy period of two months when zygotes are deposited on 0.5 percent Bold's Basal Medium agar, then transferred to fresh agar medium and illuminated for 30 to 40 days. Early germination is characterized by migration of the four plastids. Four cell products develop by successive bipartition within the spore wall. Germlings rupture the spore wall as they emerge. Photo from Biebel (1973), with permission.

degeneration of two, or even three, of the daughter nuclei (Bold and Wynne, 1985). The adaptive value of the loss of two or three meiotic products in these algae is unclear. One possible explanation might be that the survivors reap the benefit of increased resources in the form of zygote nutrient storages, but there is little empirical evidence in support of this hypothesis.

A similar rationale might be used to explain the apparent survival of a single meiotic product at germination of charalean zygotes. In comparison with other algae, charalean algae produce large zygotes, on the order of 2 mm in length, which are packed with photosynthetic storage materials (Pickett-Heaps, 1975) (Fig. 8-28). When zygotes germinate, a colorless or pale green filament known as the primary protonema is produced. This filament, presumed to represent the single product of meiosis, gives rise to a young axial shoot. The energetic requirements for such relatively complex early development, coupled with the potential for light limitation in benthic habitats, may have acted as selective force favoring maximal resource provision to the germling. Large charalean thalli may produce many zygotes in a season, with the result that reduction in numbers of meiotic products per zygote need not limit fecundity.

It should be noted, however, that charalean meiosis is poorly understood, because investigators have experienced difficulty in preserving the cellular features of germinating zygotes. Although some observations indicate that the zygote nucleus divides meiotically (see Grant, 1990, for references), the evidence is inconclusive (Bold and Wynne, 1985). Shen (1966; 1967) used microspectrophotometry to establish that sperm of *Chara zeylanica* contain the same amount of nuclear DNA as nodal vegetative cells (internodal cells, however, exhibited higher DNA levels, indicating endoduplication). Similar studies, perhaps utilizing DAPI

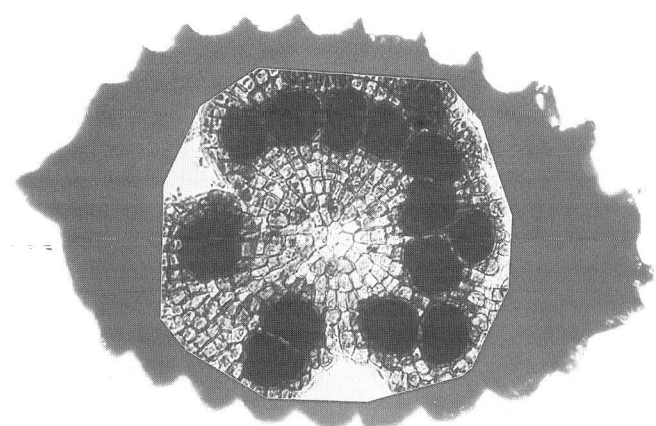

Fig. 8-28. A photo of a thallus of *Coleochaete orbicularis* has been superimposed upon a photo of a *Chara zeylanica* zygote at the same magnification, to demonstrate relative sizes of zygotes.

or other DNA-binding fluorochromes, need to be extended to germinating zygotes of charalean taxa and other charophycean algae to more accurately establish the occurrence and timing of meiosis.

Meiosis, zygote germination, and meiospore development have been more thoroughly studied in *Coleochaete* than in any other charophyte. Although zygotes of *Coleochaete* are small (on the order of 0.125 mm) in comparison with those of charalean algae (Fig. 8-28), 8 to 32 meiospores are generated per *Coleochaete* zygote (Figs. 8-29 and 8-30). A possible adaptive explanation for the production of so many meiotic products is that thalli of *Coleochaete* are much smaller than those of *Chara, Nitella,* and relatives and thus cannot rely on production of numerous gametes (zygotes) to achieve necessary fecundity levels. Disadvantageous aspects of small thallus size are possibly offset by the opportunity to occupy shallow, turbulent (CO_2-rich), and brightly illuminated waters as epiphytes. When zygotes germinate, meiospores are released into highly suitable environments for growth of new thalli and subsequent asexual reproduction by means of zoospores (Graham et al., 1986). Because germlings of *Coleochaete* are not so likely as those of charalean algae to experience light (and possibly also CO_2) limitation, more meiotic progeny can be generated per unit of zygote storage. Production of more than four meiospores per zygote thus appears to represent adaptation to life in shallow waters. Resource demand resulting from the production of comparatively large number of zygotes per thallus may have selectively influenced the appearance of placental transfer cells in one modern species of *Coleochaete* (Graham and Wilcox, 1983). Because there is some support for the hypothesis that the algal ancestors of plants inhabited shallow fresh waters (Chapter 5), perhaps similar selective pressures influenced the evolution of placentae, embryos, and production of increased numbers of meiospores per zygote 450 to 470 million years ago.

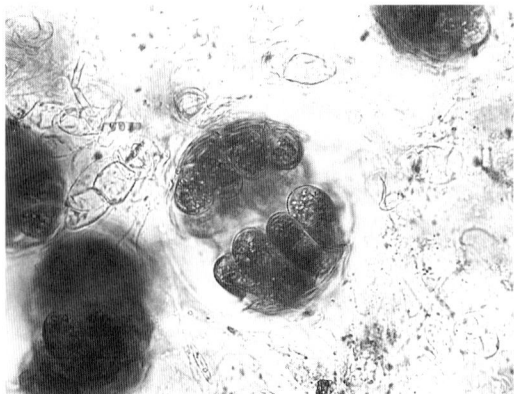

Fig. 8-29. Germination of *Coleochaete pulvinata* zygotes produces 8 to 32 meiospores; 16 have been produced from this thallus. Photo from Graham and Taylor (1986), *Journal of Phycology.*

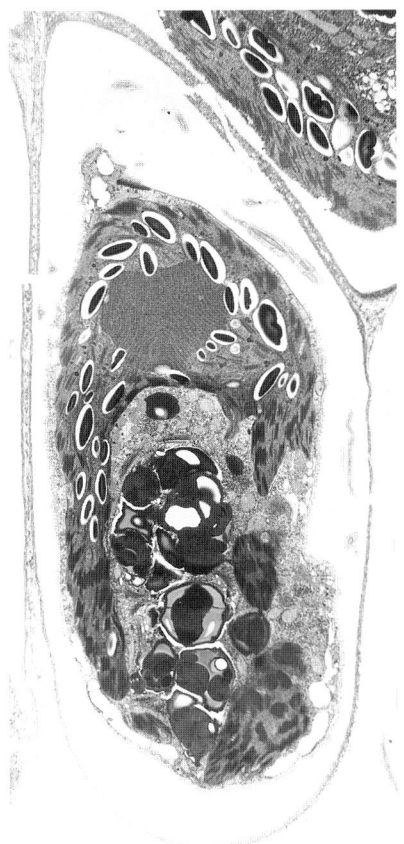

Fig. 8-30. Meiospores of *Coleochaete pulvinata* are about the same size as zoospores of the same species, but are produced within chamber walls lacking in plasmodesmata. During early stages of development, meiospores are enveloped by callose. Disappearance of the characteristic aniline blue fluorescence and resorcinol blue staining from material surrounding mature meiospores suggests developmentally controlled callose degradation. Meiospore release follows chamber wall degradation and breakage. Flagellate meiospores swim only a short distance prior to settling onto substrates, attaching firmly, and dividing to form multicellular thalli. Photo from Graham and Taylor (1986), *Journal of Phycology*.

Zygote germination in *Coleochaete* has been shown to commence with meiotic division of the nucleus (Allen, 1905). This work was substantiated by comparative Feulgen microspectrophotometric measurements of DNA in zygotes and thallus cells (Hopkins and McBride, 1976). Thus, the multicellular appearance of *Coleochaete* zygotes that have undergone meiosporogenesis (Fig. 8-28) is not homologous to small land plant sporophytes, as was once thought. Interestingly, however, Feulgen microspectrophotometric studies of *Coleochaete scutata* have revealed that prior to meiosis the nuclear DNA level reaches 8C or higher

(Hopkins and McBride, 1976) (Fig. 8-31), representing a form of endoduplication. Homologous chromosomes thus probably consist of more than two daughter chromatids. Meiosis I appears to be followed by two, three, or four meiosis II events (Hopkins and McBride, 1976), so that each meiotic daughter cell generates as many as four additional cellular products. If crossing over were to occur among chromatids at synapsis in meiosis I, it is theoretically possible for the 8 to 32 resulting meiospores to exhibit considerable genetic diversity. Although genetic diversity has not been experimentally assessed, this endoduplication and segregation process results in a population of meiospores that are potentially equivalent, in terms of genetic diversity and fecundity, to the products of a very small sporophyte. *Coleochaete*'s method for producing more than four meiospores per zygote is apparently limited to generation of no more than about 32 meiospores and is clearly not homologous or ancestral to that used by embryophytes. It represents an alternative solution to the problem of amplifying the genetic results of sexual reproduction. The direct charophycean ancestors of plants found a better way: delay of meiosis, mitotic amplification of the diploid nucleus resulting from fertilization (sporophyte production), followed by meiosporogenesis. The advantage of delay in meiosis is that sporophyte size can increase greatly, resulting in much

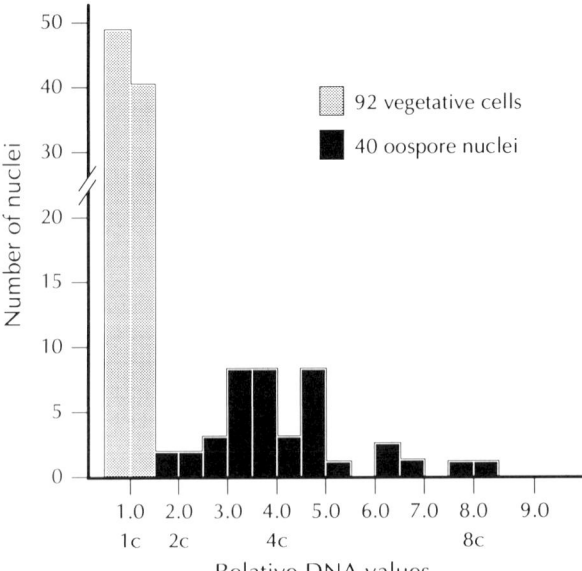

Fig. 8-31. Feulgen microspectrophotometric analysis of nuclear DNA levels in vegetative cells and oospores (zygotes) of *Coleochaete scutata* (from Hopkins and McBride [1976], used with permission of the *Journal of Phycology*). The data indicate that zygote nuclei achieve DNA levels of at least 8c prior to meiosis. This suggests that endoduplication of DNA occurs in zygote nuclei. Most likely, the number of meiospores produced by individual zygotes is determined by the number of replication cycles achieved by zygote nuclear DNA, which in turn is probably regulated by cellular resource levels.

greater amplification of the genetic recombination process and greater dispersal capability. The importance of *Coleochaete*'s comparatively feeble attempt to increase meiospore numbers lies in its illustration of the relationship between selective pressures operating in shallow freshwater habitats and adaptive reproductive responses. The example of *Coleochaete* is strong evidence that the charophycean ancestors of plants acquired reproductive innovations in shallow water that later proved indispensable in the permanent colonization and floristic domination of terrestrial habitats. As noted in earlier discussions, many groups of algae, including some charophytes, have colonized land. It is the embryophytes, however, that dominate the landscape, and the ultimate secret of their success is the invention of sporophyte generation by delay in meiosis. The evidence presently available suggests that this critical innovation was accomplished in the aquatic environment, and that spore adaptations that can be related to terrestrial selection pressures (such as resistant walls) were acquired after transition to land had occurred.

Meiospore development in *Coleochaete* is characterized by isolation of individual cells within chambers whose walls resemble those of the vegetative thallus, but which lack plasmodesmata (Fig. 8-30). Developing meiospores are also enveloped by callose, identified by a characteristic aniline blue fluorescence and resourcinol blue staining (Graham and Taylor, 1986a, b), as are developing spores of most land plants. A callosic sheath also surrounds young zygospores of *Closterium* (Dubois-Tylski, 1981) and zygotes of various flowering plants. In *Rhododendron* this callose is regarded as an important developmental marker in embryogenesis, and its function may be the isolation of zygotes from surrounding chemical influences (Williams et al., 1984). It is possible that the developmental role of callose in plant sporogenesis and embryogenesis is as ancient as sexual reproduction in charophycean algae.

The meiospores of some *Coleochaete* species are covered with a layer of small, square scales (Fig. 8-32) similar to scales ornamenting *Coleochaete* sperm (Graham and McBride, 1979), zoospores of some *Coleochaete* species, sperm of

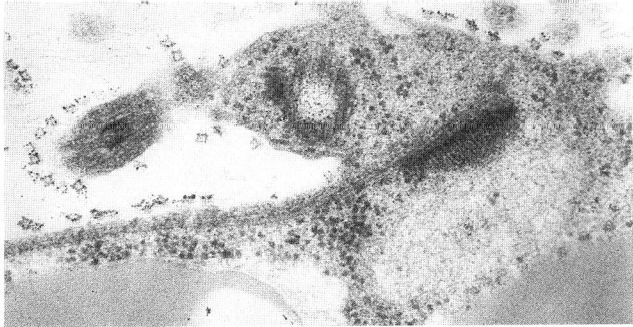

Fig. 8-32. Meiospores of *Coleochaete circularis* (shown here) contain an MLS, and associated anterior mitochondrion, and are covered with a layer of flat, diamond-shaped scales similar to those on flagellar membrane surfaces. Photo by L. Graham and C. Taylor, III.

charalean algae, and zoospores of *Chaetosphaeridium* and *Chlorokybus*. As previously mentioned (Chapter 4), similar small, square scales have been observed to occur on prasinophyte flagellates and are probably plesiomorphic for charophycean algae. In contrast, meiospores of *Coleochaete pulvinata* (Fig. 8-33) and zoospores of *Coleochaete scutata* (Fig. 8-34) are coated with a layer of distinctive, electron-dense pyramidal scales quite unlike anything observed elsewhere among green algae. The significance of differences in scale morphology with respect to relationships among the species of *Coleochaete* is unclear. Because pyramidal scales appear to be restricted to occurrence on spores, it has been suggested that they might represent an evolutionary precursor of the land plant sporoderm (Graham, 1990). However, the biochemical composition of *Coleochaete* scales has not been determined.

Blackmore and Barnes (1987) proposed that delay in the timing of deposition of sporopollenin was instrumental in the transition from deposition on zygote walls, which occurs in charophycean algae, to deposition around individual spores, as occurs in embryophytes. To date this remains a reasonable hypothesis for the origin of walled spores in early land plants. Although the earliest and most common spore tetrads lacked enclosures, some fossil spore tetrads were enclosed with a resistant envelope, as described by Norma Johnson (1985) and others. Enclosed spore tetrads could represent intermediate evolutionary stages between charophycean algae, where sporopollenin occurs only in zygote walls, and land plants, in which sporopollenin is deposited in spore walls, but not in those of spore mother cells or zygotes. In other words, the developmental timing of sporopollenin production might have switched from the zygote stage to both zygotes and meiospores, or to both spore mother cells and meiospores, and finally, to spores only. Alternatively, early spore tetrads enclosed by resistant envelopes may not represent such transi-

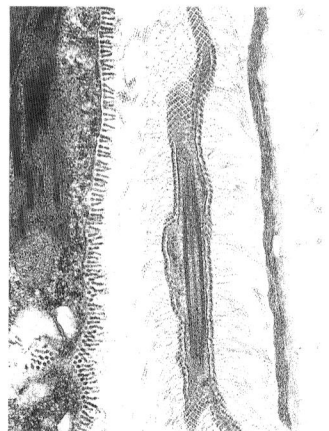

Fig. 8-33. Meiospores of *Coleochaete pulvinata* are coated with distinctive pyramidal scales which are unlike flattened flagellar scales. Photo from Graham and Taylor (1986), *Journal of Phycology*.

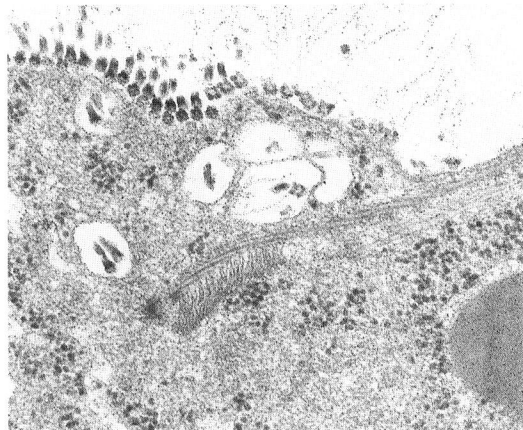

Fig. 8-34. Pyramidal body scales similar to those of *C. pulvinata* also occur on zoospores of *C. scutata*. These scales are produced in dictyosome vesicles and transported to the cell surface. An MLS is also visible in this photo from Graham and McBride (1979), *American Journal of Botany*. The difference in body scale structure between *C. circularis*, with putatively primitive flat scales on the one hand, and presumably derived pyramidal scales of other *Coleochaete* species on the other, suggests possible divergence of *Coleochaete* into two distinct lineages.

tion stages at all. Little is known regarding regulation of sporopollenin deposition in plants, and the biochemical composition of sporopollenin is poorly resolved (Southworth, 1974). When additional progress has been made toward understanding the biosynthesis of sporopollenin, it may be possible to devise strategies for tracing evolutionary changes in site of deposition. Such studies may also help in interpretation of fossil spore structures.

Niklas (1976a) discussed similarities between extant *Coleochaete* and the enigmatic Devonian fossil *Parka decipiens*. The chemical composition of *Parka* resembled that of green algae, and thallus morphology and reproductive structures were somewhat similar to those of *Coleochaete*. For example, *Parka* was a flat, circular disk that grew radially, as does *Coleochaete*. In addition, the spore masses of *Parka* were covered by polygonal cell layers that resemble the cortical layers around zygotes of *Coleochaete* (Fig. 8-35). However, *Parka* differs from *Coleochaete* in larger thallus size, production of two or more layers of thallus cells, and the fact that spore masses may contain as many as 35,000 spores (Hemsley, 1989). The spores of *Parka* were somewhat larger than those of *Coleochaete*: 25 to 45 μm in diameter for *Parka* (Hemsley, 1989), as compared with about 20 μm for *Coleochaete pulvinata* (Graham and Taylor, 1986b). More important is the fact that *Parka*'s spores were encased in a resistant, lamellate wall which, however, lacked trilete marks. Hemsley (1989) examined walls of these alete spores ultrastructurally, and concluded that they bore similarities to lamellate exines of liverwort spores, which may lack trilete marks as the result of early separation.

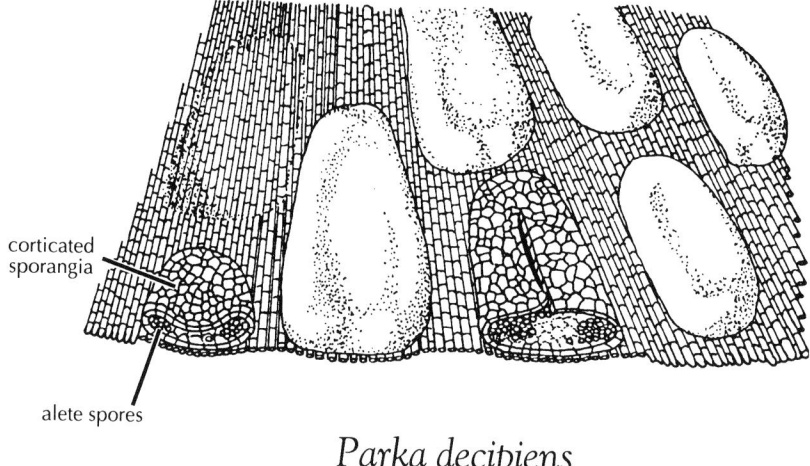

Parka decipiens

Fig. 8-35. *Parka decipiens* is an Upper Silurian–Middle Devonian compression-impression fossil that has been linked with the extant genus *Coleochaete* on the basis of morphological and reproductive similarities and some chemical evidence. This diagram illustrates the dorsiventrally flattened thallus of *Parka* and the large sporangia that were apparently covered by a polygonal cell layer. Although these features are reminiscent of *Coleochaete, Parka* thalli differed in having more than one cell layer; larger size; occurrence of a holdfast; and production of alete, but resistant-walled, spores. Diagram reproduced by permission of the Royal Society of Edinburgh and K. J. Niklas from *Transactions: Earth Sciences,* Vol. 69 (1976), p. 487.

An alternative explanation for absence of trilete marks might be that spores of *Parka* were not generated directly by meiosis, but were instead mitotic products, rather like the meiospores of *Coleochaete*. *Parka* is probably too recent to represent the earliest colonizers of land, and fossils are comparatively rare and geographically restricted. It may be important, however, as an example of the diversity of early spore producers. The fact that placental regions of *Coleochaete* and seedless plants are impregnated with resistant compounds so that wall ingrowth morphology is retained after acetolysis suggests that placental morphology might be retained in appropriate fossil remains (Delwiche et al., 1989). It would be interesting to determine if cells surrounding spore masses of *Parka* show evidence of occurrence of placental wall ingrowths. This would be suggestive, but not conclusive evidence, of embryophytic affinities. Did resistant spores evolve in plantlike organisms that had not yet acquired embryos or sporophytes (i.e., nutritional and developmental interactions between generations or sporic meiosis)? Further paleobotanical work on early spores and organisms such as *Parka* appears to be the best strategy for attempting to answer such questions regarding the evolutionary history of walled spores.

9

THE ORIGIN OF PLANT SIGNAL TRANSDUCTION SYSTEMS, PHYTOHORMONES, PHOTOMORPHOGENESIS, AND SECONDARY METABOLISM

One of the most dramatic ways in which plants differ from animals is the much greater degree to which postembryonic development is related to the physical environment. The organizational plan of animal organs and tissues (including reproductive organs) is almost completely established during embryogenesis. In contrast, plant embryos are structurally very simple, and morphogenesis occurs throughout the life of plants. Higher plant stems, roots, leaves, and reproductive organs are modular structures that can be repeatedly generated from localized meristems. Extrinsic cues such as light, water, temperature, gravity, and touch trigger specific developmental responses that affect the activity of plant meristems and, hence, plant form. Plants are thus able to deal with environmental fluctuation by modification of developmental patterns (Palme, 1992). However, the mechanisms by which plants detect environmental signals and convert them into altered structure via changes in patterns of gene transcription or other mechanisms, such as posttranslational modifications, are incompletely understood (Goldberg, 1988). The process involves sensory receptors, transduction systems that convert signals into chemical messages, and effector molecules. Some components of plant signaling systems are similar to those of animals and eucaryotes in general, whereas others, such as phytohormones, are more restricted in distribution. Because charophycean algae provide the evolutionary link between land plants and protists likely to be related to other eukaryotic lineages, charophytes may prove to be useful model systems for molecular dissection of fundamental plant-environment relationships. In this chapter, we explore current knowledge regarding signal transduction pathways in charophytes. The evidence for occurrence of phytohormones (plant growth substances) in algae is also discussed. Evidence provided by charophycean algae related to the evolutionary history of plant photomorphogenesis is reviewed. Finally, the link between environment and plant secondary metabolism is examined from an evolutionary perspective. The origin of the biosynthetic pathways leading to production by early plants of compounds such as lignin, flavonoids, cutin (and/or cutan), and sporopollenin is an important and much discussed component of concepts related to the origin of land plants. These compounds contribute significantly to the generation of plant form and plant response to environmental change. Early forms of pathways leading to secondary compounds are believed to have arisen from preexisting elements of primary metabolism in the ancestors of land plants (Stafford, 1991a). Investigations directed toward understanding the evolutionary origin of plant signal reception, transduction, and translation into structural and chemical variation may shed considerable light on the operation of these processes in higher plants.

EUKARYOTIC SIGNAL PATHWAYS

A common pattern of signal transduction elements appears to occur widely among eukaryotic organisms. Although signal transduction pathways are currently better understood in yeast and animal systems than in protists and plants, there is increasing evidence that the same fundamental elements also occur in algae and plants. The ability of eukaryotic cells to respond to external signals depends on the presence of specific receptor molecule systems located at the cell periphery, commonly the cell membrane. Eukaryotic receptor systems appear to be related to, and are probably derived from, the sensory systems of prokaryotes. Receptors are able to translate environmental information into specific intracellular responses, such as production of second messenger molecules which convey information from the cell surface to the interior (Fig. 9-1). Examples of second messengers include cyclic-AMP and derivatives of phosphoinositides, and these are found in plants as well as in other eukaryotes. Xu et al. (1992) used detergent extraction procedures to demonstrate association of a significant proportion of cellular phosphatidylinositol 4-kinase (which forms phosphatidylinositol-4-phosphate) with the cytoskeleton of carrot protoplasts. These workers suggest that phosphatidylinositol-4-phosphate may play an important role in modulation of the plant cytoskeleton, particularly actin-binding proteins (McCurdy and Williamson, 1991), in response to cellular signal transduction. Several second messenger systems relay information by causing changes in intracellular calcium localization. This results in structural changes in Ca^{2+}-binding proteins such as centrin and calmodulin and enables their interaction with other proteins. As previously discussed in Chapter 4, calcium-binding proteins are found in algae and plants and are thought to regulate cytoskeletal components, cell division, and exocytosis. Calmodulin m-RNA levels are characteristically elevated in plant meristems and developing fruits, and plant calmodulin m-RNA levels respond to flooding, touch, gravity, and pressure. Calcium ion concentration changes may also link light and other environmental signals to gene expression (Palme, 1992).

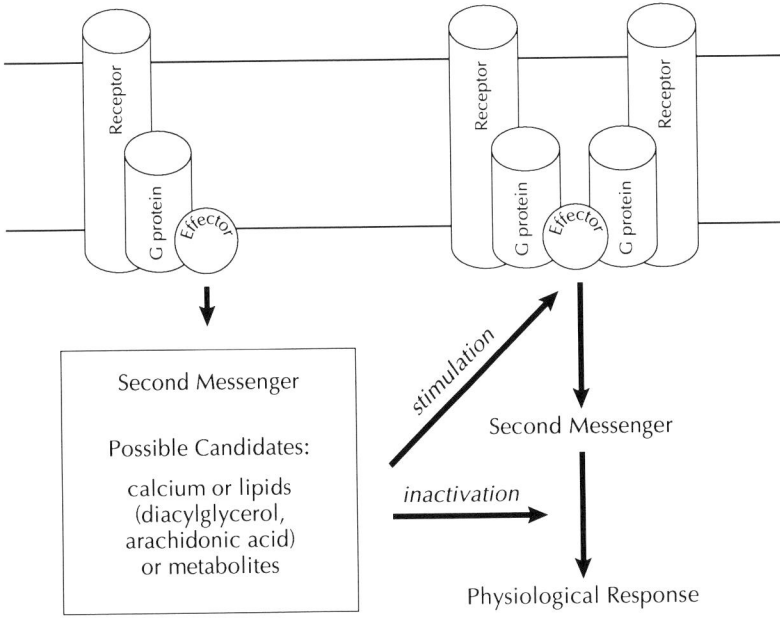

Fig. 9-1. Environmental signals are received and processed by cell membrane-bound receptors and other proteins. Second messenger molecules link the reception system with cytoplasmic and nuclear processes. Diagram modified from Palme (1992), Molecular analysis of plant signaling elements: Relevance of eukaryotic signal transduction models. *Int. Rev. Cytology* 132:236, with permission.

Eukaryotic receptor systems often include G-proteins, which have become widely diversified for different sensory functions in evolution. These proteins typically contain several transmembrane domains, of which certain regions are involved in the binding of various signal molecules, such as hormones. Guanine nucleotide-binding proteins then interact with effector enzymes inside cells (Palme, 1992). Plant G-proteins share coding sequences with known oncogenes (cancer related genes in animals), such as *ras,* and thus are implicated in control of cell proliferation. G-proteins are also important as regulators of membrane vesicle and protein transport to the cell surface. Receptor proteins, including G-proteins, can be inactivated by the action of specific protein kinases; phosphorylation of G-proteins blocks binding with external signal molecules or prevents coupling with effectors (Palme, 1992).

Protein kinases generally function as major regulatory switches in cells in controlling activation, deactivation, or amplification of enzymes; ion channel opening and closing; desensitization of receptors; and hormone production. Protein kinases themselves are "turned on or off" by protein phosphatases, which play a central role in various facets of cell physiology, including control of the cell division cycle. For example, the widely occurring eukaryotic cell division control

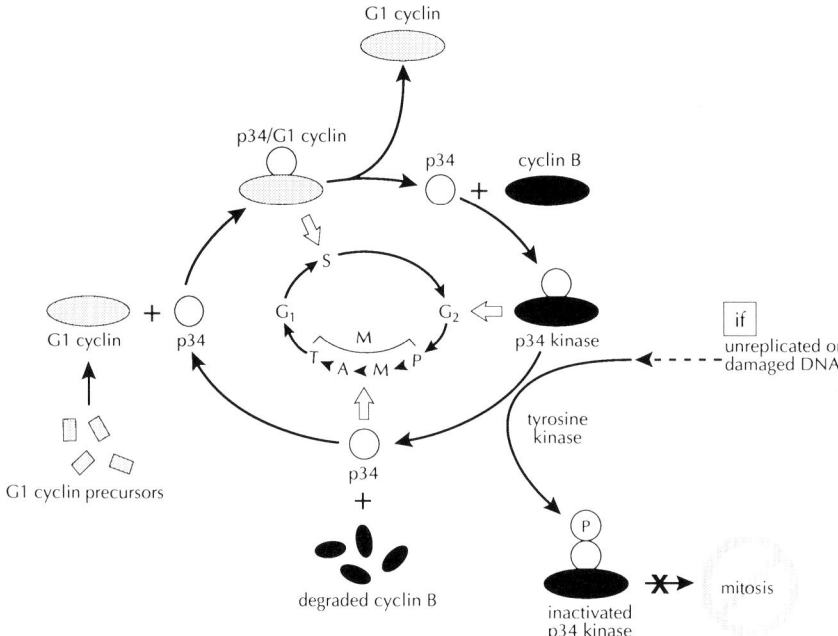

Fig. 9-2. Activation of various steps of eukaryotic mitosis is known to involve tyrosine kinases and cdc2 kinase (composed of a cyclin protein and p34^{cdc2}). This model for protein interactions relating DNA synthesis to subsequent mitotic steps is partially based on Smythe and Newport (1992), Coupling of mitosis to the completion of S phase in *Xenopus* occurs via modulation of the tyrosine kinase that phosphorylates p34^{cdc2}, *Cell* 68:787–797.

gene *cdc2* encodes a protein kinase which together with the regulatory protein cyclin, comprises MPF (maturation promoting factor). Mitosis can be blocked by the action of tyrosine kinase, which phosphorylates the *cdc2* gene product (Fig. 9-2), and the metaphase/anaphase transition results in abrupt degradation of the cyclin component of MPF (Smythe and Newport, 1992). Some evidence points to the occurrence of protein kinases in plants (and charophytes), but to date they are poorly characterized. Protein phosphatases having a high level of coding sequence similarity to other eukaryotic protein phosphatases occur in plants, but their precise role in signal transduction has yet to be elucidated (Palme, 1992).

PHYTOHORMONES

The five classes of phytohormones that affect higher plant growth and development are auxins, gibberellins, cytokinins, abscisic acid, and ethylene. Although they are transported from site of production to target site in plants and are active at low concentration, as are animal hormones, animal hormones typically differ in structure from the phytohormones listed above, and there are many more classes of animal hormones. However, some evidence is accumulating for the existence of signaling peptides in plants, similar to those occurring in animal systems. Oligosaccharides may also have signaling roles in plants (Palme, 1992).

There are many documented examples of phytohormone effects on the regulation of gene transcription, and some results suggest posttranscriptional control of plant development as well. It is possible that plant hormone systems have been evolutionarily added to signal reception, transduction, and effector systems inherited from protistan ancestors. This would explain the evidence for the similar occurrence of elements of signal transduction such as calcium-binding proteins and G-proteins in plants and other eukaryotes. However, very little is known regarding the nature of phytohormone receptor systems. Anthony Bleecker and coworkers have recently found that a higher plant ethylene receptor appears to be a transmembrane protein that, in a portion of the protein located at the cytoplasmic face, shares sequence similarity with a protein kinase (Palme, 1992). Cloning of genes encoding phytohormone binding sites is an increasingly active area of research in higher plant systems.

Growth substances such as lunularic acid (possibly related to flavonoids (Stafford, 1991a), ethylene (see Basile and Basile, 1983, for example), and other hormones (Doonan, 1991) influence morphogenesis in bryophytes. However, little is known about the origin and evolution of phytohormone production and function in embryophytes. The first phylogenetic occurrence of all five major classes of phytohormones in the plant lineage is not known. Do algae produce phytohormones? The answer to this question continues to be hotly debated. Bradley (1991) contends that there is GC/MS (gas chromatographic/mass spectrometric) and HPLC (high-pressure liquid chromatographic) evidence that auxin and cytokinin occur in seaweeds, and that application of plant hormones to axenic algal cultures affects growth and development. Evans and Trewavas (1991), however, argue that microorganisms synthesize representatives of all five classes of phytohormones as secondary metabolites, and that some reports of occurrence of phytohormones in algae could be the result of contamination. They conclude that the evidence for occurrence of phytohormones in algae remains equivocal. Isopentenyladenosine has been identified in *Chara globularis* by GC/MS (Zhang et al., 1989). These authors note, however, that the cytokinin might have resulted from the breakdown or turnover of tRNA. In addition, the algal specimens were obtained from field collection, so microbial contamination cannot be ruled out. Clearly, further investigations are warranted related to occurrence of phytohormones in algae, and especially in charophytes. When sufficient numbers of genes encoding various plant hormones and their receptors, such as an auxin-binding protein gene identified by Yu and Lazarus (1991), have been cloned, and conserved sequences identified, it may be possible to use heterologous hybridization probes to examine bryophyte and algal genomic DNA directly for evidence of occurrence of phytohormones or receptors. This approach might obviate technical concerns regarding variations in level of gene expression, production of low levels of hormones, and, possibly, microbial contamination.

PHYTOCHROME AND PHOTOMORPHOGENESIS

One aspect of higher plant morphogenesis clearly exhibited by charophycean algae is a well-documented phytochrome-mediated response. Higher plant phytochrome has been well characterized; it is a soluble blue-green chromoprotein, 116 kDa in size, which undergoes reversible conformational changes in response to visible light. Phytochrome is rapidly converted from an inactive state, Pr, which absorbs red light, to the active form, Pfr, which absorbs far-red light. Although phytochrome is known to influence gene transcription, links between the signaling network and changes in gene expression are not understood (Palme, 1992).

Phytochrome from *Mesotaenium* has been purified and partially characterized by Kidd and Lagarias (1990). The full length polypeptide is 120 kDa, similar in size to higher plant phytochrome, and is primarily localized to the cytoplasm. An estimated 150,000 to 250,000 molecules of phytochrome occur in each cell. Morand et al. (1993) found that phytochrome levels in *Mesotaenium* are light regulated. When compared to higher plant photoreceptors, *Mesotaenium* phytochrome appears to be more closely related to *phy*B gene products than to *phy*A products (Morand et al., 1993). In *Mesotaenium* and *Mougeotia,* low-irradiance white or red light results in orientation of the band-shaped chloroplast for maximal light exposure. High irradiances cause plastids to orient for minimal exposure. Orientational changes occur within a minimum of 10 minutes. Photoconversion of the phytochrome molecule results in a change in the transition moment of the chromophore from surface parallel for the Pr form to surface normal for the Pfr form. Molecular dichroism allows cells to detect changes in the direction of light as well as the intensity. The cellular response, chloroplast rotation for optimal irradiation, is probably accomplished by means of actomyosin (Wagner and Grolig, 1992). A model for the mechanism of plastid rotation invokes the occurrence of a tetrapolar gradient of active Pfr molecules, with highest levels at front and rear of an irradiated cell, and lowest levels at the flanks. This gradient is proposed to direct the formation of a corresponding gradient of actin anchorage sites along the cell membrane. Grolig (1992) speculates that myosin at the chloroplast envelope binds to microfilaments, thus connecting the edges of the plastid to the cell periphery and forming a mechanical basis for rotation of this large organelle. Evidence for a motive role of actomyosin is provided by reversible cytochalasin inhibition of chloroplast movement (but not light perception). According to Wagner and Grolig (1992), calcium channel fluxes are not likely to be directly involved in plastid rotation, which occurs at a velocity two to three times more slowly than actin-based cytoplasmic streaming phenomena observed in charalean algae (see Chapter 10). *Mougeotia* cells contain and release phenolic compounds to the external medium, probably as a chemical defense against herbivores or microbial attack. Interestingly, the phenolic compounds are localized in calcium-binding vesicles, which are abundant at chloroplast edges. Whether these vesicles act as a calcium buffer, maintaining calcium homeostasis within cells, is unclear (Wagner and Grolig, 1992).

Although there are other reports of phytochrome-mediated responses in charophytes, these are not as well documented, and the evolutionary origin of phytochrome has not been elucidated. The possible occurrence of phytochrome in charophytes below the level of Zygnematales or in putative flagellate ancestors of charophytes has not been explored. It may be possible to use conserved, heterologous DNA sequences to probe algal DNA for genes homologous to those encoding higher plant phytochrome and thus trace the ancestry of this important molecule.

CHAROPHYCEAN PHENOLIC COMPOUNDS: POSSIBLE RELEVANCE TO THE ORIGIN OF LIGNIN, FLAVONOIDS, CUTIN, CUTAN, AND SPOROPOLLENIN OF EMBRYOPHYTES

As previously discussed (Chapter 3), there is evidence that charophycean algae produce phenolic compounds (Gunnison and Alexander, 1975; Delwiche et al., 1989; Wagner and Grolig, 1992). It is possible that these phenolics are related to the evolution of various secondary products that occur in land plants. In land plants phenolic compounds are utilized in the biosynthesis of lignin (Sarkanen and Ludwig, 1971), flavonoids (Stafford, 1991a), cutin, and suberin (Kolattukudy, 1980) and have also been reported to occur in sporopollenin (Southworth, 1990) and extensin (see discussion in McCann and Roberts, 1991). In this section, the characteristics of charophycean phenolic materials are summarized. Then follows a consideration of the biosynthetic pathways leading to production of lignin and flavonoids in higher plants, and a discussion of strategies by which the evolutionary history of these pathways might be elucidated. The characteristics of material occurring on upper surfaces of thalloid *Coleochaete* species are described, and the possible relevance of this material to the origin of plant cuticle is discussed. The relevance of the biosynthesis of various plant polymers to the possible solution of difficult problems related to composition and synthesis of sporopollenin is also considered.

Charophycean Phenolics

As discussed earlier in this chapter, calcium-binding vesicles in *Mougeotia* cells contain phenolics, which are hypothesized to have a defense function. The vesicles absorb vital dyes such as neutral red and rhodamine B in vivo. Dye absorption stabilizes the vesicles so that they can be isolated and characterized. Vesicles were found to contain a protonated group with $pK_8 = 9.9$, characteristic of phenolic hydroxyl groups. Titration resulted in release of phenolic compounds and vital dye from vesicles (Wagner and Grolig, 1992). Interestingly, in a study of the hair cells of *Coleochaete* Geitler (1961) reported that vacuoles of seta cells were selectively stained with neutral red, whereas vacuoles of other vegetative cells were not. Whether this staining reaction indicates hydroxy-substituted phenolics is not known.

Gunnison and Alexander (1975) reported the occurrence of "ligninlike" compounds associated with cell walls of *Staurastrum* cells and suggested that the

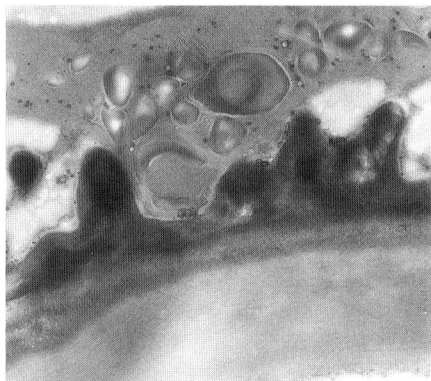

Fig. 9-3. Acetolysis resistance and specific autofluorescence in fertile thalli of *Coleochaete orbicularis* correlate with the occurrence of cortical cell wall ingrowths, shown here, as well as other cell walls within the sphere of influence of zygotes. Photo from Graham and Wilcox (1983), *American Journal of Botany*.

compounds conferred resistance to microbial degradation. There is fossil evidence that ancient charophytes were vulnerable to microbial attack (Taylor et al., 1992), and land plants commonly produce phenolic compounds in response to microbial elicitors (Lamb et al., 1989). Delwiche et al. (1989) noted the occurrence of acetolysis-resistant material in fertile, but not vegetative, thalli of *Coleochaete* species. Acetolysis-resistant material occurred in wall ingrowths characterizing cortical cells associated with *Coleochaete orbicularis* zygotes (Figs. 9-3 and 9-4), walls of cells adjacent to cortical cells, and cells associated with zygotes of other

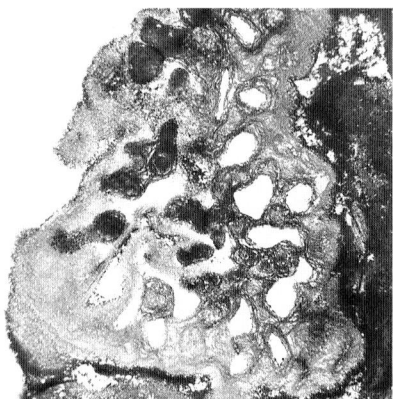

Fig. 9-4. Cortical cell ingrowth structure is preserved when fertile thalli of *Coleochaete orbicularis* are subjected to standard acetolysis procedures. Acetolysis-resistant material appears relatively electron dense in transmission electron micrographs. Photo by L. Graham and C. Delwiche.

Coleochaete species. The resistant material was proposed to have an antimicrobial function.

Infrared (IR) spectra obtained from individual acetolyzed fertile thalli by Fourier transform infrared (FTIR)microscopy were consistent with those obtained by Gunnison and Alexander (1975) for "ligninlike compounds" of *Staurastrum*. The IR spectra were distinguishable from those obtained by FTIR analysis of angiosperm and *Coleochaete* sporopollenin. In addition, the ligninlike compounds of *Coleochaete* were solubilized by moderate base treatments which left sporopollenin intact. The acetolysis-resistant material of *Coleochaete* was localized within the walls of nonreproductive cells in the vicinity of zygotes, whereas sporopollenin was localized to the inner zygote wall. When large numbers of zygotes were present, nearly the entire thallus was resistant to acetolysis. The distribution of acetolysis-resistant material was highly correlated with the occurrence of autofluorescence in ultraviolet (365 nm) and violet (395 to 440 nm) light that could be distinguished from sporopollenin autofluorescence (Delwiche et al., 1989). Pyrolysis GC/MS analysis also indicates the occurrence of phenolic compounds in *Coleochaete*. The pattern exhibited by acetolysis resistance and specific autofluorescence was interpreted as resulting from phenolic production by vegetative cells in response to unknown chemical inducers emanating from individual zygotes, within overlapping spheres of influence. This pattern is also reflected in the appearance of *Coleochaete* thalli that have been stained with conventional stains used in preparation of plant tissues for light microscopy (Fig. 9-5).

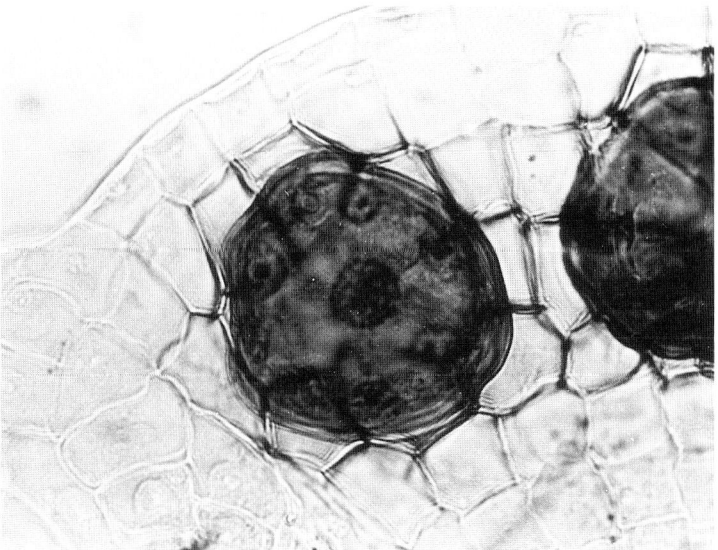

Fig. 9-5. Acetolysis resistance and specific autofluorescence also correlate with the appearance of *Coleochaete orbicularis* thalli which have been fixed and stained according to standard procedures for light microscopy. Note the darker appearance of cell walls within the sphere of influence of zygotes.

Interestingly, acetolysis-resistant material that was similar in autofluorescence properties to *Coleochaete* polyphenolic material was also reported to occur in walls of cells at the placental junction of bryophytes. The polyphenolic material in *Coleochaete* appears to deter microbial decomposition of parental thalli, so that attached zygotes germinate in a favorable environment for growth of progeny (Delwiche et al., 1989). However, function of the putatively homologous material at the bryophyte placental region is obscure. It was suggested that resistant, fluorescent compounds in bryophyte placental cells represent inheritance of the trait from ancestral charophytes (Delwiche et al., 1989). In embryophytes, placental location of these compounds may be analogous to the human appendix: essentially functionless, perhaps even detrimental—the result of patterns of evolutionary descent. Alternatively, as previously discussed in Chapter 6, regulated production of phenolic compounds may function to shut off the flow of photosynthate from gametophytes to sporophytes at particular developmental stages. However, this hypothesis has not been tested.

A deep brown coloration resulted upon application of the Maule reagent series ($KMnO_4/HCl/NH_3$) to acetolyzed *Coleochaete* (Delwiche et al., 1989). The Maule test is thought to stain syringylpropane moieties in syringyl lignins (characteristic of angiosperms), typically yielding a red color. Although brown Maule reactions are sometimes reported to occur in land plants, the significance of this color reaction is unknown. Strong Maule staining has also been attributed to lignans (phenylpropanoid dimers) that are also characterized by syringyl moieties (Lewis and Yamamoto, 1990). Lignans occur in vascular plants and (surprisingly) in hornworts, but the lignans so far described in hornworts do not contain syringyl groups (Takeda et al., 1990), and the Maule staining characteristics of hornwort lignans are not known. Thus, it remains difficult to interpret the Maule reaction results in *Coleochaete*.

In view of the phylogenetic distance separating charophycean algae from seed plants, it is not surprising that histochemical staining reactions characteristic of higher plant lignins were not observed in charophycean algae. Lewis and Yamamoto (1990) suggest that existing histochemical staining procedures lack sufficient specificity for definitive use in identification of lignins. These authors reviewed the major methods for chemical analysis of lignin and concluded that careful analysis of ^{13}C NMR spectra may be particularly useful in studies of lignin. However, application of this technique to nonvascular plants and algae may be limited by biomass availability. In addition, resistant, autofluorescent materials of *Coleochaete* are present only after sexual reproduction has occurred, and thus are developmentally regulated. Ligninlike compounds of desmids may be detectable only in the walls of older cells; young populations may lack these materials. Moreover, unialgal (or preferably axenic) cultures must be used in order to preclude the possibility of contamination with lignin breakdown products from the natural environment.

The studies described above provide evidence for the possible occurrence in charophycean algae of hydroxyphenols, as well as regulated production of resistant phenolic polymers that can be distinguished from sporopollenin. It is for this reason that they were described as "ligninlike." The alternative would have been to hypothesize the existence of an entirely new class of resistant biopolymers. The considerable body of evidence linking charophycean algae to the ancestry of embryophytes suggests that phenolic polymers produced by charophycean algae are somehow related to the evolutionary origin of resistant compounds in plants. Similarly, there is considerable evidence for the occurrence of cell wall-associated hydroxyphenyl derivatives and polyphenols in bryophytes, although unequivocal evidence is lacking for the occurrence of higher plant lignins (Grisebach, 1981; Lewis and Yamamoto, 1990). The study of bryophyte phenolics by techniques in common use in the study of vascular plant lignins are also limited by biomass availability and the specter of environmental contamination. Phenolic materials in charophytes and/or bryophytes possibly related to higher plant lignins may be present in such small amounts as to preclude study by methods used in the analysis of more abundant higher plant lignins. It is clear that alternate strategies are required to elucidate the nature of phenolic and polyphenolic compounds present in charophytes and bryophytes. However, discussion of such alternatives requires us to consider the properties of lignin in more detail and to examine closely the biosynthetic pathways by which lignin is generated. This venture will also shed light upon methods for approaching the phylogenetic origin of flavonoids.

Lignin: Properties and Biosynthesis

In terms of dry weight, lignin constitutes 25 percent of the estimated 100 billion metric tons of photosynthetic biomass produced annually and is second only to cellulose in amount. Lignin, containing a higher proportion of carbon than do cellulose and other plant polysaccharides, is thus more important as fixed carbon than is indicated by relative weight. Lignin is extremely important in the earth's carbon cycle, not only because of its abundance, but also because of its resistance to microbial degradation. Biodegradation resistance is conferred by high molecular weight, complex three-dimensional structure (Fig. 9-6), and nonrepeating types of interunit linkages such as ether bonds and C-C bonds that are difficult to hydrolyze (Obst, 1990). Lignin contributes to the persistence of peat and humus in the soil and to the production of coalified plant remains. It also protects the more vulnerable constituents of woody plants from microbial action. Lignin can thus be said to be of critical significance in the generation of important long-term carbon sinks (fossil fuels and forests) and therefore pivotal in global biogeochemical cycling of carbon and maintenance of geologically recent atmospheric homeostasis (Robinson, 1990). By inhibiting the oxygen-requiring respiration activities of decomposers, phenolic compounds may also have been historically influential in establishment of atmospheric oxygen levels. Jennifer Robinson (1990) has noted that the burial of organic carbon is the most important factor controlling the level of atmospheric oxygen, that burial on land is strongly influenced by the abundance of decay-resistant compounds in plants, that lignin is decay resistant, and that it also inhibits the decay of associated carbohydrates, both on land and in aquatic sediments. She has suggested that the advent of land plants

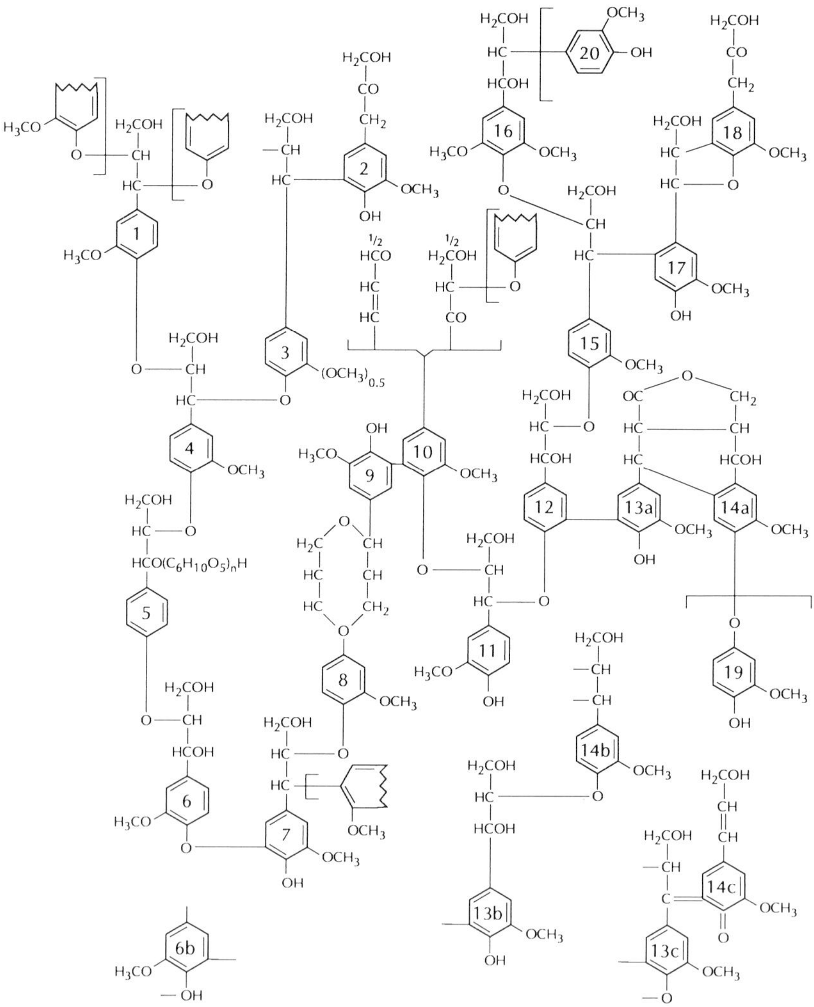

Fig. 9-6. A diagrammatic representation of a portion of higher plant lignin, illustrating the complexity of this resistant biopolymer. Cinnamyl alcohol derivatives (p-hydroxyphenyl-propane monomers) are linked together by the action of a specific peroxidase that generates phenoxy radicals. C-C and C-O-C linkages that are not directly hydrolyzable are formed, yielding a complex, three-dimensional natural plastic. Diagram adapted from Reddy (1984), Physiology and biochemistry of lignin degradation, In: Klug and Reddy (eds.), *Current Perspectives in Microbial Ecology*, American Society for Microbiology, Washington, D.C., p. 561.

destabilized the global redox balance, inasmuch as lignified organic carbon resists oxidation, and that this resistance culminated in all-time maxima of buried organic carbon and atmospheric oxygen during the Carboniferous period. The reduction in rates of organic carbon burial since that time is attributed to the evolution of lignolytic fungi, which function best in conditions of relatively high oxygen, and angiospermous plants, which are comparatively sparing in the generation of lignified tissues (Robinson, 1990).

In vascular plants, lignin is primarily found in secondary cell walls, where it is closely associated with hemicelluloses, and in middle lamellae, where it helps cement plant cells together. In addition to inhibition of microbial invasion and growth, lignin confers structural rigidity and resistance to impact, compression, and bending, and reduces the movement of water out of xylem tissues. Some ancient plants, for example, the lower Devonian *Gosslingia,* apparently produced material resembling lignin in its decay resistance, not only in vascular tissues, but also in peripheral regions (Kenrick and Edwards, 1988). Similar lignification of peripheral stem tissues is also reported to occur in the extant pteridophyte *Psilotum* (Kenrick and Edwards, 1988), but its significance is not understood. This evidence suggests the possibility that early land plant lignin may have been localized at the thallus surface, serving as a peripheral support system or as protection from radiation or microbes. Unfortunately, degradative processes remove diagnostic chemical markers of lignin, so it is difficult to test this hypothesis by chemical examination of putative lignin remains in the fossil record.

Lignin has not yet been isolated in its unaltered (natural) state (Lewis and Yamamoto, 1990), but its deduced structure is a highly complex polymer of three p-hydroxyphenylpropane derivatives: coniferyl, sinapyl, and coumaryl alcohols. Considerable variation in the occurrence and proportions of these monomers is found among plant taxa and within individual plants. The polymerization process involves dehydrogenation activity by specific wall-associated peroxidases (Graham and Graham, 1991), which utilize cytoplasmically produced monomers as substrates. Involvement of the endoplasmic reticulum with monomer production is likely, and the secretory system may package monomers into vesicles for transport through the cell membrane. Regulation of lignin deposition is poorly understood (Lewis and Yamamoto, 1990), but deposition sites may be defined by the localization of glycine-rich proteins (GRPs) in cell walls (Roberts, 1990; Ye and Varner, 1991). GRPs are encoded by genes whose transcription is influenced by light, abscisic acid, and ethylene (Linthorst, 1991). Localized cell membrane-associated microtubules are also correlated with lignin deposition sites (Pickett-Heaps and Northcote, 1966). Lignin deposition in xylem development and defense reactions likely involves highly coordinated interactions between signal transduction (including gene expression) and secretory and cytoskeletal systems of plant cells. When details of the interactions are worked out, the lignin deposition process will no doubt prove to be a dramatic example of plant developmental and regulatory capabilities.

At this point it is necessary to focus on the biosynthetic origins of lignin monomers, for if charophytes and bryophytes are incapable of generating these monomers or homologous molecules, ligninlike polymers could not be formed. Lignin biosynthesis begins with the shikimic acid pathway, which produces the three aromatic amino acids: phenylalanine, tyrosine, and tryptophan. Duplicate pathways appear to generate aromatic amino acids in plastids and the cytoplasm. The amino acids that enter into lignin (and flavonoid) synthesis are generally believed to originate via the cytoplasmic pathway, but additional work on localization is needed (Lewis and Yamamoto, 1990). In a second biosynthetic phase, the phenylpropanoid segment converts phenylalanine into substrates that are used in the formation of lignin and flavonoids. The first enzyme in the phenylpropanoid pathway, phenylalanine ammonia lyase (PAL), deaminates phenylalanine to cinnamic acid (Fig. 9-7). PAL appears to localize with the next pathway enzyme, cinnamate 4-hydroxylase, which is known to occur in the endoplasmic reticulum

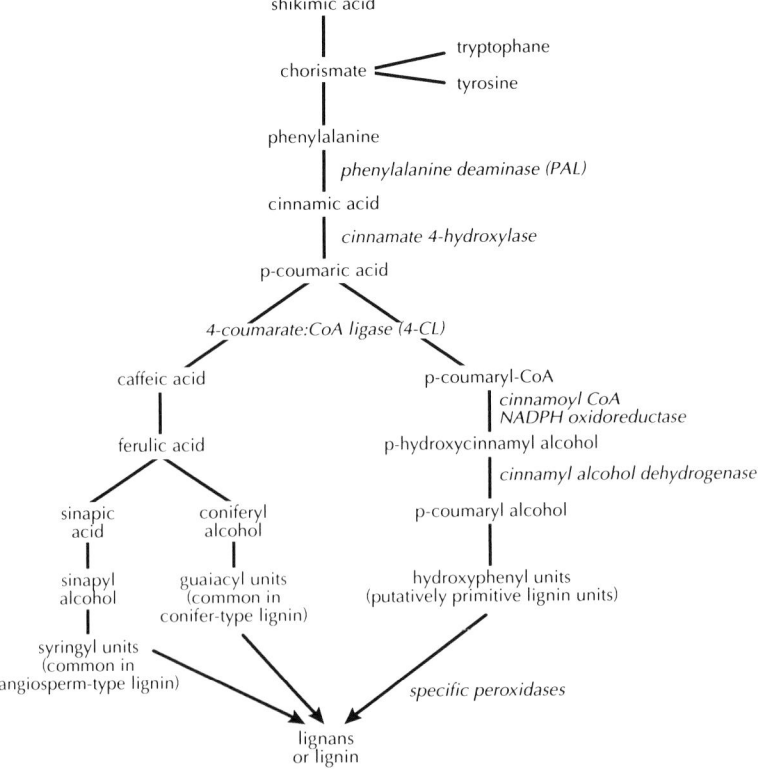

Fig. 9-7. The pathway leading to lignin monomers and precursors to flavonoids begins with phenylalanine ammonia lyase (PAL), the enzyme that deaminates phenylalanine to cinnamic acid. Higher plant lignins contain variable proportions of hydroxyphenyl, guiacyl, and syringyl units, but early lignins may have been simpler in structure, perhaps composed primarily of hydroxyphenyl moieties.

(Hrazdina and Zobel, 1991). Whether these enzymes occur in charophycean algae is not known. Lewis and Yamamoto (1990) note that PAL has been found in the green flagellate *Dunaliella,* but not in *Chlorella* or *Ulva.* They suggest that the absence of PAL from *Chlorella* and *Ulva* indicates that PAL is not likely to be generally present in green algae. It must be observed, however, that these green algae are representatives of the chlorophycean and ulvophycean lineages, respectively, and that there is substantial evidence for significant divergence of charophycean cytoplasmic enzyme systems from those of other green algal lineages (see Chapter 3). Absence of PAL from *Chlorella* and *Ulva* does not necessarily indicate that charophycean algae lack this enzyme, but charophytes have apparently not yet been examined.

Cinnamate 4-hydroxylase is an important enzyme with respect to lignin formation because it generates p-coumaric acid. This acid and its derivatives are characterized by the p-hydroxyl group that is ultimately dehydrogenated by peroxidases, leading to formation of highly reactive oxygen radicals and crosslinkage of lignin monomers (Fig. 9-7). Although there is some evidence that hydroxylated phenols are produced by charophycean algae (Wagner and Grolig, 1992), details of structure and biogenesis are unknown. The final enzyme in the land plant phenylpropanoid pathway is 4-coumarate:CoA ligase (4-CL). This enzyme forms CoA-thioesters from cinnamic acid and its derivatives, which include p-coumaric acid, caffeic acid, ferulic acid, and sinapic acid. However, of the naturally occurring hydroxycinnamic acids, only p-coumaric, ferulic, and sinapic acids are considered to be genuine lignin precursors (Lewis and Yamamoto, 1990). The formation of p-coumaryl-CoA is especially important because it is a precursor for flavonoid biosynthesis, which is discussed in more detail below. Two additional enzymes, cinnamoyl CoA NADPH oxidoreductase and cinnamyl alcohol dehydrogenase, catalyze the final steps in production of substituted cinnamyl alcohol derivatives (monolignols), which are the building blocks of lignin. Reduction of the CoA thioesters to alcohols is regarded as specific to the lignin biosynthetic pathway (Grisebach, 1981). Interestingly, soybean cinnamyl alcohol dehydrogenase is remarkably similar in molecular weight, cofactor, kinetics, stereospecificity, and inhibition characteristics to alcohol dehydrogenases from yeast and animals. These similarities are consistent with derivation from a common ancestral gene (Grisebach, 1981). This increases the chance that charophycean algae are likely to possess an homologous enzyme.

The activities of cinnamoyl CoA NADPH oxidoreductase and cinnamyl alcohol dehydrogenase generate p-coumaryl alcohol from p-coumaric acid, coniferyl alcohol from ferulic acid, and sinapyl alcohol from synapic acid. As previously mentioned, lignins from different plants and different parts of the same plant may vary in the proportions of the three monolignol constituents. Coniferyl alcohol forms the basis of guaiacyl lignins, which are abundant in gymnosperm woods, whereas sinapyl alcohols are the basis for lignins with a high proportion of syringyl groups, which are more common in angiosperm woods. Coumaryl alcohol-derived lignin subunits (hydroxyphenyl units) may be regarded as the least derived monolignols. Theoretically speaking, the earliest lignins might have been primarily polymers of p-coumaryl alcohols, but as previously mentioned, direct testing of this hypothesis by analysis of fossil remains has not proven possible. Hydroxyphenyl lignins are generally present in small and variable amounts in seed plants and pteridophytes, and there is typically a high proportion of hydroxyphenyl moieties in the compression wood that forms in stressed regions of gymnosperms (Timell, 1980). There is also some evidence for occurrence of p-hydroxyphenyl units in bryophytes, but in the latter group questions remain regarding possible contamination (Lewis and Yamamoto, 1990).

Niklas (1982) suggested that elaboration of vascular tissues and formation of true lignin might actually have been disadvantageous for bryophytes, where the dominant generation is the gametophyte, because if gametophytes became too tall, gamete transfer would have become more difficult. Gametophytes of some bryophytes such as the relatively tall mosses *Dawsonia* and *Dendroligotrichum,* and other mosses, exhibit well-defined conduction systems that are more or less comparable to xylem and phloem (Hebant, 1977). It is clear, however, that elaboration of lignified vascular tissues would be more useful to plants having independent sporophytes (Niklas, 1982). Plants below the grade exemplified by modern pteridophytes would not be expected to produce large amounts of phenylpropanoids for structural purposes. However, it has been suggested that phenylpropanoid phenolics could serve as effective UV filters (Lowry et al., 1980), and that they probably functioned as UV filters prior to the origin of flavonoids (Stafford, 1991a). In this regard it is interesting that only about half of the mosses and liverworts tested produce flavonoids, and that flavonoids appear to be absent from hornworts (Markham, 1990). The epidermal cells of hornwort sporophytes, however, exhibit characteristic staining reactions (Fig. 9-8) and autofluorescence suggestive of polyphenolics. As previously mentioned earlier in this chapter, hornworts produce the phenolic polymers known as lignans (Takeda et al., 1990), but whether these compounds function as radiation screens is not known. Although flavonoids were earlier reported to occur in *Chara* and *Nitella* (Markham and Porter, 1978), more recently this finding has been questioned (Markham, 1988). The possibility that low levels of phenylpropanoid phenolics might function as primitive UV filters in charophytes, hornworts, and other bryophytes should be explored.

Flavonoid Properties and Biosynthesis

The UV absorption coefficients of flavonoids are higher than those of phenylpropanoid phenolics on the basis of molarity or weight. Thus, although flavonoids may have originally functioned in low concentrations as intracellular regulators or chemical messengers, production of larger amounts targeted to vacuoles could have resulted in a new role as UV filters in some bryophytes and the progenitors of vascular plants (Stafford, 1991a). Flavonoids function in higher plants in production of fruit and flower colors, in defense, as UV protectants, in regulation of auxin transport, in induction of nodulation, and during

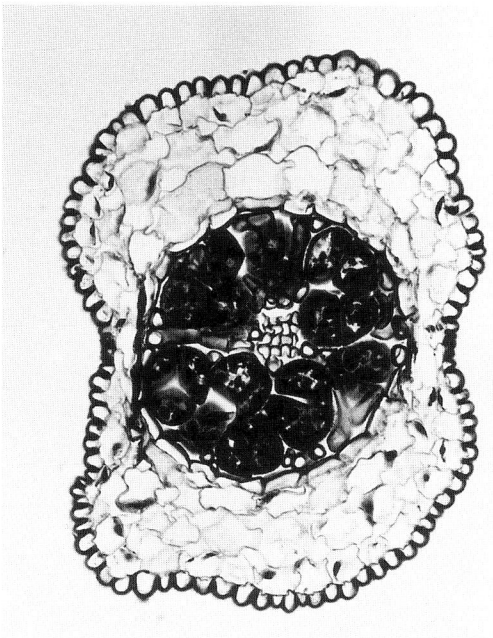

Fig. 9-8. Cross-sections of hornwort sporophytes reveal heavy staining reactions of sporopollenin-impregnated spore walls (center) and outer walls of epidermal cells. The chemical composition of epidermal walls has not been elucidated, so possible relationship to cutin or cutan polyesters or other compounds is not known.

male gametophyte development (van der Meer et al., 1992). In plants, flavonoid synthesis commences with the condensation of p-coumaryl CoA with three malonyl CoA units from the acetate pathway to form naringenin chalcone, the precursor of all flavonoids (Fig. 9-9). These reactions are catalyzed by the enzymes chalcone synthase and chalcone isomerase. The result is a three-ring compound (C6-C3-C6). It has been suggested that because the stepwise addition of C-2 units to form the A-ring resembles the activity of the beta-ketoacyl-acyl carrier protein of fatty acid synthases, this enzyme might be the precursor of chalcone synthase (Stafford, 1991a). Moreover, it has been postulated that formation of the central C-ring to form a flavanone by isomerization might originally have been nonenzymatic. Addition of chalcone isomerase would have ensured production of only one isomer, necessary for subsequent enzymatic steps that generate flavanone derivatives (e.g., flavones, biflavones, proanthocyanidins, anthocyanidins, flavanones) (Stafford, 1991a). Thus, given a source of p-coumaryl CoA, the addition of a single enzyme, chalcone synthase, would have been sufficient for production of primitive flavonoids.

It is apparent that higher plant lignin and flavonoid biosynthesis is phylogenetically rooted in ancestral pathways. For example, Helen Stafford (1991a) has

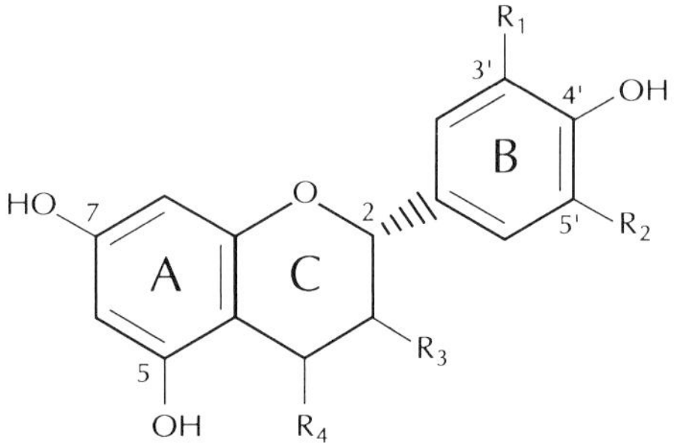

Fig. 9-9. Flavonoids are based in part on subunits derived from the phenylpropanoid pathway—these form the B ring. The A ring is formed by stepwise addition of C-2 units from the malonyl-CoA pathway, catalyzed by chalcone synthase. The C ring is generated by the activity of an isomerase. Diagram from Stafford (1991), Flavonoid evolution: an enzymic approach, *Plant Physiology* 96:681, with permission.

provided evidence for progressive phylogenetic elaboration of flavonoid pathways, from bryophytes through pteridophytes to seed plants. She argues that enzymes of the ancestral pathways were likely to have been derived from those of primary metabolism, and that because the first enzymes capable of generating flavonoids were not as abundant or efficient as those of modern plants, cellular concentrations would be low. Therefore, if charophytes are able to produce phenypropanoid phenolics and/or flavonoids at all, these substances are likely to be present only in low quantities and production may be developmentally or environmentally regulated. Charophytes may possess some, but not all, of the enzymes necessary to generate phenolic compounds similar to those of embryophytes, or ancestral charophycean enzymes might use alternative substrates. If so, procedures currently used for chemical analysis of lignin and flavonoids from higher plant extracts may not provide the resolution necessary for detection of ancestral pathways.

The genes encoding several key enzymes of the phenylpropanoid and flavonoid pathways have now been cloned from flowering plants (Linthorst, 1991). It may be possible to utilize heterologous sequences to probe charophyte and bryophyte genomes for these genes. It will be important to use highly conserved sequences, because plant phenylpropanoid pathway genes can be highly divergent. There are, for example, at least three PAL genes that encode proteins exhibiting different affinities for the substrate phenylalanine, as well as very different patterns of regulation during development (Lamb et al., 1989). This approach, recommended earlier for other purposes, has the advantage that gene expression is not required. It is not necessary to generate material in particular developmental stages, as in the case of resistant materials produced in *Coleochaete* only by fertile thalli. In the case of flavonoids, irradiation of plants with full spectrum sunlight (including UV) is often necessary to ensure that flavonoid expression occurs. Alternately, conserved gene sequences related to key enzymes can be used to design oligonucleotide primers for use in polymerase chain reaction amplification of putatively homologous sequences from very small amounts of charophyte or bryophyte DNA. Antisense RNA inhibition of expression of genes in the flavonoid pathway could be used to explore flavonoid function in pteridophytes, bryophytes, and, possibly, charophytes, as has been done in flowering plants (van der Meer et al., 1992). It is apparent that the tools of molecular biology will allow a more efficient approach to questions regarding the evolution of biosynthetic pathways leading to lignin and flavonoids than has been possible in the past. The prospects for eventual elucidation of the early evolutionary history of lignin and flavonoid biosynthesis appear excellent.

Cutin and Cutan

Charophytes and bryophytes may also illuminate the evolutionary history of the plant cuticle, generally regarded as a critical adaptation acquired by early land plants. The cuticle of vascular plants is composed of a structural biopolyester matrix and waxes, which function in waterproofing (Kolattukudy, 1980). In addition to reducing water loss, plant cuticles are valuable in resisting microbial invasion, and they may also reflect UV radiation (Raven, 1984). Chemical and microscopic evidence suggests that cutin is present on the epidermal surfaces of bryophytes and also occurs on submerged seagrasses and freshwater macrophytes (see references in Holloway, 1982), but is generally regarded as being absent from algae, on the assumption that a cuticular layer would be detrimental in the aquatic habitat. Certainly it is true that surficial wax layers would retard absorption of water, CO_2, and minerals into algal cells, but the presence of cutin in bryophytes suggests that the polyester by itself may not present a major barrier to water and carbon dioxide movement into cells. This raises the possibility that cutin, or an evolutionary precursor, might not be restricted to embryophytes. Surficial polyester layers might be useful in retarding microbial attack, or they might play a role in UV absorption in the aquatic environment. Because plant polyesters are weak cation exchangers, they might also function in nutrient uptake (Kolattukudy, 1980). Therefore, it is not completely surprising that thalloid species of the aquatic, charophycean alga *Coleochaete* have been observed to produce a specialized surface layer that is analogous (at least in position) to cutin layers of bryophytes (Figs. 9-10, 9-11, 9-12, and 9-13). The *Coleochaete* surface material can be separated from the thallus surface by mild acid treatment, lacks detectable waxes, as indicated by the absence of Sudan IV staining, and can be partially degraded by treatment with a putative bacterial cutinase (Graham and Delwiche, 1992). Chemical characterization is otherwise unknown. Because future work may make it possible to trace the evolutionary history of biosynthetic pathways leading to cutin, it is useful to consider cutin structure and synthesis in greater detail.

The cutin polymer is composed of interesterified monomers consisting of C16 and C18 fatty acids, most of which are hydroxy fatty acids and some of which are epoxy fatty acids. Phenolic moieties may also be present (Kolattukudy, 1980;

Fig. 9-10. Concentric rings of surface material occur on the dorsal surfaces of *Coleochaete scutata* thalli. SEM photo by Sonia Cook (University of California-Davis).

Holloway, 1982) (Fig. 9-14). The hydroxylating enzymes are located in the endoplasmic reticulum. Hydroxylated monomers are subsequently transported to the extracellular location, probably as activated thioesters (cf. lignin monomers), and incorporated into the growing polymer. Transport through the extracellular matrix to the site of cutin synthesis is probably mediated by a lipid transfer protein, encoded by the EP2 gene (Sterk et al., 1991). However, not much else is known about the monomer transport process. Epoxidation of fatty acids is believed to occur within the cuticular matrix. Polymerization is accomplished by the action of

Fig. 9-11. Surface material is continuous across adjacent cell lineages, and also covers zygotes of *Coleochaete scutata.* Photo by Sonia Cook (University of California-Davis).

Fig. 9-12. Absence of surface material from older, central areas of *Coleochaete* thalli and presence at younger, peripheral regions indicates that the "cuticular" layer does not represent accumulation of environmental depositions, because such deposits would likely be heaviest on oldest portions of thalli. Photo courtesy of Sonia Cook (University of California-Davis).

hydroxyacyl-CoA:cutin transacylase, which transfers the hydroxyacyl moiety to free hydroxyl groups present in the growing polymer (Kolattukudy, 1980). Cutin is a highly amorphous polymer (Kolattukudy, 1980; Holloway, 1982) whose composition could vary greatly depending on the types and proportions of monomers incorporated. Strategies for unraveling the evolutionary history of cutin biosynthesis might well focus on enzymes critical to the pathway. Niklas (1982) has

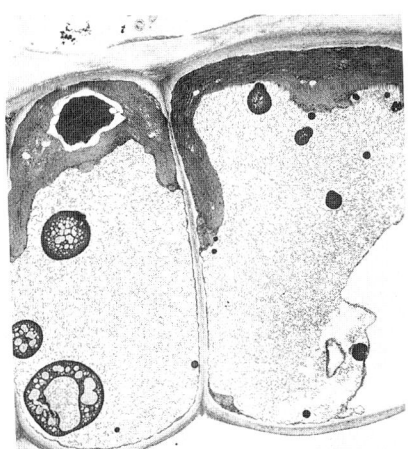

Fig. 9-13. Transmission electron micrograph of *Coleochaete orbicularis* showing dorsal surface layer with thickness variation consistent with ridged appearance in SEM.

CUTIN

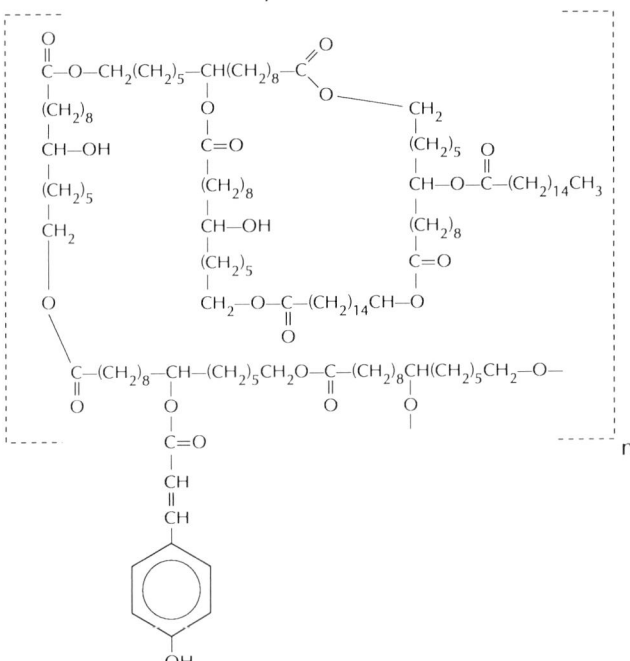

Fig. 9-14. Plant cutin is a biopolymer whose structure is poorly understood. Cutin monomers are various C_{16} and C_{18} hydroxyalkanoic acids. However, structural and phylogenetic relationships between cutin and a more resistant plant cuticular polyester, cutan, are unclear. Reproduced from Kolattukudy (1980), Biopolyester membranes of plants: cutin and suberin, *Science* 208:994. Copyright 1980 by the American Association for the Advancement of Science.

suggested that cutin evolution might have involved only a single innovative step, hydroxylation of fatty acids, and that there is some evidence that polymerization might be at least partially nonenzymatic. When more is known about the enzymes that catalyze fatty acid hydroxylation, this information might be used to devise a strategy for tracing evolution of cutin biosynthesis pathways.

Cutin is not the only matrix polymer found in plant cuticles. The cuticular matrix of higher plants may also contain the insoluble, nonhydrolyzable, polymethylenic biopolymer cutan. The molecular structure and biosynthesis of cutan is not well understood, but cutan is much more stable to biochemical degradation than is cutin (Tegelaar et al., 1991). In fact, cutan formed the major matrix component of all fossil gymnosperm and angiosperm cuticles examined in one study (Tegelaar et al., 1991). This suggests that plant cuticles dominated by cutin, rather than cutan, have a low fossilization potential, and that the fossil record is biased toward taxa that originally had significant amounts of cutan in the cuticular matrix (Tegelaar et al., 1991). The acetolysis-resistant sheets (putative cuticles) found in late Ordovician and early Silurian deposits are thus perhaps more likely to be made of cutan than cutin. This suggests that cutan may be of ancient origin, and that studies of its possible occurrence in seedless plants are in order.

Sporopollenin Composition and Synthesis

As briefly discussed in Chapter 8, the composition of sporopollenin is controversial. Sporopollenin is regarded as the most resistant biopolymer on earth. Known solvents include fused KOH; oxidizing mixtures such as hypochlorite/hydrochloric acid, potassium dichromate/sulfuric acid, and hydrogen peroxide/sulfuric acid; and organic bases such as 2-aminoethanol, 3-aminopropanol, 2,2,2-nitriloethanol, and 4-methylmorpholene-N-oxide. Sporopollenin is insoluble in nonoxidizing acids and dilute bases and is resistant to hydrolytic enzymes. Acetolysis is a common procedure for purifying sporopollenin, but it acetylates sporopollenin and may modify it by inducing condensation reactions. In some situations, it is possible that acetolysis may create sporopollenin where it was not originally present (Southworth, 1990).

Sporopollenin is a cross-linked polymer with saturated and unsaturated hydrocarbons and phenolics. One hypothesis suggests that sporopollenin is an oxidative polymer of carotenoids and carotenoid esters. This proposal is supported by the fact that artificial polymers similar in IR spectra to naturally occurring sporopollenin can be produced by oxidizing carotenoids and carotenoid esters. Breakdown products of the synthetic polymers include saturated fatty acids, dicarboxylic acids, and phenolic acids. These constituents could explain the UV absorbance, autofluorescence, and toluidine blue staining characteristics of sporopollenin (Southworth, 1990).

Alternatively, sporopollenin might be a product of phenylpropanoid metabolism. Some evidence for this is the occurrence of phenylalanine ammonia lyase (PAL) in anthers. In addition, changes in toluidine blue staining characteristics indicate that the phenolic content of sporopollenin changes during pollen wall development (Southworth, 1990). If phenylpropanoid metabolism turns out to be a fundamental aspect of sporopollenin biosynthesis, the well-established presence of sporopollenin in charophytes (and other algae) (Graham, 1990) suggests that at least some aspects of phenylpropanoid metabolism are quite ancient. Phenolic residues might, however, represent derived characteristics of higher plant sporopollenins or perhaps result from the application of harsh chemical analytic procedures. It is possible that sporopollenin polymerization processes are related to those that generate cutin and lignin from esterified monomers. Perhaps progress in deciphering the evolutionary history of phenylpropanoid metabolism, and in the phylogeny of biosynthesis of other plant polymers, will also shed light on aspects of sporopollenin regulation, biosynthesis, and deposition in plants.

10

LAND PLANT ORIGINS—
A SUMMARY

We cannot physically travel back in time to observe firsthand the appearance of the earliest land plants. In view of this constraint, some would no doubt argue that hypotheses regarding early plant evolution are not directly testable and therefore must be merely conjectural. In contrast, it is quite clear to many modern evolutionary biologists that the past history of modern organisms is encoded in their genes, biochemistry, cellular structure, and body construction, and that this history can be extracted if appropriate attention is paid to analysis of homology. It is possible to develop testable phylogenetic hypotheses, and various examples have been cited in this book.

A robust picture of early events in the history of plant life on earth can now be painted. No doubt future developments will render portions of this view obsolete, inaccurate, incomplete, or downright wrong. However, it is possible to identify certain aspects of modern knowledge regarding early plant evolution that are supported by sufficient evidence to allow wide acceptance for the time being. This chapter begins with a survey of these well-substantiated concepts in the hope that they will be used by the writers of introductory biology and botany texts. Many of the descriptions of plants origins found in recent textbooks are sadly outdated and poorly representative of exciting modern ideas. It is no wonder that many biology students lack appreciation for plants and the impact they have had on the evolution of planet Earth's biosphere.

Although much has been learned about the origin of plants, many important questions remain. The second portion of this chapter is concerned with identification of significant topics that require further research, together with suggestions for approaches for clarifying the issues. The third and final section of this chapter focuses on the use of model systems among algal protists and seedless plants in the extraction of fundamental information about the higher plants on which we all depend. Although *Arabidopsis* is a valuable experimental system in many ways, there is no doubt that this model plant system is highly derived and thus incapable of answering all our questions regarding structure and function of plants. By virtue of phylogenetic divergence at particular points in time, various plant and protist groups are positioned so as to provide insight into the solution of important and intractable problems regarding plant development and reproduction. No single model system will be sufficient, and the proponents of individual plant or protist systems should be open-minded about the value of other systems. Moreover, the results of studies obtained with the use of plant and protist model systems *must* be considered within an evolutionary context in order to avoid misinterpretation and unwarranted extrapolation.

Seedless plants and protists harbor secrets regarding coevolution of cell organelles that are likely to be as exciting as coevolutionary relationships occurring on a more conspicuous level between terrestrial plants and animals. Protists and seedless plants can also tell us much about how cells perceive their environments and communicate among themselves, and how these signals are translated into developmental changes. When these relationships have been uncovered at the molecular level, we will have the power to solve some of the most fundamental and perplexing mysteries of plant biology.

WHAT WE KNOW ABOUT THE ORIGIN OF THE PLANT KINGDOM

Microfossil evidence indicates that land-adapted plants originated sometime between the mid-late Ordovician and the early Silurian periods in the early Paleozoic era, possibly about 450 to 470 million years ago. Colonization of land by plants thus occurred millions of years earlier than appearance of the benchmark vascular plant *Cooksonia,* at about 430 million years ago. Origin of land plants coincided with a period of glaciation and mass extinction rivaling that of the more famous K/T event. Atmospheric carbon dioxide levels were probably higher then than they are today, but oxygen and ozone levels may have been nearly equivalent to modern concentrations. An older concept, that the timing of plant colonization of land was tied to achievement of a particular level of oxygen (and ozone), is less strongly supported now than in the past.

There is conclusive evidence that land plants are derived from progenitors that would be classified with modern charophycean algae. The algal ancestors of land plants were distinct in many ways from other protists, including noncharophycean green algae, and it is uncertain whether the group of organisms encompassing green algae and land plants is monophyletic or polyphyletic. The closest marine relatives of the plant lineage were most likely scaly, unicellular flagellates. Land plants almost certainly did not arise from multicellular seaweeds that washed up onto beaches and sprouted there—a concept often presented in introductory textbooks.

The balance of evidence indicates that land plants arose most directly from freshwater algae that were constructed of branched filaments or perhaps an early form of parenchyma (coherent cell lineages that form tissue). It is quite possible that morphologically simple charophycean algae passed through one or more phases of life on land, but as yet there is no concrete evidence that terrestrial selection pressures influenced the development of advanced morphologies and reproductive features of charophytes. However, there is empirical evidence that features of freshwater environments correlate well with the occurrence of derived charophyte morphologies and reproduction.

The algal ancestors of plants and the earliest plants were most likely monoplastidic, with Rubisco aggregated into a well-defined pyrenoid. A phytochrome and actin system had developed in unbranched charophytes as a mechanism for orienting the plastid with respect to the incidence of solar radiation. This cytoskeletal system may later have functioned to reorient plastids prior to divisions occurring at right angles to the main axis (branching). It has been speculated that the need to move a relatively massive plastid to a new position perpendicular to the previous orientation may have resulted in elaboration of earlier, simple phragmoplasts, and in the origin of centrifugally developing cell plates (in place of previous centripetal furrowing). The evolution of cell plates may be related to the origin of plasmodesmata and symplastic communication in charophytes. The origin of branching in charophycean algae is probably related to development of the cytoskeletal apparatus necessary for subsequent evolution of plant tissues and more complex morphological development. Ecological studies show that morphological variation in modern charophytes (from relatively delicate, branched filaments to more robust, tissuelike thalloid constructions) is correlated with habitat differences (deeper versus shallow, turbulent waters). Low surface area/volume ratio conferred by thalloid structure would also provide selective advantage on land, in terms of reduced water loss.

Ecological evidence and laboratory culture experiments also suggest that most charophytes may be less able to utilize the bicarbonate ion as a carbon source in photosynthesis than many other algae. This suggests that reliance on dissolved carbon dioxide may be the plesiomorphic condition for charophytes. Growth of CO_2 users should be favored in turbulent, shallow waters where diffusion boundary layers are reduced. Occupants of the shallow littoral region would be exposed to high irradiance and full-spectrum sunlight. Availability of red light would promote elaboration of phytochrome-mediated cell processes, and exposure to UV may have promoted adaptive responses such as phenolic, phenylpropanoid, or flavonoid sunscreens, or cuticular polyesters. Development of surface UV protectants may also have provided protection against microbial attack and eventually led to evolution of a desiccation-resistant cuticle useful in the terrestrial habitat.

The charophycean ancestor of plants was likely oogamous, with eggs retained upon the parental thallus during fertilization by biflagellate sperm. Zygotes were probably retained too, as a mechanism for ensuring that meiospores resulting from zygote germination would be produced in an environment abundant in light and carbon dioxide, suitable for germination and growth. Production of greater numbers of meiotic products would have been possible if retained zygotes were provided with photosynthate exported from parental thallus cells. Some evidence suggests that charophycean algae may be primitively capable of importing organic carbon (specifically glucose) for vegetative growth. Although further study is needed, a capacity for cellular uptake of sugars may have served as a preadaptation for the nutritive support of zygotes at placental regions (and later, phloem loading) in land plants. The capacity of early plants to produce sporopollenin-encased spores is almost certainly derived from charophycean production of sporopollenin in zygote walls. In monoplastidic charophytes, the zygote plastid anticipates meiosis by undergoing preprophase division into four daughter plastids and subsequent tetrahedral orientation. This behavior, characteristic of sporocytes of various seedless plant groups, is probably derived from the algal condition. Involvement of callose in land plant gametogenesis, embryogenesis, and sporogenesis is also likely homologous to use of callose in charophycean reproductive systems.

The relationship of charophytes, which are characterized exclusively by zygotic meiosis, to embryophytes, featuring sporic meiosis, strongly argues for origin of the sporophyte generation and alternation of generations in plants by delay in meiosis. The alternative hypothesis, that embryophytes arose from green algae having alternation of generations, and that free-living sporophytes assumed a parasitic association with gametophytes, is not well supported by modern data. Presumably the first sporophytes were small, the products of only a few intercalated mitotic divisions prior to meiosis, but gradually increased in size as selection for greater amplification of the sexual process occurred. The example of modern *Coleochaete* shows that selection for more than four meiotic products occurs in shallow water habitats. Various authors have argued that archegonia must have preceded the evolutionary origin of embryos, because modern seedless plants are all characterized by archegonia. Yet charophycean algae provide substantial evidence that the origin of the embryophytic habit (nutritional and developmental interactions between haploid and diploid generations) probably evolved *prior* to the origin of recognizable archegonia. Gametangial elaboration, however, is more likely to be correlated with terrestrial selection pressures. Thus, there is basis for speculation that the first embryophytes were inhabitants of shallow water, rather than terrestrial habitats. But the first archegoniates are more likely to have been land dwellers. Little evidence is available on which to base speculation as to whether multicellular sporophytes (delay in meiosis) evolved before, after, or concurrently with recognizable archegonia.

Molecular evidence indicates that the charophycean ancestors of plants moved the *tuf*A gene from a plastid to a nuclear genomic location, with the addition of targeting sequences for importation of cytoplasmically translated protein back to the plastid. The selective advantage of this change in genomic location is unknown, but it emphasizes the importance of genetic interactions between cellular organelles that have occurred during the evolutionary history of plants. There is some evidence that ancestors of the charophycean/embryophyte clade may have been distinct from other protists since the origin of eukaryotic lineages. In other words, the cells of charophytes may have been formed by different endosymbiotic organelle/host combinations than even those giving rise to other green algae. Hypothetically, incorporation of a chlorophyll a and b-containing endosymbiont, or modification of a chlorophyll a and phycobilin-containing endosymbiont could have occurred multiple times in response to the spectral characteristics of shallow water habitats.

The features ascribed to the algal ancestors of land plants that have been described above suggest that shallow freshwater environments played a strong

About 500 million years ago green algae decide to invade the land.

selective role in shaping the characteristics of embryophytes, preadapting their algal ancestors for success on land. Critical adaptations which appear to have originated in response to aquatic selection pressures that later proved useful in terrestrial habitats include glycolate oxidase (scavenges carbon that would otherwise be lost from the cell), sporopollenin-encased (hence, desiccation and microbe resistant) reproductive cells, and transport of photosynthate from haploid to diploid generation (the nutritional basis for genetic amplification of sexual reproduction). The alternative concept, that embryophytes evolved from unicellular terrestrial algae in response to terrestrial selection pressures alone, is not so strongly supported by empirical evidence.

The immediate algal ancestors of plants were mostly likely rather small, attached forms highly adapted to life in shallow water, similar to modern filamentous species of *Coleochaete,* which also bear resemblances to some of the less morphologically derived charalean taxa (i.e., *Nitella*), rather than more derived charalean genera such as *Chara.* At least some of the modern Charales are also known to utilize bicarbonate in photosynthesis and would be less likely to profit by migration to the CO_2-rich, but significantly drier, terrestrial environment than would nonbicarbonate users. Thallus structure, physiology, and reproductive biology indicate that most modern charalean taxa are well adapted for success in benthic environments. Consequently, to date the characteristics of Charales have proven useful in documenting examples of parallel evolution of traits (e.g., polyplastidy, loss of pyrenoids, barrel-shaped spindles) that were probably not features of the earliest embryophytes, but that are characteristic of most modern land plants.

Molecular and cytological evidence demonstrates that charalean algae are definitely closely related to the ancestors of land plants, and some recent gene trees indicate that Charales represent the charophyte group most recently diverged from the line leading to embryophytes. These results might be explained by postulating rapid radiation of lines leading to modern *Coleochaete,* Charales, and embryophytes from an archetypal form that was an oogamous, branched filament similar to certain putatively primitive *Coleochaete* species, followed by rapid morphological and reproductive specialization in all three lineages. Plesiomorphic and apomorphic characters are thus variously distributed among charophytes in a mosaic pattern of evolution. For example, loss of the chloroplast *tuf*A gene occurred in both Zygnematales and land plants, whereas *Coleochaete* and Charales still maintain a chloroplast homologue (Baldauf et al., 1990). However, other evidence (for example, the distribution of cp tRNA introns [Manhart and Palmer, 1990]) suggests that *tuf*A loss occurred independently in the two groups and by itself does not necessarily indicate that Zygnematales are the closest charophycean relatives of plants. As another example, *Coleochaete scutata* exhibits derived cellulose synthesizing complexes as compared with those of *Nitella* and land plants (Okuda and Brown, 1992), but also has the plesiomorphic characteristic, single plastids with pyrenoid. It is clear that single characters, even if these are molecular variations, cannot be used to define relationships between organisms of such ancient divergence. Although particular gene sequences may

By James Graham

Members of the Charophycean algae gather for a group photo before sending off their first "terranaut" to explore the land.

exhibit multiple character state changes, all of these may not be completely independent, i.e., modification in one region of a sequence might influence selection effects on other regions of the same sequence, depending on the cumulative effect on the shape/function of the encoded molecule.

In addition, particular gene sequences may have been influenced by differential selective effects operating on related taxa that have come to occupy different habitats, after their divergence from a common ancestor. For example, charophycean algae occupy habitats that differ in the extent to which carbon is limiting to photosynthesis, and it is possible that this has influenced sequences of genes such as *rbc*L. Charophytes that do not use bicarbonate, but which have remained for hundreds of millions of years in aquatic environments where the level of dissolved CO_2 limits photosynthesis, might be expected to exhibit adaptive, divergent Rubisco structure. In contrast, the fossil record indicates that aquatic, calcified *Chara* species escaped carbon limitation hundreds of millions of years ago by acquiring the ability to use bicarbonate (which rarely limits modern aquatic photosynthesis). These charalean algae, terrestrial charophytes, and plants, whose ancestors long ago migrated to the land (where CO_2 diffusivity is 10^4 greater than in water) might display less *rbc*L sequence divergence because the evolutionary history of these organisms has been less strongly influenced by carbon limitation. Comparative gene sequencing may indicate whether such variation occurs. The challenge will be to distinguish changes resulting from evolutionary divergence from those reflecting differential environmental influences. Further, even if ribosomal RNA, *rbc*L, or other specific gene sequences indicate that a particular taxon of modern charophytes is most closely related to land plants, it would be a mistake to conclude that the modern taxon resembled algal ancestors of plants in all aspects of cell biology, biochemistry, morphology, reproduction, physiology, and ecology. Modern charophytes have been diverging in various ways from the ancestral type for perhaps 450 to 470 million years and during this time have retained various plesiomorphic features while undergoing change in others.

Morphological and reproductive characteristics of modern Coleochaetales correlate well with environmental conditions in shallow fresh waters, and as argued above, the charophycean ancestors of embryophytes were probably littoral forms. There is also strong evidence that the direct algal ancestors of land plants were monoplastidic (Brown and Lemmon, 1990c). Therefore, even if modern coleochaetalean algae turn out not to reflect the organization of particular embryophyte genes as well as modern charalean forms, they may better model the ecophysiology, morphology, and reproduction of the transmigrants. By way of analogy, *Ceratophyllum* (which, as noted in Chapter 4, is rather similar in structure to Charales), although the earliest diverging modern angiosperm taxon known, is not regarded to reflect accurately the morphology of early angiosperms in general. Rather, *Ceratophyllum*'s morphology may be correlated with its freshwater benthic habitat, as is *Chara*'s.

Regardless of the nature of the immediate charophycean ancestors it is probable that the first terrestrial plants quickly acquired the following autapomorphies of modern seedless plants: loss of zoospores; absence of centrioles except during spermatogenesis; bicentriolar centrosomes; changes in sperm microarchitecture, including modification of MLS lamellar angle; simple preprophase microtubule bands; sporic meiosis; sporopollenin-coated spores; localized meristematic cells with four cutting faces; complete loss of the chloroplast *tuf*A gene and stabilization of chloroplast genome size and architecture; elaboration of the cuticular matrix; and development toward a nascent phenylpropanoid/flavonoid pathway. As previously discussed (Chapter 4), this character assemblage forms the basis for tentative conclusion that modern embryophytes are monophyletic, that is, derived from a single common ancestor. It is possible that this view may be altered as more information becomes available regarding the evolution of these characters. However, the hypothesis of monophyly of embryophytes strongly suggests that the various bryophyte groups are not the results of separate, dead-end colonizations of land, but rather hold the secrets of vascular plant origins.

Gray and Shear (1992) have written convincingly that the paleobotanical evidence suggests that the earliest land plants looked and reproduced very much like modern thalloid liverworts. Their conclusion is supported by molecular systematics (Waters et al., 1992), immunofluorescence studies (Brown and Lemmon, 1990c), and phylogenetic (cladistic) analysis of morphological, developmental, and biochemical characteristics of bryophytes (Mishler and Churchill, 1984; 1985). The fact that modern liverworts (and other bryophytes) exhibit dependent sporophytes adds to the evidence provided by charophycean algae that alternation of generations in plants arose by elaboration of the diploid generation (the antithetic hypothesis), as espoused by Bower (1908).

Bryophytes illustrate various character state changes required for transition from poikilohydry (variability in water content), which was most likely characteristic of the first land plants, to homeohydry (maintenance of stable tissue water content), the hallmark of vascular plants. Raven (1984) concluded that the probable sequence of acquisition of vascular plant characteristics was as follows: origin of heteromorphic alternation of generations with a multicellular, erect sporophyte; cuticularization of the sporophyte generation; evolution of xylem; development of intercellular air spaces with pores in the epidermis; and origin of stomatal activity of the pores, with evolution of the endodermis and phloem long-distance photosynthate transport occurring at the time of xylem evolution or later. He also suggested that lignin's first role was antimicrobial, a conclusion supported by the occurrence of ligninlike substances (i.e., lignans) in nonvascular plants, and that this preadaptation later formed the basis for compression resistance and structural support. Although vascular plant investment in nonphotosynthetic structural tissues and commitment to lignin biosynthesis is expensive, consequent achievement of homeohydry is postulated to lead to greater growth rates (Raven, 1977; 1984).

Further illustration of the relevance of bryophytes to the evolution of vascular plants is provided by comparative structural studies of water-conducting cells of the Lower Devonian axes known as *Sennicaulis hippocrepiformis* and *Gosslingia breconensis* (Kenrick et al., 1991). *Gosslingia*-type (G-type) water-conducting cells resemble protoxylem elements of some modern pteridophytes. In contrast, water-conducting cells of *Sennicaulis,* called S-type, combine helical thickenings characteristic of tracheids with a thin lignified wall penetrated by numerous micropores. Continuous microporate sheets of lignin are not known to form on the inner walls of conducting tissues of extant vascular plants. However, numerous plasmodesmata-derived pores occur in walls of conducting cells of certain liverworts and the moss *Takakia* (Hebant, 1977). Whether these are homologous to the microporate walls of S-type conducting cells is not known, but S-type cells may represent an evolutionary intermediate between more primitive conducting cells and tracheids. *Sennicaulis,* then, exhibits a level of conduction system organization that falls between those of modern bryophytes and vascular plants (Kenrick et al., 1991). Hebant's (1977) elegant review of the structure of bryophyte conducting tissues provides a substantial basis for further comparative study of conducting cells produced by modern bryophytes and those of early vascular plants.

Bryophytes may even be useful in elucidating the phylogenetic origins of leaves and roots. Hagemann has interpreted structural features of thalloid bryophytes as possible evolutionary intermediates in root evolution (1991a). He has also challenged Zimmermann's longstanding telome "theory," suggesting instead that primitive megaphyllous leaves and their marginal meristems are derived from thallose, ribbonlike plants with midribs, wings, and dichotomous branching (1991b). It may be possible to test Hagemann's ideas when more is known regarding the gene expression characteristics of various plant meristems. For example, there appears to be a specific gene expression pattern characteristic of the root meristem which differentiates it from shoot meristems (Yamamoto et al., 1991). Whether Hagemann's iconoclastic ideas will withstand further scrutiny is questionable, but his (1991b) call for wholesale revolution of the theoretical framework of structural botany is a worthy challenge.

MAJOR GAPS IN OUR UNDERSTANDING OF PLANT ORIGINS

What was the origin of plant sexual reproduction? As we have noted, land plants inherited many reproductive characteristics from ancestral charophytes. But we do not know whether syngamy and meiosis evolved more than once in the charophycean clade. A strong argument could be made that sexual reproduction in the charophyte-embryophyte clade is monophyletic, because there are a number of autapomorphic characteristics of reproduction in this group of taxa (association of sporopollenin, callose, and diffusible substances with sexual reproduction). However, even if monophyly of sex in the charophyte-embryophyte clade is accepted, it is not known whether syngamy and meiosis evolved separately in various protist groups (including charophytes) or whether syngamy and meiosis were inherited from a common protistan ancestor. To answer this important question more resources must be directed toward fundamental studies of protist diversity, reproduction, and genetic systems.

Why did charophycean algae or their early embryophyte descendants invade the land? This question is related to the timing of the transition to terrestrial environments, and several authors have speculated regarding environmental correlates. Land plant origins have been associated by some with late Ordovician glaciation and eustatic sea-level drop. However, Gray and Boucot (1978) suggest that sea-level changes occur too slowly to be correlated directly with the timing of plant colonization of land, and that availability of unoccupied habitats was the primary stimulus.

The timing of land plant origins has also been related to the accumulation of a sufficient ozone screen or the development of internal UV screens such as phenylpropanoids or derived flavonoids. Increase in oxygen to a threshold level (generating protective ozone levels or sufficient oxygen to allow phenylpropanoid metabolism) is thus suggested as the mechanism constraining permanent adaptation of plants to land (Lowry et al., 1980; Chapman, 1985). This hypothesis is based on the assumption of a relatively late rise in oxygen to critical levels (Chapman, 1985), and as previously discussed, other authors have argued that evidence for earlier oxygen level increase renders the oxygen level control hypothesis less attractive.

An alternative hypothesis invokes the timely association of a green alga with a fungal partner in a mutualistic partnership that allowed both organisms to survive desiccation and nutritional stresses of life on land (Pirozynski and Malloch, 1975). Evidence for this hypothesis includes the frequency of endophytic fungi in gametophytes of modern seedless plants. The presence of fungal associates in heterotrophic gametophytes of the pteridophyte *Psilotum* is often cited as an example. Recently, Taylor et al. (1992) reported the occurrence of a parasitic member of the Plasmodiophoromycetes in cells of *Paleonitella* from 400-million-year-old Lower Devonian Rhynie Chert deposits. The permineralized remains showed enlargement of the algal cells, indicating host responses to parasitism similar to those exhibited by modern *Chara*. Presumably, microbial parasites of charophycean algae might have evolved into mutualistic partners, but as Taylor (1982) points out, it is often difficult to determine whether fungal remains in fossilized plant tissues represent symbiotic associations or evidence of fungal penetration of tissue after plant death. Thus, the evidence that acquisition of fungal partners represented a turning point in land plant evolution remains suggestive at best. It is, however, relevant that modification in a single gene locus was sufficient to transform the filamentous fungal parasite *Colletotrichum magna* into a nonpathogenic, endophytic mutualist (Freeman and Rodriguez, 1993). This evidence suggests that symbiotic associations between early land plants and fungi might have occurred rather easily. The benefits to plants in terms of increased efficiency in nutrient and water acquisition are well known. In addition, there is evidence that formation of mutualistic association confers upon the plant partner resistance to infection by pathogenic fungal strains that are related to the fungal partner (Freeman and Rodriguez, 1993).

Atsatt (1991), recognizing that the appearance of the plant embryo was a critical event in the history of plants, suggested that the origin of transfer cells at the placental junction was related to responses to fungal parasites, because transfer cell morphology is also manifested in land plants where they interface with mycorrhizal fungi. Atsatt (1991) makes the very reasonable assumption that selection operated on gametophytes of early land plants to facilitate nutrient acquisition via mycorrhizal associations. He then suggests that the associated evolution of transfer cells preadapted gametophytes to incorporate a second symbiont, the diploid sporophyte. This latter hypothesis, however, assumes that embryophytes necessarily evolved directly from ancestors having separate, free-living generations, as described by the homologous hypothesis (see Chapter 3). The placental region of embryophytes is thus interpreted as a parasitic structure, as is also suggested by the term "haustorium," often used as a synonym for the land plant placenta. As has earlier been argued, the balance of modern evidence much more strongly supports the alternative antithetic hypothesis, that is, origin of embryophytes directly from charophycean algae having zygotic meiosis, by meiotic delay. Other hypotheses suggested by Atsatt (1991) include incorporation of fungal genes that conferred parasitic capabilities upon embryophytes and the possibility that embryophytes might share enzymatic features with fungi by virtue of common descent from ancestral protists. The latter concept is most consistent with the evidence presented in this book. In Chapters 5 and 8 it was argued that charophycean algae inherited from ancestral protists the ability to take up exogenous organic compounds, specifically sugars, and that this capability preadapted charophytes for the origin of nutritional interactions between generations. This hypothesis is supported by only a small amount of experimental evidence, but in contrast, no empirical work related to charophycean algae, hornworts, or liverworts is yet available to support the conjecture that timing of the origin of the plant embryo was directly dependent on acquisition of fungal symbionts or genes. Proposed tests of the fungal symbiosis model for the origin of the plant embryo that utilize comparative gene sequences (suggested by Atsatt, 1991) will have to differentiate between plesiomorphic gene similarities among plants, algae, and fungi resulting from common protistan ancestry, and genes acquired at later points in time. As noted earlier (Chapter 4), endosymbiosis and subsequent coevolution of cellular genomes, as exemplified by the origin of transit peptide sequences, played a very large role indeed in shaping the biology of the charophyte-plant lineage. In view of the highly integrated nature of modern biotic systems, coevolutionary and symbiotic interactions between charophytes and early embryophytes, on one hand, and fungi and nitrogen-fixing procaryotes, on the other, are quite likely to have been integral to the success of plants on land. The genetic basis for molecular integration of the *Rhizobium*-plant interaction has been elucidated (Fisher and Long, 1992). This work provides a model for investigations of analogous mutualistic interactions between seedless plants and their fungal and cyanobacterial partners that could shed light on the evolutionary origins of these relationships.

Chapter 5 presents evidence in support of the hypothesis that transmigration of charophytes to land was based on increased carbon availability. Elements of the argument include (1) evidence that many charophytes lack physiological adaptations for growth at high pH levels, that is, they appear to use primarily dissolved carbon dioxide rather than bicarbonate as a carbon source, and (2) the relatively low

By James Graham

On nights with a full moon, the little gametophytes would huddle together—fearing the coming of the vampire sporophyte.

diffusibility of carbon dioxide in water as compared with air. Additional correlates are the tendency of some charophytes to occupy terrestrial habitats and the possibility that atmospheric (and hence, dissolved) carbon dioxide concentrations had decreased substantially prior to the late Ordovician period as a result of calcium carbonate formation by cyanobacteria, marine seaweeds, and other organisms (see Chapter 2). Certainly, the timing of embryophyte origins is likely to have resulted from the interplay of many factors, including geographic and climatic change, alterations in atmospheric and water chemistry, and biotic interrelationships. However, among the possible forcing factors, carbon dioxide availability appears to be a particularly strong candidate. Evaluation of this hypothesis will require much more detailed study of the carbon economies of charophycean algae.

By James Graham

"That's one small step for an alga,
one giant step for a plant kingdom."

SEEDLESS PLANTS AND ALGAE AS MODEL SYSTEMS FOR UNDERSTANDING HIGHER PLANT PROCESSES

As Goldberg (1988) has pointed out, the lack of techniques for easy access to higher plant reproductive cells suggests that algal and seedless plant models might contribute to understanding fundamental aspects of plant reproductive development. Mosses and ferns (Raghavan, 1992) show promise as useful experimental systems for elucidation of the genetic control of higher plant structure and development. In particular, gametophyte stages of nonseed plants display the attributes of rapid growth and ease in application of mutant screening procedures. Methods such as use of high-velocity microprojectiles for transformation of chloroplast and other genomes developed for *Chlamydomonas* (Boynton et al., 1988) are being applied to seedless plants and could be applied to charophycean algae, increasing the usefulness of these organisms as model genetic systems. The following discussion focuses on the strengths of particular algal and nonvascular plant systems with regard to particular research questions.

Noncharophycean Protists

Protists are the systems of choice for elucidation of highly conserved biological features appearing in higher animals and higher plants. Obtaining the answers to questions concerning the elaboration of such fundamental biological attributes as signal transduction pathways and interactions with cytoskeletal proteins, homeotic genes, and hexose transport systems will require much more intensive study of protists. For example, a homologue of the cell cycle control protein $p34^{cdc2}$ kinase is involved in the division cycle of *Chlamydomonas* (John et al., 1989), as well as those of higher organisms. In addition, *Chlamydomonas* is of particular importance because of the immense body of knowledge accumulated about this taxon and because the occurrence of sexual reproduction allows molecular-genetic analysis (Harris, 1989). *Euglena* and its relatives offer special potential in regard to the elucidation of early signal transduction mechanisms and other fundamental eukaryotic cell processes. Representatives of nearly all groups of algal protists are significant systems for study of the evolution of cytoskeletal systems. Prasinophyte algae are of special importance in studies focused on the origin of cell walls of charophytes (and plants), as well as those of other green algal groups.

Charophycean Green Algae

On several research fronts focused on the origin of fundamental plant characteristics, charophytes are the experimental organisms of choice. The subject of these investigations include the origin of genetic relationships between higher plant cell organelles: the process of gene transfer from one organellar genome to another, the origin of chloroplast targeting sequences, the evolution of plastid envelope receptor systems, and the underpinnings of the relationship between plastids and the cytoskeletal system that occurs in seedless plants. The origin of changes in the nature and location of microtubular organizing centers related to loss of centrioles, changes in centriole biogenesis, and the origin of phragmoplasts will not become clear without comparative study of charophytes. Clues to the basis for division site establishment in plant cells, calcium ion control of plant cell wall deposition, and evolution of plant meristems also lie in wait for those using charophytes as model experimental systems.

As discussed previously (Chapter 3), charophytes are unique among green algae and similar to higher plants (and animals) in exhibiting breakdown of the nuclear envelope at meiosis. In animals, dissolution and reestablishment of the nuclear envelope is associated with changes in phosphorylation of nuclear lamins, proteins similar in sequence to intermediate filaments. It has also been speculated that intermediate filaments are important structural components of microtubular organizing centers (MTOCs). Assembly of both intermediate filaments and nuclear lamins is regulated by phosphorylation, and both of these proteins are substrates for p34^{cdc2} kinase, a component of the M-phase promoting factor system in eukaryotes (Kimble and Kuriyama, 1992). Charophycean algae are the protist links between the animal and plant kingdoms. Because plants have been awarded the prize for the least well known MTOCs (Kimble and Kuriyama, 1992), study of phosphorylation effects on MTOCs and lamins of charophytes may illuminate various poorly understood aspects of plant cell biology.

Study of charophytes may contribute to hypotheses explaining the origin of sex in general, and plant sexual reproduction in particular. Because the origin of sex in protists is so poorly understood, it cannot be assumed that information derived from study of noncharophycean protists, including green algae such as *Chlamydomonas*, necessarily applies to sexual reproduction in embryophytes. Charophycean algae also appear to exhibit evolutionary precursors to the chemical cross-talk that probably characterizes the plant embryo-gametophyte interface. Further study of chemical interchange between charophycean zygotes and parental thalli may contribute to a better understanding of embryogenesis in higher plants. The origin of enzymatic pathways leading to production of resistant plant polymers (sporopollenin, phenolic polymers, and precursors of cutin or cutan) requires examination of charophycean metabolism. Charophytes may also be useful in tracing the origin of the quadripolar microtubular system characteristic of plant sporogenesis, the role of callose in plant reproduction, the relationship of cell wall biochemistry to plant morphogenesis, and changes in the efficiency of carbon fixation that have occurred during the evolutionary history of plants.

The large cells of charalean algae have proven to be excellent systems for study of plant cell electrophysiology, cytoplasmic streaming, and gravitropism. Wayne (1993) reviewed the use of patch-clamp techniques to evaluate membrane action potentials and ionic compartmentation in charalean algae. Charales continue to be the plant systems in which cytoplasmic streaming is best understood (Williamson, 1992). Staves et al. (1992) discovered that application of hydrostatic pressure to one end of a *Chara* cell could be used to mimic the effects of gravity, and that such pressure induces a polarity in cytoplasmic streaming. This work suggests that gravity and hydrostatic pressure act at the same signal transduction point, and that characean cells respond to gravity by sensing the resulting pressure differential between cell ends (Staves et al., 1992). Kiss and Staehelin (1993) noted that *Chara* rhizoids represent good systems for graviperception studies because all steps in gravitropism occur in a single cell. They reviewed the evidence in support of the 'vesicle blockage' hypothesis, which explains differential rhizoid growth resulting in curvature as a function of sedimentation of barium- and sulfur-containing vesicles (statoliths), and their interference with distribution of wall-associated Golgi (dictyosome) vesicles. In short, it appears that the *Chara* model may provide new insight into the process of graviperception and signal transduction by plants.

Seedless Plants

Liverworts have proven to be central to illumination of the origin of acentric mitosis in plants, whereas hornworts are the plants of choice for studies focused on the origin of polyplastidy in plants. Immunofluorescence studies of mosses have been useful in investigations of preprophase band development and function. All three bryophyte groups will be useful in tracing evolutionary elaboration of the sporophyte and the phenypropanoid pathway leading to lignin in vascular plants. It is improbable that lignin suddenly appeared "de novo" in vascular plants. Enzymatic pathways ancestral to those involved in lignin biosynthesis are likely to occur in bryophytes. It may prove useful to correlate results of molecular/genetic studies of phenylpropanoid pathways in bryophytes and pteridophytes with data obtained in paleobotanical studies focused on evolution of vascular plants.

Among pteridophytes, lycopods have been identified as the earliest diverging extant vascular plants on the basis of gene sequence data (Raubesson and Jansen, 1992), and this is supported by the occurrence of plesiomorphic spermatozoid characters in lycopods (Robbins and Carothers, 1975; 1978). Comparative study of centrosomal components may illuminate evolutionary transition from bicentriolar centrosomes exhibited by *Lycopodium* (and bryophytes) to the polycentriolar aggregates involved in development of multiflagellate sperm of ferns and some gymnosperms. Such work is essential to tracing the evolutionary history of plant male gametes; evolutionary history may explain the structure and function of spermatozoids of economically important seed plants.

Pteridophyte prothalli are also useful as experimental systems for elucidation of signal transduction system relationships to morphology (Kadota and Wada, 1992a, b, c) and the relationship between cytokinetic systems and morphology (Cooke and Lu, 1992). Kadota and Wada (1992c) found that low levels of red and blue light stimulated the formation of circular structures composed of actin microfilaments upon the surfaces of chloroplasts on the side facing the plasma membrane. These workers suggested that actin plays a role in the anchorage of chloroplasts during intracellular photo-orientation. They also noted that microtubular arrays were dependent on precursor arrays of microfilaments. These findings are directly relevant to photo-orientation of plastids and preprophase plastid migration mechanisms in other seedless plants and in charophytes. Further analyses of the cytoskeletal roles and photo-responses of microfilaments and microtubules in pteridophyte thalli are likely to illuminate developmental transitions from tip growth to histogenesis in bryophytes and pteridophytes and to shed light on evolutionary transition from the filamentous morphology of charophytes to the parenchymatous organization of land plant thalli (not to mention their increasing the understanding of seed plant signal transduction systems).

Cooke and Lu (1992) compared polygonal arrays of various algae (*Ulva, Porphyra, Laminaria,* and *Coleochaete*) with those formed by land plants, including pteridophyte gametophytes. These authors observed that the type of cytokinesis (furrowing versus cell plate formation) does not affect the ability of an organism to form a polygonal array of cells, but that it does influence the mean number of walls exhibited by cells of the array. Land plant polygonal arrays are characterized by cells having an average number of anticlinal walls of about six. This is attributed to avoidance of three-way cell junctions during production of new cross walls, characteristic of the cell plate formation process (see Chapter 7 for further discussion). In contrast, cytokinetic furrowing in algae does not appear to involve avoidance of four-way junction formation. *Coleochaete scutata,* which expresses both furrowing and cell plate formation, depending on the direction of division (Marchant and Pickett-Heaps, 1973), generates thalli resembling polygonal arrays of other algae (Cooke and Lu, 1992). This work emphasizes the importance of the evolutionary transition toward cell plate formation which occurred in charophycean algae during the morphological evolution of land plants.

Knowledge gained in the pursuit of evolutionary questions like those examined in this book is not only useful to the systematist, but will almost certainly contribute to our increased ability to manipulate higher plants to human advantage. Far from merely "filling in the evolutionary gaps," the use of protists, charophytes, and seedless plants as model systems has already contributed much toward the solution of fundamental problems in plant biology. To ignore evolutionarily relevant protist, charophyte, and nonvascular plant systems in favor of exclusive focus on flowering plant experimental systems would be comparable to abandonment by animal scientists of their exceedingly valuable invertebrate systems. The widely quoted dictum that nothing in biology makes sense except in the light of evolution holds true in the plant sciences. To truly know the seed plants on which we primarily depend, it is necessary in addition to understand their simpler relatives, the seedless plants, the small geometries known as charophycean algae, and other protists and their symbionts that first harnessed the power of chlorophyll.

By James Graham

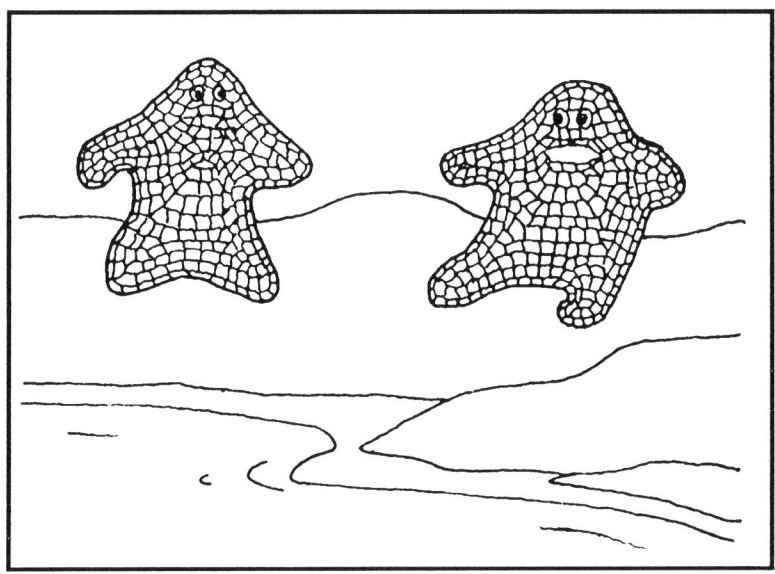

"So we made it onto the land!
In 500 million years, who will care?"

LITERATURE CITED

Allen, C. E. 1905. Die Keimung der Zygote bei *Coleochaete*. *Ber. Deutsch. Bot. Ges.* 23:285–292.

Allen, T. F. H. 1971. Multivariate approaches to the ecology of algae on terrestrial rock surfaces in north Wales. *J. Ecol.* 59:803–826.

Andrews, M., R. Box, S. McInroy, and J. A. Raven. 1984. Growth of *Chara hispida*. II. Shade adaptation. *J. Ecol.* 72:885–895.

Ariztia, E. V., R. A. Anderson, and M. L. Sogin. 1991. A new phylogeny for chromophyte algae using 16S-like rRNA sequences from *Mallomonas papillosa* (Synurophyceae) and *Tribonema aequale* (Xanthophyceae). *J. Phycol.* 27:428–436.

Atchley, W. R., and W. M. Fitch. 1991. Gene trees and the origins of inbred strains of mice. *Science* 254:554–557.

Atkinson, A. W., B. E. S. Gunning, and P. C. L. John. 1972. Sporopollenin in the cell wall of *Chlorella* and other algae: Ultrastructure, chemistry, and incorporation of ^{14}C-acetate, studied in synchronous cultures. *Planta* 107:1–32.

Atsatt, P. R. 1991. Fungi and the origin of land plants, pp. 301–315. *In:* Margulis and Fester (eds.), *Symbiosis as a Source of Evolutionary Innovation. Speciation and Morphogenesis.* The MIT Press, Cambridge, Mass.

Auer, M. T., J. M. Graham, L. E. Graham, and J. A. Kranzfelder. 1983. Factors regulating the spatial and temporal distribution of *Cladophora* and *Ulothrix* in the Laurentian Great Lakes, pp. 135–145. *In:* Wetzel (ed.), *Periphyton of Freshwater Ecosystems.* Junk, The Hague.

Baldauf, S. L., and J. D. Palmer. 1990. Evolutionary transfer of the chloroplast *tuf*A gene to the nucleus. *Nature* 344:262–265.

Baldauf, S. L., J. R. Manhart, and Palmer, J. D. 1990. Different fates of the chloroplast *tuf*A gene following its transfer to the nucleus in green algae. *Proc. Natl. Acad. Sci.* 87:5317–5321.

Basile, D. V. 1990. Morphoregulatory role of hydroxyproline-containing proteins in liverworts, pp. 225–243. *In:* Chopra and Bhatla (eds.), *Bryophyte Development: Physiology and Biochemistry.* CRC Press, Boca Raton.

Basile, D. V., and M. R. Basile. 1983. Desuppression of leaf primordia of *Plagiochila arctica* (Hepaticae) by ethylene antagonists. *Science* 220:1051–1052.

———. 1990. Hydroxyproline metabolism and hydroxyproline-containing glycoproteins in leafy liverworts, pp. 275–288. *In:* Zinsmeister and Mues (eds.), *Bryophytes. Their Chemistry and Chemical Taxonomy.* Clarendon Press, Oxford.

Baum, D. A., and A. Larson. 1991. Adaptation reviewed: a phylogenetic methodology for studying character macroevolution. *Syst. Zool.* 40:1–18.

Beardall, J. 1985. Occurrence and importance of HCO_3^- utilization in microscopic algae, pp. 83–96. *In:* Lucas and Berry (eds.), *Inorganic Carbon Uptake by Aquatic Photosynthetic Organisms.* Am. Soc. Pl. Physiol., Rockville, Md.

Becker, B., D. Becker, J. P. Kamerling, and M. Melkonian. 1991. 2-Keto-sugar acids in green flagellates: a chemical marker for prasinophycean scales. *J. Phycol.* 27:498–504.

Beerbower, R. 1985. Early development of continental ecosystems, pp. 47–91. *In:* Tiffney (ed.), *Geological Factors and the Evolution of Plants.* Yale Univ. Press, New Haven.

Berner, R. A. 1991. A model for atmospheric CO_2 over Phanerozoic time. *Am. J. Sci.* 291:339–376.

Berner, R. A., and D. E. Canfield. 1989. A new model of atmospheric O_2 over Phanerozoic time. *Am. J. Sci.* 289:336–361.

Biebel, P. 1973. Morphology and life cycles of saccoderm desmids in culture. *Beih. Nova Hedwigia* 42:39–57.

Blackmore, S., and S. H. Barnes. 1987. Embryophyte spore walls: origin, development, and homologies. *Cladistics* 3:185–195.

Bold, H. C., and M. J. Wynne. 1985. *Introduction to the Algae. Structure and Reproduction.* 2d ed. Prentice-Hall, Englewood Cliffs, N.J.

Bold, H. C., C. J. Alexopoulos, and T. Delevoryas. 1987. *Morphology of Plants and Fungi.* Harper and Row, New York.

Bower, F. O. 1908. *The Origin of a Land Flora. A Theory Based upon the Facts of Alternation.* Macmillan and Co., London.

———. 1935. *Primitive Land Plants.* Hafner, New York.

Bowes, G. 1985. Pathways of CO_2 fixation by aquatic organisms, pp. 187–210. *In:* Lucas and Berry (eds.), *Inorganic Carbon Uptake by Aquatic Photosynthetic Organisms.* Am. Soc. Pl. Physiol., Rockville, Md.

Boynton, J. E., N. W. Gillham, E. H. Harris, J. P. Hosler, A. M. Hohnson, A. R. Jones, B. L. Randolph-Anderson, D. Robertson, T. M. Klein, K. B. Shark, and J. C. Sanford. 1988. Chloroplast transformation in *Chlamydomonas* with high-velocity microprojectiles. *Science* 240:1534–1537.

Bradley, P. M. 1991. Plant hormones do have a role in controlling growth and development of algae. *J. Phycol.* 27:317–321.

Brandham, P. E. 1967. Time-lapse studies of conjugation in *Cosmarium botrytis.* I. Gamete fusion and spine formation. *Revue Algologique* 8:312–316.

Bray, A. A. 1985. The evolution of the terrestrial vertebrates: environmental and physiological considerations, pp. 289–319. *In:* Chaloner and Lawson (eds.), *Evolution and Environment in the Late Silurian and Early Devonian.* The Royal Society, London.

Bremer, K., C. J. Humphries, B. D. Mishler, and S. P. Churchill. 1987. On cladistic relationships in green plants. *Taxon* 36:339–349.

Brenchley, S. K. 1989. The late Ordovician extinction, pp. 104–132. *In:* Donovan (ed.), *Mass Extinctions: Processes and Evidence.* Columbia University Press, New York.

Brook, A. J. 1968. The discoloration of roofs in the United States and Canada by algae. *J. Phycol.* 4:250.

Brown, R. C., and B. E. Lemmon. 1984. Plastid apportionment and preprophase microtubule bands in monoplastidic root meristem cells of *Isoetes* and *Selaginella. Protoplasma* 123:95–103.

———. 1988. Preprophasic microtubule systems and development of the mitotic spindle in hornworts (Bryophyta). *Protoplasma* 143:11–21.

———. 1989. Morphogenetic plastid migration and microtubule organization during megasporogenesis in *Isoetes. Protoplasma* 152:136–147.

———. 1990a. Polar organizers mark division axis prior to preprophase band formation in mitosis of the hepatic *Reboulia hemisphaerica* (Bryophyta). *Protoplasma* 156:74–81.

———. 1990b. Sporogenesis in bryophytes, pp. 55–94. *In:* Blackmore and Knox (eds.), *Microspores: Evolution and Ontogeny.* Academic Press, London.

———. 1990c. Monoplastidic cell division in lower land plants. *Amer J. Bot.* 77:559–571.

———. 1992. Polar organizers in monoplastidic mitosis of hepatics (Bryophyta). *Cell Motility and the Cytoskeleton* 22:72–77.

Brown, R. C., B. E. Lemmon, and L. E. Graham. 1993. Morphogenetic plastid migration and microtubule arrays in mitosis and cytokinesis in the green alga *Coleochaete* (submitted, Amer. J. Bot.).

Brown, R. M., Jr. 1985. Cellulose microfibril assembly and orientation: recent developments, pp. 13–32. *In:* Robert, Johnston, Lloyd, Shaw, and Woolhouse (eds.), *The Cell Surface in Plant Growth and Development.* The Company of Biologists Limited, Cambridge.

Browning, A. J., and B. E. S. Gunning. 1979a. Structure and function of transfer cells in the sporophyte haustorium of *Funaria hygrometrica* Hedw. II. Kinetics of uptake of labeled sugars and localization of absorbed products by freeze-substitution and autoradiography. *J. Experimental Botany* 30:1265–1273.

———. 1979b. Structure and function of transfer cells in the sporophyte haustorium of *Funaria hygrometrica.* III. Translocation of assimilate into the attached sporophyte and along the seta of attached and excised sporophytes. *J. Experimental Botany* 30:1265–1273.

Buick, R. 1992. The antiquity of oxygenic photosynthesis: Evidence from stromatolites in sulphate-deficient Archaean lakes. *Science* 255:74–77.

Butz, N. D., and T. D. Sharkey. 1989. Activity ratios of ribulose-1,5-bisphosphate carboxylase accurately reflect carbamylation ratios. *Plant Physiol.* 89:735–739.

Campbell, D. H. 1912. The classification of the liverworts. Amer. Nat. 46:684–695.

Canuto, V. M., J. S. Levine, T. R. Augustsson, and C. L. Imhoff. 1982. UV radiation from the young sun and oxygen and ozone levels in the prebiological paleoatmosphere. *Nature* 296:816–820.

Caussin, C. 1983. Absorption of some amino acids by sporophytes isolated from *Polytrichum formosum* and ultrastructural characteristics of the haustorium transfer cells. Ann. Bot. 51:167–173.

Cavalier-Smith, T. 1981. Eukaryote kingdoms: seven or nine? *BioSystems* 14:461–481.

Celakovsky, L. 1874. *Bedeutung des Generationswechsels der Pflanzen.* Prague.

Cerling, T. E. 1991. Carbon dioxide in the atmosphere: evidence from Cenozoic and Mesozoic paleosols. *Am. J. Sci.* 291:377–400.

Chapman, D. J. 1985. Geological factors and biochemical aspects of the origin of land plants, pp. 23–45. *In:* Tiffney (ed.), *Geological Factors and the Evolution of Plants.* Yale Univ. Press, New Haven.

Chapman, R. L., and M. A. Buchheim. 1991. Ribosomal RNA gene sequences: analysis and significance in the phylogeny and taxonomy of green algae. *Crit. Rev. Pl. Sci.* 10:343–368.

Chapman, R. L., and M. C. Henk. 1986. Phragmoplasts in cytokinesis of *Cephaleuros parasiticus* (Chlorophyta) vegetative cells. *J. Phycol.* 22:83–88.

Chasen, R. 1992. In the realm of the homeodomain. *The Plant Cell* 4:237–240.

Chrispeels, M. J., and N. V. Raikhel. 1992. Short peptide domains target proteins to plant vacuoles. *Cell* 68:613–616.

Church, A. H. 1919. *Thalassiophyta and the Subaerial Migration.* Oxford Bot. Mem. No. 3:1–19.

Cooke, T. J., and B. Lu. 1992. The independence of cell shape and overall form in multicellular algae and land plants: cells do not act as building blocks for constructing plant organs. *Int. J. Plant Sci.* 153:57–527.

Crawford, D. J. 1990. *Plant molecular systematics. Macromolecular approaches.* Wiley, New York.

Crowley, T. J., J. G. Mengel, and D. A. Short. 1987. Gondwanaland's seasonal cycle. *Nature* 329:803–807.

Darvill, A. G., P. Albersheim, M. McNeil, J. M. Lau, W. S. York, T. T. Stevenson, J. Thomas, S. Doares, D. J. Gollin, P. Chelf, and K. Davis. 1985. Structure and function of plant cell wall polysaccharides, pp. 203–217. *In:* Roberts, Johnston, Lloyd, Shaw, and Woolhouse (eds.), *The Cell Surface in Plant Growth and Development.* The Company of Biologists, Ltd., Cambridge.

deJesus, M. D., F. Tabatabai, and D. J. Chapman. 1989. Taxonomic distribution of copper-zinc superoxide dismutase in green algae and its phylogenetic importance. *J. Phycol.* 25:767–772.

Delmer, D. P. 1991. The biochemistry of cellulose synthesis, pp. 102–107. *In:* Lloyd (ed.), *The Cytoskeletal Basis of Plant Growth and Form.* Academic Press, London.

Delwiche, C. F., L. E. Graham, and N. Thomson. 1989. Lignin-like compounds and sporopollenin in *Coleochaete,* an algal model for land plant ancestry. *Science* 245:399–401.

DeVries, P. J. R., J. Simons, and A. P. VanBeem. 1983. Sporopollenin in the spore wall of *Spirogyra* (Zygnemataceae, Chlorophyceae). *Acta Botanica Neerlandica* 32:25–28.

DeWitt, N. D., J. F. Harper, and M. R. Sussman. 1991. Evidence for a plasma membrane proton pump in phloem walls of higher plants. *The Plant Journal* 1:121–128.

Ding, B., R. Turgeon, and M. V. Parthasarathy. 1991. Microfilaments in the preprophase band of freeze substituted tobacco root cells. *Protoplasma* 165:209–211.

Domozych, D. S., K. D. Stewart, and K. R. Mattox. 1980. The comparative aspects of cell wall chemistry in the green algae (Chlorophyta). *J. Mol. Evol.* 15:1–12.

Doonan, J. H. 1991. The cytoskeleton and moss morphogenesis, pp. 289–301. *In* Lloyd (ed.), *The Cytoskeletal Basis of Plant Growth and Form.* Academic Press, London.

Doonan, J. H., and J. G. Duckett. 1988. The bryophyte cytoskeleton: experimental and immunofluorescence studies of morphogenesis. *Adv. in Bryol.* 3:1–31.

Doonan, J. H., D. J. Cove, F. M. K. Corke, and C. W. Lloyd. 1987. Pre-prophase band of microtubules, absent from tip-growing moss filaments, arises in leafy shoots during transition to intercalary growth. *Cell Motility and the Cytoskeleton* 7:138–153.

Douglas, S. E., and D. G. Durnford. 1989. The small subunit of ribulose -1,5-bisphosphate carboxylase is plastid encoded in the chlorophyll c-containg alga *Cryptomonas* Φ. *Plant Molec. Biol.* 13:13–20.

Dubois-Tylski, T. 1981. Utilisation de fluorochromes pour l'observation des parois cellulaires chez trois especes de *Closterium* (Desmidiales) au cours de leur reproduction sexée. *Cryptogamie: Algologie II* 4:277–287.

Duckett, J. G., and K. S. Renzaglia. 1988. Ultrastructure and development of plastids in the bryophytes. *Adv. in Bryol.* 3:33–93.

Dutcher, S. K. 1989. Linkage group XIX in *Chlamydomonas reinhardtii* (Chlorophyceae): genetic analysis of basal body function and assembly, pp. 39–53. *In:* Coleman, Goff, and Stein-Taylor (eds.), *Algae as Experimental Systems.* Liss, New York.

Edwards, D., and U. Fanning. 1985. Evolution and environment in the Late Silurian–Early Devonian: the rise of the pteridophytes, pp. 147–165. *In:* Chaloner and Lawson (eds.), *Evolution and Environment in the Late Silurian and Early Devonian.* The Royal Society, London.

Edwards, D., K. L. Davies, and L. Axe. 1992. A vascular conducting strand in the early land plant *Cooksonia. Nature* 357:683–685.

Edwards, D., U. Fanning, and J. B. Richardson. 1986. Stomata and sterome in early land plants. *Nature* 323:438–440.

Ellis, J. R. 1991. Chaperone function: cracking the second half of the genetic code. *The Plant Journal* 1:9–13.

Espie, G. S., and B. Coleman. 1985. Is extracellular HCO_3^- a direct source of inorganic carbon for terrestrial plant cell photosynthesis? pp. 169–186. *In:* Lucas and Berry (eds.), *Inorganic Carbon Uptake by Aquatic Photosynthetic Organisms.* Am. Soc. Pl. Physiol., Rockville, Md.

Evans, L. V., and A. J. Trewavas. 1991. Is algal development controlled by plant growth substances? *J. Phycol.* 27:322–326.

Fawley, M. W., and C. M. Lee. 1990. Pigment composition of the scaly green flagellate *Mesostigma viride* (Micromonadophyceae) is similar to that of the siphonous green alga *Bryopsis plumosa* (Ulvophyceae). *J. Phycol.* 26:666–670.

Feist, M., and N. Grambast-Fessard. 1991. The genus concept in Charophyta: evidence from Palaeozoic to Recent, pp. 189–203. *In:* Riding (ed.), *Calcareous Algae and Stromatolites.* Springer-Verlag, New York.

Fisher, R. F., and S. R. Long. 1992. *Rhizobium*-plant signal exchange. *Nature* 357:655–660.

Floyd, G. L., and Salisbury, J. L. 1977. Glycolate dehydrogenase in primitive green algae. *Amer. J. Bot.* 64:1294–1296.

Floyd, G. L., K. D. Stewart, and K. R. Mattox. 1972. Cellular organization, mitosis, and cytokinesis in the ulotrichalean alga, *Klebsormidium. J. Phycol.* 8:176–184.

Foissner, I. 1990. Wall appositions induced by ionophore A 23187, $CaCl_2$, $LaCl_3$, and nifedipine in characean cells. *Protoplasma* 154:80–90.

Forsberg, C. 1965. Nutritional studies of *Chara* in axenic cultures. *Physiologia Plantarum* 18:275–290.

Fowke, L. C., and J. D. Pickett-Heaps. 1969. Cell division in *Spirogyra.* II. Cytokinesis. *J. Phycol.* 5:273–281.

Frantz, T. C., and A. J. Cordone. 1967. Observations on deepwater plants in Lake Tahoe, California and Nevada. *Ecology* 48:709–714.

Franzen, L.-G., J.-D. Rochaix, and G. von Heijne. 1990. Chloroplast transit peptides from the green alga *Chlamydomonas reinhardtii* share features with both mitochondrial and higher plant chloroplast presequences. *FEBS Letters* 260:165–168.

Frederick, S. E., P. J. Gruber, and N. E. Tolbert. 1973. The occurrence of glycolate dehydrogenase and glycolate oxidase in green plants. *Plant Physiol.* 52:318–323.

Freeman, S., and R. J. Rodriguez. 1993. Genetic conversion of a fungal plant pathogen to a nonpathogenic endophytic mutualist. *Science* 260:75–78.

Friedman, W. E. 1990a. Double fertilization in *Ephedra,* a nonflowering seed plant: Its bearing on the origin of angiosperms. *Science* 247:951–954.

———. 1990b. Sexual reproduction in *Ephedra nevadensis* (Ephedraceae): Further evidence of double fertilization in a nonflowering seed plant. *Amer. J. Bot.* 77:1582–1598.

———. 1992. Evidence of a pre-angiosperm origin of endosperm: implications for the evolution of flowering plants. *Science* 255:336–339.

Fritsch, F. E. 1916. The algal ancestry of the higher plants. *The New Phytologist* 15:233–250.

Galway, M. E., and A. R. Hardham. 1991. Immunofluorescent localization of microtubules throughout the cell cycle in the green alga *Mougeotia* (Zygnemataceae). *Amer. J. Bot.* 78:451–461.

Gambardella, R., and F. Alfano. 1990. Monoplastidic mitosis in the moss *Timmiella barbuloides* (Bryophyta). *Protoplasma* 156:29–38.

Garatt, M. J., J. D. Tims, R. B. Rickards, T. C. Chambers, and J. G. Douglas. 1984. The appearance of *Baragwanathia* (Lycophytina) in the Silurian. *Botanical Journal of the Linnean Society* 89:355–358.

Garcia-Pichel, F., and R. W. Castenholz. 1991. Characterization and biological implications of scytonemin, a cyanobacterial sheath pigment. *J. Phycol.* 27:395–409.

Geitler, L. 1942. Morphologie, Entwicklungsgeschichte und Systematik neuer bemerkens werter atmophytischer Algen aus Wien. *Flora NF* 136:1–29.

———. 1955. Über die cytologisch bemerkenswerte Chlorophycee *Chlorokybus atmophyticus* Osterr. *Bot. Z.* 102:20–29.

———. 1961. Spontaneous partial rotation and oscillation of the protoplasm in *Coleochaete* and other Chlorophyceae. *Am. J. Bot.* 48:738–741.

Gensel, P. A., and H. N. Andrews. 1984. *Plant Life in the Devonian.* Praeger, New York.

———. 1987. The evolution of early land plants. *Amer. Sci.* 75:478–489.

Giddings, T. H., Jr., and S. H. Staehelin. 1991. Microtubule-mediated control of microfibril deposition: A re-examination of the hypothesis, *In:* Lloyd (ed.), *The Cytoskeletal Basis of Plant Growth and Form.* Academic Press, London.

Gifford, E. M., and A. S. Foster. 1989. *Morphology and Evolution of Vascular Plants.* Freeman, New York. Chapter 5.

Giovannoni, S. J., S. Turner, G. J. Olsen, S. Barns, D. J. Lane, and N. R. Pace. 1988. Evolutionary relationships among cyanobacteria and green chloroplasts. *J. Bacteriol.* 170:3584–3592.

Glover, H. E. 1989. Ribulosebisphosphate carboxylase/oxygenase in marine organisms. *Int. Rev. Cytol.* 115:67–138.

Goldberg, R. B. 1988. Plants: Novel developmental processes. *Science* 240:1460–1467.

Good, B. H., and R. L. Chapman. 1978. The ultrastructure of *Phycopeltis* (Chroolepidaceae: Chlorophyta). I. Sporopollenin in the cell walls. *Amer. J. Bot.* 65:27–33.

Goto, Y., and K. Ueda. 1988. Microfilament bundles of F-actin in *Spirogyra* observed by fluorescence microscopy. *Planta* 173:442–446.

Gould, S. J., and E. Vrba. 1982. Exaptation—a missing term in the science of form. *Paleobiology* 8:4–15.

Graham, L. E. 1982a. Cytology, ultrastructure, taxonomy, and phylogenetic relationships of Great Lakes filamentous algae. *J. Great Lakes Res.* 8:3–9.

———. 1982b. The occurrence, evolution, and phylogenetic significance of parenchyma in *Coleochaete* Breb. (Chlorophyta). *Amer. J. Bot.* 69:447–454.

———. 1984. *Coleochaete* and the origin of land plants. *Amer. J. Bot.* 71:603–608.

———. 1985. The origin of the life cycle of land plants. *Amer. Sci.* 73:178–186.

———. 1990. Meiospore formation in charophycean algae, pp. 43–54. *In:* Blackmore and Knox (eds.), *Microspores: Evolution and Ontogeny.* Academic Press, London.

Graham, L. E., and C. F. Delwiche. 1992. The occurrence and phylogenetic significance of a surface layer on thalli of *Coleochaete* (Charophyceae). *Amer. J. Bot.* 79:102.

Graham, L. E., and Y. Kaneko. 1991. Subcellular structures of relevance to the origin of land plants (embryophytes) from green algae. *Crit. Rev. in Plant Sci.* 10:323–342.

Graham, L. E., and G. E. McBride. 1979. The occurrence and phylogenetic significance of a multilayered structure in *Coleochaete* spermatozoids. *Amer. J. Bot.* 66:887–894.

Graham, L. E., and W. M. Repavich. 1989. Spermatogenesis in *Coleochaete pulvinata* (Charophyceae): early blepharoplast development. *Amer. J. Bot.* 76:1266–1278.

Graham, L. E., and C. Taylor, III. 1986a. Occurrence and phylogenetic significance of "special walls" at meiosporogenesis in *Coleochaete*. *Amer. J. Bot.* 73:597–601.

———. 1986b. The ultrastructure of meiospores of *Coleochaete pulvinata* (Charophyceae). *J. Phycol.* 22:299–307.

Graham, L. E., and G. J. Wedemayer. 1984. Spermatogenesis in *Coleochaete pulvinata* (Charophyceae): sperm maturation. *J. Phycol.* 20:301–309.

Graham, L. E., and L. W. Wilcox. 1983. The occurrence and phylogenetic significance of putative placental transfer cells in the green alga *Coleochaete*. *Amer. J. Bot.* 70:113–120.

Graham, L. E., C. F. Delwiche, and B. Mishler. 1991. Phylogenetic connections between the 'green algae' and the 'bryophytes'. *Adv. in Bryol.* 4:213–244.

Graham, L. E., J. M. Graham, and J. A. Kranzfelder. 1986. Irradiance, daylength and temperature effects on zoosporogenesis in *Coleochaete scutata* (Charophyceae). *J. Phycol.* 22:35–39.

Graham, L. E., F. J. Macentee, and H. C. Bold. 1981. An investigation of some subaerial green algae. *Texas J. Sci.* 33:13–16.

Graham, L. E., J. M. Graham, J. Downs, and J. Gerwing. 1992. *Coleochaete* and the periphyton ecology of a northern oligotrophic lake. *J. Phycol.* 28:8.

Graham, M. Y., and T. L. Graham. 1991. Rapid accumulation of anionic peroxidase and phenolic polymers in soybean cotyledon tissues following treatment with *Phytophthora megasperma* f. sp. *Glycinea* wall glucan. *Plant. Physiol.* 97:1445–1455.

Grant, M. C. 1990. Phylum Chlorophyta. Class Charophyceae. Order Charales, pp. 641–648. *In:* Margulis, Corliss, Melkonian, and Chapman (eds.), *Handbook of Protoctista.* Jones and Bartlett, Boston.

Grambast, L. J. 1974. Phylogeny of the Charophytes. *Taxon* 23:463–481.

Gray, J. 1984. Ordovician-Silurian land plants: The interdependence of ecology and evolution. *In:* Bassett and Lawson (eds.), *Autecology of Silurian Organisms.* The Paleontological Association.

———. 1985. The microfossil record of early land plants: advances in understanding of early terrestrialization, 1970–1984. *Phil. Trans. R. Soc. Lond. B* 309:167–195.

Gray, J., and A. J. Boucot. 1977. Early vascular land plants: proof and conjecture. *Lethaia* 10:145–174.

———. 1978. The advent of land plant life. *Geology* 6:489–492.

Gray, J., and W. Shear. 1992. Early life on land. *Am. Sci.* 80:444–456.

Gray, J., D. Massa, and A. J. Boucot. 1982. Caradocian land plant microfossils from Libya. *Geology* 10:197–201.

Gray, M. W., and P. H. Boer. 1988. Organization and expression of algal (*Chlamydomonas reinhardtii*) mitochondrial DNA. *Proc. Roy. Soc. Lond. B* 319:135–147.

Gray, M. W., R. Cedergren, Y. Abel, and D. Sankoff. 1989. On the evolutionary origin of the plant mitochondrion and its genome. *Proc. Natl. Acad. Sci.* 86:2267–2271.

Grisebach, H. 1981. Lignins, pp. 457–477. *In:* Stumpf and Conn (eds.), *The Biochemistry of Plants, a Comprehensive Treatise.* Vol. 7. *Secondary Plant Products.* Academic Press, London.

———. 1992. The cytoskeleton of the Zygnemataceae, pp. 165–193. *In:* Menzel (ed.), *The Cytoskeleton of the Algae.* CRC Press, Boca Raton, Fl.

Grolig, F., and G. Wagner. 1988. Light-dependent chloroplast reorientation in *Mougeotia* and *Mesotaenium*: biased by pigment-regulated plasmalemma anchorage sites to actin filaments? *Bot. Acta* 101:2–6.

Gunning, B. E. S., and R. L. Overall. 1983. Plasmodesmata and cell-to-cell transport in plants. *BioScience* 33:260–265.

Gunning, B. E. S., and S. M. Wick. 1985. Preprophase bands, phragmoplasts, and spatial control of cytokinesis, pp.157–179. *In:* Roberts, Johnston, Lloyd, Shaw, and Woolhouse (eds.), *The Cell Surface in Plant Growth and Development.* The Company of Biologists Limited, Cambridge.

Gunnison, D., and M. Alexander. 1975. Basis for the resistance of several algae to microbial decomposition. *Applied Microbiol.* 29:729–738.

Hagemann, W. 1978. Zur phylogenese der terminalen Sprossmeristeme. *Ber. Deutsch. Bot. Ges.* 91:699–716.

———. 1991a. What is a root? *In: Root Ecology and Its Practical Application—A Contribution to the Investigation of the Whole Plant.* Verein fur Wurzelforschung, Klagelfurt.

———. 1991b. The evolution of pteridophytes, new ideas based on the comparative evaluation of the construction of plants. *In:* Bhardwaj and Gena (eds.), *Aspects of Plant Sciences. Perspectives in Pteridology: Present and Future.* 13:1–20. Today & Tomorrow's Printers & Publishers, New Delhi.

Halstead, L. B. 1985. The vertebrate invasion of fresh water, pp. 243–258. *In:* Chaloner and Lawson (eds.), *Evolution and Environment in the Late Silurian and Early Devonian.* The Royal Society, London.

Harris, E. H. 1989. *The Chlamydomonas Sourcebook.* Academic Press, New York.

Hartwig, J. H., M. Thelin, A. Rosen, P. A. Janmey, A. C. Nairn, and A. Aderem. 1992. MARCKS is an actin filament crosslinking protein regulated by protein kinase C and calcium-calmodulin. *Nature* 356:618–622.

Hashimoto, H. 1992. Involvement of actin filaments in chloroplast division of the alga *Closterium ehrenbergii. Protoplasma* 167:88–96.

Hearst, E. 1991. Psychology and nothing. *Am. Sci.* 79:432–443.

Hebant, C. 1977. *The Conducting Tissues of Bryophytes.* J. Cramer, Vaduz.

Hemsley, A. R. 1989. The ultrastructure of the spores of the Devonian plant *Parka decipiens. Ann. Bot.* 64:359–367.

Hennig, W. 1966. *Phylogenetic Systematics.* Univ. of Illinois Press, Urbana.

Herendeen, P. S., D. H. Les, and D. L. Dilcher. 1990. Fossil *Ceratophyllum* (Ceratophyllaceae) from the Tertiary of North America. *Amer. J. Bot.* 77:7–16.

Hofmeister, W. 1851. *Vergleichende Untersuchungen der Keimung, Entfaltung und Fruchtbildung höherer Kryptogamen.* Leipzig.

Holland, C. H. 1985. Synchronology, pp. 11–24. *In:* Chaloner and Lawson (eds.), *Evolution and Environment in the Late Silurian and Early Devonian.* The Royal Society, London.

Holland, H. D., B. Lazar, and M. McCaffrey. 1986. Evolution of the atmosphere and oceans. *Nature* 320:27–33.

Holloway, P. J. 1982. The chemical constitution of plant cutins. *In:* Cutler, Alvin, and Price (Eds.), *The plant cuticle.* Academic Press, London. Pages 45–85.

Hopkins, A. W., and G. E. McBride. 1976. The life history of *Coleochaete scutata* (Chlorophyceae) studied by a Feulgen microspectrophotometric analysis. *J. Phycol.* 12:29–35.

Hori, T., I. Inouye, T. Horiguchi, and G. T. Boalch. 1985. Observations on the motile stage of *Halosphaera minor* Ostenfeld (Prasinophyceae) with special reference to the cell structure. *Bot. Mar.* 28:529–537.

Hoshaw, R. W., R. M. McCourt, and J.-C. Wang. 1990. Phylum Conjugaphyta, pp. 119–131. *In:* Margulis, Corliss, Melkonian, and Chapman (eds.), *Handbook of Protoctista.* Jones and Bartlett, Boston.

Hotchkiss, A. T., and R. M. Brown, Jr. 1987. The association of rosette and globule terminal complexes with cellulose microfibril assembly in *Nitella translucens* var. *axillaris* (Charophyceae). *J. Phycol.* 23:229–237.

Hotchkiss, A. T., M. R. Gretz, K. B. Hicks, and R. M. Brown, Jr. 1989. The composition and phylogenetic significance of the *Mougeotia* (Charophyceae) cell wall. *J. Phycol.* 25:646–654.

Hotta, Y., S. Tabata, L. Stubbs, and H. Stern. 1985. Meiosis-specific transcripts of a DNA component replicated during chromosome pairing: homology across the phylogenetic spectrum. *Cell* 40:785–793.

Hrazdina, G., and A. M. Zobel. 1991. Cytochemical localization of enzymes in plant cells. *Int. Rev. Cytol.* 129:269–322.

Hu, H., and Y. Wei. 1991. *Radioramus*—a new genus of Coleochaetaceae (Chlorophyceae). *Oceanologia et Limnologia Sinica* 22:422–426.

Inouye, I., T. Hori, and M. Chihara. 1990. Absolute configuration analysis of the flagellar apparatus of *Pterosperma cristatum* (Prasinophyceae) and consideration of its phylogenetic position. *J. Phycol.* 26:329–344.

Jacobshagen, S., and C. Schnarrenberger. 1990. Two class I aldolases in *Klebsormidium flaccidum* (Charophyceae): an evolutionary link from chlorophytes to higher plants. *J. Phycol.* 26:312–317.

Jeffrey, C. 1962. The origin and differentiation of the archegoniate land-plants. *Botaniska Notiser* 115:446–454.

―――. 1968. The origin and differentiation of the archegoniate land plants: a second contribution. *Kew Bull.* 21:335–349.

―――. 1982. Kingdoms, codes and classification. *Kew Bull.* 37:403–416.

Jeram, A. J., P. A. Selden, and D. Edwards. 1990. Land animals in the Silurian: arachnids and myriapods from Shropshire, England. *Science* 250:658–661.

John, P. C. L., F. J. Sek, and M. G. Lee. 1989. A homolog of the cell cycle control protein p34^{cdc2} participates in the division cycle of *Chlamydomonas* and a similar protein is detectable in higher plants and remote taxa. *Plant Cell* 1:1185–1193.

Johnson, N. J. 1985. Early Silurian palynomorphs from the Tuscarora formation in central Pennsylvania and their paleobotanical and geological significance. *Review of Palaeobotany and Palynology* 45:307–360.

Johnson, W. J. and R. H. Goldstein. 1993. Cambrian sea water preserved as inclusions in marine low-magnesium calcite cement. *Nature* 362:335–337.

Jost, L. 1895. Beiträge zur Kenntniss der Coleochaeteen. *Ber. d. Deutsch. Bot. Ges.* 13:433–452.

Kadota, A., and M. Wada. 1992a. The circular arrangement of cortical microtubules around the subapex of tip-growing fern protonemata is sensitive to cytochalasin B. *Plant Cell Physiol.* 33:99–102.

———. 1992b. Reorganization of the cortical cytoskeleton in tip-growing fern protonemal cells during phytochrome-mediated phototropism and blue-light-induced apical swelling. *Protoplasma* 166:35–41.

———. 1992c. Photoinduction of formation of circular structures by microfilaments on chloroplasts during intracellular orientation in protonemal cells of the fern *Adiantum capillus-veneris. Protoplasma* 167:97–107.

Kaplan, D. R., and W. Hagemann. 1991. The relationship of cell and organism in vascular plants. *BioScience* 41:693–703.

Kasting, J. F. 1987. Theoretical constraints on oxygen and carbon dioxide concentrations in the Precambrian atmosphere. *Precambrian Research* 34:205–229.

Katasaros, C. and B. Galatis. 1992. Immunofluorescence and electron miscroscopic studies of microtubule organization during the cell cycle of *Dictyota dichtoma* (Phaeophyta, Dictyotales). *Protoplasma* 169:75–84.

Kauss, H. 1985. Callose biosynthesis as a Ca^{2+}-regulated process and possible relations to the induction of other metabolic changes, pp. 89–103. *In:* Roberts, Johnston, Lloyd, Shaw, and Woolhouse (eds.), *The Cell Surface in Plant Growth and Development.* The Company of Biologists Limited, Cambridge.

Keller, G.-A., S. Krisans, S. J. Gould, J. M. Sommer, C. C. Wang, W. Schliebs, W. Kunau, S. Brody, and S. Subramani. 1991. Evolutionary conservation of a microbody targeting signal that targets proteins to peroxisomes, glyoxysomes, and glycosomes. *J. Cell Biol.* 114:893–904.

Kenrick, P., and P. R. Crane. 1991. Water-conducting cells in early fossil land plants: implications for the early evolution of tracheophytes. *Bot. Gaz.* 152:335–356.

Kenrick, P., and D. Edwards. 1988. The anatomy of Lower Devonian *Gosslingia breconensis* Heard based on pyritized axes, with some comments on the permineralization process. *Bot. J. Linn. Soc.* 97:95–123.

Kenrick, P., D. Edwards, and R. C. Dales. 1991. Novel ultrastructure in water-conducting cells of the Lower Devonian plant *Sennicaulis hippocrepiformis. Paleontology* 34:751–766.

Khan, M., and Y. S. R. K. Sarma. 1984. Cytogeography and cytosystematics of Charophyta, pp. 303–330. *In:* Irvine and John (eds.), *Systematics of the Green Algae.* Academic Press, New York.

Kidd, D. G., and J. C. Lagarias. 1990. Phytochrome from the green alga *Mesotaenium caldariorum.* Purification and preliminary characterization. *J. Biol. Chem.* 265:7029–7035.

Kiermayer, O., and U. Meindl. 1989. Cellular morphogenesis: the desmid (Chlorophyceae) system, pp. 149–167. *In:* Coleman, Goff, and Stein-Taylor (eds.), *Algae as Experimental Systems.* Liss, New York.

Kimble, M., and R. Kuriyama. 1992. Functional components of microtubule-organizing centers. *Int. Rev. Cyt.* 136:1–50.

Kiss, J. Z., and L. A. Staehelin. 1993. Structural polarity in the *Chara* rhizoid: A reevaluation. *Amer. J. Bot.* 80:273–282.
Knoll, A. H. 1992. The early evolution of eukaryotes: a geological perspective. *Science* 256:622–627.
Knoll, A. H., and K. J. Niklas. 1987. Adaptation, plant evolution, and the fossil record. *Rev. Paleobotany and Palynology* 50:127–149.
Kolattukudy, P. E. 1980. Biopolyester membranes of plants: cutin and suberin. *Science* 208:990–1000.
Kondrashov, A. S. 1988. Deleterious mutations and the evolution of sexual reproduction. *Nature* 336:435–440.
Kondrashov, A. S., and J. F. Crow. 1991. Haploidy or diploidy: which is better? *Nature* 351:314–315.
Kranz, E., J. Bautor, and H. Lörz. 1990. In vitro fertilization of single, isolated gametes, transmission of cytoplasmic organelles and cell recognition of maize (*Zea mays* L.), pp. 252–257. *In:* Nijkamp, Plas, and vanAatrijk (eds.), *Progress in Plant Cellular and Molecular Biology.* Kluwer Academic Publishers, Dordrecht.
Kubitzki, K. 1987. Phenylpropanoid metabolism in land plant evolution. *J. Plant Physiol.* 131:17–24.
Kwiatkowska, M., and J. Maszewski. 1986. Changes in the occurrence and ultrastructure of plasmodesmata in antheridia of *Chara vulgaris* L. during different stages of spermatogenesis. *Protoplasma* 132:179–188.
Kwiatkowska, M., K. Poplonska, and K. Zylinska. 1990. Biological role of endoreplication in the process of spermatogenesis in *Chara vulgaris* L. *Protoplasma* 155:176–187.
Lamb, C. J., M. A. Lawton, M. Dron, and R. A. Dixon. 1989. Signals and transduction mehanisms for activation of plant defenses against microbial attack. *Cell* 56:215–224.
Lancelle, S. A. and P. K. Hepler. 1991. Association of actin with cortical microtubules revealed by immunogold localization in *Nicotiana* pollen tubes. *Protoplasma* 165:167–172.
Les, D. H., D. K. Garvin, and C. F. Wimpee. 1991. Molecular evolutionary history of ancient aquatic angiosperms. *PNAS* 88:10119–10123.
Lewis, N. G., and E. Yamamoto. 1990. Lignin: occurrence, biogenesis and biodegradation. *Ann. Rev. Plant Physiol. Plant Mol. Bio.* 41:455–496.
Li, W.-H., and D. Grauer. 1991. *Fundamentals of Molecular Evolution.* Sinauer, Sunderland, Mass.
Ligrone, R., and R. Gambardella. 1988. The sporophyte-gametophyte junction in bryophytes. *Adv. in Bryol.* 3:225–274.
Linthorst, H. J. M. 1991. Pathogenesis-related proteins of plants. *Crit. Rev. Plant Sci.* 10:123–150.
Liu, B., J. Marc, H. C. Joshi, and B. A. Palevitz. 1993. A γ-tubulin-related protein associated with the microtubule arrays of higher plants in a cell cycle-dependent manner. *Journal of Cell Science* 104:1–12.
Livermore, R. A., A. G. Smith, and J. C. Briden. 1985. Palaeomagnetic constraints on the distribution of continents in the Late Silurian and Early Devonian, pp. 29–56. *In:* Chaloner and Lawson (eds.), *Evolution and Environment in the Late Silurian and Early Devonian.* The Royal Society, London.
Lloyd, C. W. 1991. Cytoskeletal elements of the phragmosome establish the division plane in vacuolated higher plant cells, pp. 245–257. *In:* Lloyd (ed.), *The Cytoskeletal Basis of Plant Growth and Form.* Academic Press, New York.

Lockhart, P. J., T. J. Beanland, C. J. Howe, and A. W. D. Larkum. 1992. Sequence of *Prochloron didemni atpBE* and the inference of chloroplast origins. *PNAS* 89:2742–2746.

Lokhorst, G. W., and W. Star. 1985. Ultrastructure of mitosis and cytokinesis in *Klebsormidium mucosum* nov. comb., formerly *Ulothrix verrucosa* (Chlorophyta). *J. Phycol.* 21:466–476.

Lokhorst, G. M., H. J. Sluiman., and W. Star. 1988. The ultrastructure of mitosis and cytokinesis in the sarcinoid *Chlorokybus atmosphyticus* (Chlorophyta, Charophyceae) revealed by rapid freeze fixation and freeze substitution. *J. Phycol.* 24:237–248.

Lowry, B., D. Lee, and C. Hebant. 1980. The origin of land plants: a new look at an old problem. *Taxon* 29:183–197.

Lucas, W. J. 1985. Bicarbonate utilization by *Chara*: a re-analysis, pp. 229–254. *In:* Lucas and Berry (eds.), *Inorganic Carbon Uptake by Aquatic Photosynthetic Organisms.* Am. Soc. Pl. Physiol., Rockville, Md.

Lucas, W. J., F. Brechignac, T. Mimura, and J. W. Oross. 1989. Charasomes are not essential for photosynthetic utilization of exogenous HCO_3^- in *Chara corallina. Protoplasma* 151:106–114.

Lyon, F. 1904. The evolution of the sex organs of plants. *Bot. Gaz.* 37:280–299.

McBride, G. E. 1967. Cytokinesis in the green alga *Fritschiella. Nature* 216:939.

———. 1974. The seta-bearing cells of *Coleochaete scutata* (Chlorophyceae, Chaetophorales). *Phycologia* 13:271–285.

McCann, M. C., and K. Roberts. 1991. Architecture of the primary cell wall, pp. 109–129. *In:* Lloyd (ed.), *The Cytoskeletal Basis of Plant Growth and Form.* Academic Press, New York.

McCurdy, D. W., and R. E. Williamson. 1991. Actin and actin-associated proteins, pp. 4–14. *In:* Lloyd (ed.), *The Cytoskeletal Basis of Plant Growth and Form.* Academic Press, London.

McKay, R. M. L., and S. P. Gibbs. 1991. Composition and function of pyrenoids: cytochemical and immunocytochemical approaches. *Can. J. Bot.* 69:1040–1052.

McKay, R. M. L., S. P. Gibbs, and K. C. Vaughn. 1991. RuBisCo activase is present in the pyrenoid of green algae. *Protoplasma* 162:38–45.

Maguire, M. P. 1992. The evolution of meiosis. *J. Theor. Biol.* 154:43–55.

Manhart, J. R., and J. D. Palmer. 1988. Variation in charophyte chloroplast DNAs. *J. Phycol.* 24:24.

———. 1990. The gain of 2 chloroplast tRNA introns marks the green algal ancestors of land plants. *Nature* 345:268–270.

Manhart, J. R., K. Kelly, B. S. Dudock, and J. D. Palmer. 1989. Unusual characteristics of *Codium fragile* DNA revealed by physical and gene mapping. *Mol. Gen. Genet.* 216:417–421.

Marchant, H. J. 1976a. Actin in the green algae *Coleochaete* and *Mougeotia. Planta* 131:119–120.

———. 1976b. Plasmodesmata in algae and fungi. *In:* Gunning and Robards (Eds.), *Intercellular communication in plants:* Studies on plasmodesmata. Springer-Verlag, Berlin. Pages 59–80.

———. 1977. Ultrastructure, development and cytoplasmic rotation of seta-bearing cells of *Coleochaete scutata* (Chlorophyceae). *J. Phycol.* 13:28–36.

Marchant, H. J., and J. D. Pickett-Heaps. 1973. Mitosis and cytokinesis in *Coleochaete scutata. J. Phycol.* 9:461–471.

Marchant, H. J., J. D. Pickett-Heaps, and K. Jacobs. 1973. An ultrastructural study of zoosporogenesis and the mature zoospore of *Klebsormidium flaccidum. Cytobios* 8:95–107.

Marger, M. D. and M. H. Saier, Jr. 1993. A major superfamily of transmembrane facilitators that catalyse uniport, symport and antiport. *Trends in Biological Science* 18: 13–20.

Markham, K. R. 1988. Distribution of flavonoids in the lower plants and its evolutionary significance, pp. 427–468. *In:* Harborne (ed.), *The Flavonoids: Advances in Research Since 1980.* Academic Press, New York.

———. 1990. Bryophyte flavonoids, their structure, distribution and evolutionary significance. *In:* Zinsmeister and Mues (eds.), *Bryophytes. Their Chemistry and Chemical Taxonomy.* Clarendon Press, Oxford.

Markham, K. R., and L. J. Porter. 1969. Flavonoids in the green algae. *Phytochemistry* 8:1777–1781.

———. 1978. Chemical constituents of the bryophytes. *Prog. Phytochem.* 5:181–272.

Maszewski, J., and P. Kolodziejczyk. 1991. Cell cycle duration in antheridial filaments of *Chara* spp. (Characeae) with different genome size and heterochromatin content. *Plant Systematics and Evolution* 175:23–38.

Mattox, K. R., and K. D. Stewart. 1984. Classification of the green algae: a concept based on comparative cytology, pp. 29–72. *In:* Irvine and John (eds.), *Systematics of the Green Algae.* Academic Press, London.

Meeuse, A. D. J. 1966. The early evolution of the archegoniatae: a reappraisal. *Acta. Bot. Neerl.* 15:162–177.

Meiners, S., A. Xu, and M. Schindler. 1991. Gap junction protein homologue from *Arabidopsis thaliana*: evidence for connexins in plants. *PNAS* 88:4119–4122.

Melkonian, M. 1975. The fine structure of the zoospores of *Fritschiella tuberosa* Iyeng. (Chaetophorineae, Chlorophyceae) with special reference to the flagellar apparatus. *Protoplasma* 86:391–404.

———. 1989. Flagellar apparatus ultrastructure in *Mesostigma viride* (Prasinophyceae). *Plant System. Evol.* 164:93–122.

Melkonian, M., B. L. Beech, C. Katsaros, and D. Schulze. 1992. Centrin-mediated cell motility in algae. *In:* Melkonian (ed.), *Algal Motility.* Chapman and Hall, New York. Pages 179–220.

Mimura, T., S. A. Frost-Shartzer, and W. L. Lucas. 1991. Influence of protein synthesis inhibitors on the extracellular current pattern and electrophysiological properties of internodal cells of *Chara corallina. Can. J. Bot.* 69:1109–1115.

Mineyuki, Y., and B. E. S. Gunning. 1990. A role for preprophase bands of microtubules in maturation of new cell walls, and a general proposal on the function of preprophase band sites in cell division in higher plants. *J. Cell Sci.* 97:527–537.

Mishler, B. D. 1986. A Hennigian approach to bryophyte phylogeny. *J. Bryol.* 14:71–81.

Mishler, B. D., and S. P. Churchill. 1984. A cladistic approach to the phylogeny of the "bryophytes." *Brittonia* 36:406–434.

———. 1985. Transition to a land flora: Phylogenetic relationships of the green algae and bryophytes. *Cladistics* 1:305–328.

Misonou, T., J. Ishihara, J. Y. Pak, and T. Nitta. 1989. Restriction endonuclease analysis of chloroplast and mitochondrial DNAs from *Bryopsis* (Derbesiales, Chlorophyta). *Phycologia* 28:422–428.

Moestrup, O. 1970. The fine structure of mature spermatozoids of *Chara corallina*, with special reference to microtubules and scales. *Planta* 93:295–308.

———. 1974. Ultrastructure of the scale-covered zoospores of the green alga *Chaetosphaeridium,* a possible ancestor of the higher plants and bryophytes. *Biol. J. Linn. Soc.* 6:111–125.

———. 1978. On the phylogenetic validity of the flagellar apparatus in green algae and other chlorophyll a and b containing plants. *BioSystems* 10:117–144.
Moore, L. J., and A. W. Coleman. 1989. The linear 20 Kb mitochondrial genome of *Pandorina morum* (Volvocaceae, Chlorophyta). *Plant Mol. Biol.* 13:459.
Mora, C. I., S. G. Driese, and P. G. Seager. 1991. Carbon dioxide in the Paleozoic atmosphere: evidence from carbon-isotope compositions of pedogenic carbonate. *Geology* 19:1017–1020.
Morland, Z., D. G. Kidd, and J. C. Lagarias. 1993. Phytochrome levels in the green alga *Mesotaenium caldariorum* are light regulated. *Plant Physiol.* 101:97–103.
Moser, J. W., J. G. Duckett, and Z. B. Carothers. 1977. Ultrastructural studies of spermatogenesis in the Anthocerotales. I. The blepharoplast and anterior mitochondrion in *Phaeoceros laevis*: early development. *Amer. J. Bot.* 1097–1106.
Murphy, J. B., and R. D. Nance. 1992. Mountain belts and the supercontinent cycle. *Sci. Am.* (April):84–91.
Nakahara, H., and T. Ichimura. 1992. Convergent evolution of gametangiogamy both in the zygnematalean green algae and in the pennate diatoms. *Japanese J. Phycol.* 40:161–166.
Newman, S. M., and R. A. Cattolico. 1990. Ribulose bisphosphate carboxylase in algae: synthesis, enzymology and evolution. *Photosynthesis Research* 26:69–85.
Niklas, K. J. 1976a. Morphological and ontogenetic reconstruction of *Parka decipiens* Fleming and *Pachytheca* Hooker from the Lower Old Red Sandstone, Scotland. *The Royal Society of Edinburgh Transactions* 69:483–499.
———. 1976b. The chemotaxonomy of *Parka decipiens* from the Lower Old Red Sandstone, Scotland (U.K.). *Rev. Paleobot. & Palynol.* 21:205–217.
———. 1982. Chemical diversification and evolution of plants as inferred from paleobiochemical studies, pp. 29–91. *In:* Nitecki (ed.), *Biochemical Aspects of Evolutionary Biology.* Univ. of Chicago Press, Chicago.
Niklas, K. J., and L. M. Pratt. 1980. Evidence for lignin-like constituents in Early Silurian (Llandoverian) plant fossils. *Science* 209:396–398.
Northcote, D. H. 1989. Control of cell wall formation during growth. *In:* Brett and Hillman (eds.), *Biochemistry of Plant Cell Walls.* Cambridge Univ. Press, New York.
Obst, J. R. 1990. Lignins: structure and distribution in wood and pulp. *In:* Caulfield, Passaretii, Sobczynski (eds.), *Materials Interactions Relevant to the Pulp, Paper, and Wood Industries.* Materials Research Society 197:11–20.
Ohyama, K., H. Fukuzawa, T. Kohchi, H. Shirai, T. Sano, S. Sano, K. Umesono, Y. Shiki, M. Takeuchi, Z. Chang, S.-I. Aoto, H. Inokuchi, and H. Ozeki. 1986. Chloroplast gene organization deduced from complete sequence of liverwort *Marchantia polymorpha* chloroplast DNA. *Nature* 322:572–579.
Ohyama, K., Y. Ogura, K. Oda, K. Yamato, E. Ohta, Y. Nakamura, M. Takemura, N. Nozato, K. Akashi, T. Kanegae, and Y. Yamada. 1991. Evolution of organellar genomes, pp. 187–198. *In:* Osawa and Honjo (eds.), *Evolution of Life. Fossils, Molecules, and Culture.* Springer-Verlag, Tokyo.
O'Kelly, C. J. 1992. Flagellar apparatus architecture and the phylogeny of green algae: Chlorophytes, Euglenoids, Glaucophytes, pp. 315–345. *In:* Menzel (ed.), *The Cytoskeleton of the Algae.* CRC Press, Boca Raton, Fl.
Okuda, K., and R. M. Brown, Jr. 1992. A new putative cellulose-synthesizing complex of *Coleochaete scutata. Protoplasma* 168:51–63.
Oltmanns, F. 1898. Die Entwicklung der Sexualorgane bei *Coleochaete pulvinata. Flora* 85:1–14.
Palenik, B., and R. Haselkorn. 1992. Multiple evolutionary origins of prochlorophytes, the chlorophyll b-containing prokaryotes. *Nature* 355:265–267.

Palme, K. 1992. Molecular analysis of plant signaling elements: relevance of eukaryotic signal transduction models. *Int. Rev. Cytology* 132:223–283.

Palmer, J. D. 1985. Evolution of chloroplast and mitochondrial DNA in plants and algae, pp. 131–240. *In:* MacIntyre (ed.), *Molecular Evolutionary Genetics.* Plenum, New York.

Pentecost, A. 1985. Photosynthetic plants as intermediary agents between environmental HCO_3^- and carbonate deposition, pp. 459–480. *In:* Lucas and Berry (eds.), *Inorganic Carbon Uptake by Aquatic Photosynthetic Organisms.* Am. Soc. Pl. Physiol., Rockville, Md.

Perasso, R., A. Baroin, L. H. Qu, J. P. Bachellerie, and A. Adouette. 1989. Origin of the algae. *Nature* 339:142–144.

Pickett-Heaps, J. D. 1967a. Ultrastructure and differentiation in *Chara* sp. I. Vegetative cells. *Aust. J. Biol. Sci.* 20:539–551.

———. 1967b. Ultrastructure and differentiation in *Chara* sp. II. Mitosis. *Aust. J. Biol Sci.* 20:883–894.

———. 1968. Ultrastructure and differentiation in *Chara (fibrosa).* IV. Spermatogenesis. *Aust. J. Biol Sci.* 21:655–690.

———. 1975. *Green Algae. Structure, Reproduction and Evolution in Selected Genera.* Sinauer, Sunderland, Mass.

———. 1976. Cell division in eukaryotic algae. *BioScience* 26:445–450.

Pickett-Heaps, J. D., and H. J. Marchant. 1972. The phylogeny of the green algae: a new proposal. *Cytobios* 6:255–264.

Pickett-Heaps, J. D., and D. H. Northcote. 1966. Organization of microtubules and endoplasmic reticulum during mitosis and cytokinesis in wheat meristems. *J. Cell Sci.* 1:109–120.

Pickett-Heaps, J. D., and R. Wetherbee. 1987. Spindle function in the green algae *Mougeotia*: absence of anaphase A correlates with postmitotic nuclear migration. *Cell Motil. & the Cytoskeleton* 7:68–77.

Pirozynski, K. A., and D. W. Malloch. 1975. The origin of land plants: a matter of mycotrophism. *BioSystems* 6:153–164.

Poethig, R. S. 1987. Clonal analysis of cell lineage patterns in plant development. *Am. J. Bot.* 74:581–594.

Pratt, L. M., T. L. Phillips, and J. M. Dennison. 1978. Evidence of non-vascular land plants from the Early Silurian (Llandoverian) of Virginia, U.S.A. *Rev. Palaeobot. Palynol.* 25:121–149.

Prescott, G. W. 1951. *Algae of the Western Great Lakes Area.* Cranbrook Institute of Science, Bloomfield Hills, Mich.

Pringsheim, N. 1860. Beiträge zur Morphologie and Systematik der Algen. III. Die Coleochaeteen. *Jahr. fr wissen. Bot.* 2:1–38.

———. 1863. Ueber die Vorkeim und die nacktfssigen Zweige der Charen. *Jahr. für wissen. Bot.* 3:294–324.

———. 1878. Ueber Sprossung der Moosfruchte und den Generationswechsel der Thallophyten. *Jahr. für wissen. Bot.* 11:1–46.

Printz, H. 1964. Die Chaetophoralean der Binnengewässer. Eine systematische Ubersicht. *Hydrobiol.* 24:1–376.

Quader, H., and E. Schnepf. 1989. Actin filament array during side branch initiation in protonema cells of the moss *Funaria hygrometrica*: an actin organization center at the plasma membrane. *Protoplasma* 151:167–170.

Raghavan, V. 1992. Germination of fern spores. *Am. Sci.* 80:176–185.

Raubeson, L. A., and R. K. Jansen. 1992. Chloroplast DNA evidence on the ancient evolutionary split in vascular land plants. *Science* 255:1697–1699.

Raven, J. A. 1977. The evolution of vascular plants in relation to supracellular transport processes. *Adv. Bot. Res.* 5:153–219.

———. 1984. Physiological correlates of the morphology of early vascular plants. *Bot. J. Linn. Soc.* 88:105–126.

Raven, J. A., F. A. Smith, and S. M. Glidewell. 1979. Photosynthetic capacities and biological strategies of giant-celled and small-celled macroalgae. *The New Phytologist* 83:299–309.

Rawitscher-Kunkel, E., and L. Machlis. 1962. The hormonal integration of sexual reproduction in *Oedogonium*. *Am. J. Bot.* 49:177–183.

Remy, W., and H. Hass. 1986. Das Ur-Landpflanzen-Konzept—Unter Besonderer Berucksichtingung der Organization Altdevonischer Gametophyten. *Argumenta Palaeobotanica* 7:173–214.

———. 1988. Neue Befunde zur Gestalt der Ur-Landpflanzen. *Biologie in unserer Zeit* 3:77–80.

Remy, W., and R. Remy. 1980a. Devonian gametophytes, with anatomically preserved gametophytes. *Science* 208:295–296.

———. 1980b. *Lyonophyton rhyniensis* nov. gen. et nov. spec., Ein Gametophyt aus dem Chert von Rhynie (Unterdevon, Schottland). *Argumenta Palaeobotanica* 6:37–72.

Renault, S., J. L. Bonnemain, L. Faye, and J. P. Caudillere. 1992. Physiological aspects of sugar exchange between the gametophyte and the sporophyte of *Polytrichum formosum*. *Plant Physiol.* 100:1815–1822.

Renzaglia, K. S. 1978. A comparative morphology and developmental anatomy of the Anthocerotophyta. *J. Hattori Bot. Lab.* 44:31–90.

Renzaglia, K. S., and Z. B. Carothers. 1986. Ultrastructural studies of spermatogenesis in the Anthocerotales. IV. The blepharoplast and mid-stage spermatid of *Notothylas*. *J. Hattori Bot. Lab.* 60:97–104.

Renzaglia, K. S., and J. G. Duckett. 1987. Spermatogenesis in *Blasia pusilla*: From young antheridium through mature spermatozoid. *The Bryologist* 90:419–449.

———. 1988. Different developmental processes underlie similar spermatozoid architecture in mosses, liverworts and hornworts. *J. Hattori Bot. Lab.* 64:219–236.

———. 1991. Towards an understanding of the differences between the blepharoplasts of mosses and liverworts, and comparisons with hornworts, biflagellate lycopods and charophytes: a numerical analysis. *New Phytol.* 117:187–208.

Retallack, G. J. 1985. Fossil soils as grounds for interpreting the advent of large plants and animals on land, p. 105–140. *In:* Chaloner and Lawson (eds.), *Evolution and Environment in the Late Silurian and Early Devonian*. The Royal Society, London.

Rethy, R. 1968. Red (R), far-red (FR) photoreversible effects on the growth of *Chara* sporelings. *Zeitschrift fur Pflanzen Physiol.* 59:100–102.

Robards, A. W., and W. J. Lucas. 1990. Plasmodesmata. *Ann. Rev. Physiol. Plant Mol. Biol.* 41:369–419.

Robbins, R. R. 1984. Origin and behavior of bicentriolar centrosomes in the bryophyte *Riella americana*. *Protoplasma* 121:114–119.

Robbins, R. R., and Z. B. Carothers. 1975. The occurrence and structure of centrosomes in *Lycopodium complanatum*. *Protoplasma* 86:279–284.

———. 1978. Spermatogenesis in *Lycopodium*: the mature spermatozoid. *Amer. J. Bot.* 65:433–440.

Roberts, K. 1990. Structures at the plant cell surface. *Current Opinion in Cell Biology* 2:920–928.

Roberts, K., C. Grief, G. J. Hills, and P. J. Shaw. 1985. Cell wall glycoproteins: structure and function, pp. 105–127. *In:* Roberts, Johnston, Lloyd, Shaw, and Woolhouse (eds.), *The Cell Surface in Plant Growth and Development.* The Company of Biologists Limited, Cambridge.

Robinson, D. G. 1991. What is a plant cell? the last word. *The Plant Cell* 3:1145–1146.

Robinson, J. M. 1990. Lignin, land plants, and fungi: biological evolution affecting Phanerozoic oxygen balance. *Geology* 15:607–610.

Rogers, C. E., K. R. Mattox, and K. D. Stewart. 1980. The zoospore of *Chlorokybus atmophyticus*, a charophyte with sarcinoid growth habit. *Amer. J. Bot.* 67:774–783.

Rogers, C. E., D. S. Domozych, K. D. Stewart, and K. R. Mattox. 1981. The flagellar apparatus of *Mesostigma viride* (Prasinophyceae): multilayered structures in a scaly green flagellate. *Pl. Syst. Evol.* 138:247–258.

Rolfe, W. D. I. 1985. Early terrestrial arthropods: a fragmentary record, pp. 207–218. *In:* Chaloner and Lawson (eds.), *Evolution and Environment in the Late Silurian and Early Devonian.* The Royal Society, London.

Rothman, J. E., and L. Orci. 1992. Molecular dissection of the secretory pathway. *Nature* 355:409–415.

Sack, F. 1991. What is a plant cell? continued . . . *The Plant Cell* 3:844.

Salisbury, J. L. 1989. Algal centrin: calcium-sensitive contractile organelles, pp. 19–37. *In:* Coleman, Goff, and Stein-Taylor (eds.), *Algae as Experimental Systems.* Liss, New York.

Salisbury, J. L. 1992. Centrin-based calcium-sensitive contractile fibers of the green algae. *In:* Menzel (ed.), *The Cytoskeleton of the Algae.* CRC Press, Boca Raton.

Salisbury, J. L., A. T. Baron, and M. A. Sanders. 1988. The centrin-based cytoskeleton of *Chlamydomonas reinhardtii*: distribution in interphase and mitotic cells. *J. Cell Biol.* 197:635–641.

Sarkanen, K. V., and C. H. Ludwig. 1971. *Lignins. Occurrence, Formation, Structure and Reactions.* Wiley, New York.

Sauer, N., K. Friedländer, and U. Gräml-Wicke. 1990. Primary structure, genomic organization and heterologous expression of a glucose transporter from *Arabidopsis thaliana*. *The EMBO Journal* 9:3045–3050.

Sauer, N. and W. Tanner. 1989. The hexose carrier from *Chlorella*. *FEBS Letters* 259:43–46.

Sawidis, T., H. Quader, M. Bopp, and E. Schnepf. 1991. Presence and absence of the preprophase band of microtubules in moss protonemata: a clue to understanding its function? *Protoplasma* 163:156–169.

Schindler, D. W., K. H. Mills, D. F. Malley, D. L. Findlay, J. A. Shearer, I. J. Davies, M. A. Turner, G. A. Lindsay, and D. R. Cruickshank. 1980. Long-term ecosystem stress: the effects of years of experimental acidification on a small lake. *Science* 228:1395–1401.

Schlesinger, W. H. 1991. *Biogeochemistry. An Analysis of Global Change.* Academic Press, San Diego.

Schmid, V. H. R. and U. Meindl. 1992. Microtubules do not control orientation of secondary cell wall microfibril deposition in *Micrasterias*. *Protoplasma* 169:148–154.

Schuster, R. M. 1984a. Evolution, phylogeny and classification of the Hepaticae, Chap. 15. *In:* Schuster (ed.), *The New Manual of Bryology,* Vol. II. The Hattori Botanical Laboratory, Nichinan-shi, Miyazaki-Ken, Japan.

Schuster, R. M. 1984b. Morphology, phylogeny and classification of the Anthocerotae, Chap. 16. *In:* Schuster (ed.), *The New Manual of Bryology,* Vol. II. The Hattori Botanical Laboratory, Nichinan-shi, Miyazaki-ken, Japan.

Scotese, C. R., R. Van derVoo, and S. F. Barrett. 1985. Silurian and Devonian base maps, pp. 57–76. *In:* Chaloner and Lawson (eds.), *Evolution and Environment in the Late Silurian and Early Devonian.* The Royal Society, London.

Searles, R. B. 1980. The strategy of the red algal life history. *Am. Nat.* 115:113–120.

Selden, P. A. 1985. Eurypterid respiration, pp. 219–224. *In:* Chaloner and Lawson (eds.), *Evolution and Environment in the Late Silurian and Early Devonian.* The Royal Society, London.

Selden, P. A., and D. Edwards. 1989. Colonisation of the land, pp. 122–152. *In:* Allen and Briggs (eds.), *Evolution and the Fossil Record.* Smithsonian Institution Press, Washington, D.C.

Shaw, P. J., D. J. Fairbairn, and C. W. Lloyd. 1991. Cytoplasmic and nuclear intermediate filament antigens in higher plant cells, pp. 69–81. *In:* Lloyd (ed.), *The Cytoskeletal Basis of Plant Growth and Form.* Academic Press, London.

Sheath, R. G., and K. M. Cole. 1980. Distribution and salinity adaptations of *Bangia atropurpurea* (Rhodophyta), a putative migrant into the Laurentian Great Lakes. *J. Phycol.* 16:412–420.

Shen, E. Y. F. 1966. Oospore germination in two species of *Chara*. *Taiwania* 12:39–46.

———. 1967. Microspectrophotometric analysis of nuclear DNA in *Chara zeylanica*. *J. Cell Biol.* 35:377–384.

Silverman, M. 1991. Structure and function of hexose transporters. *Ann. Rev. Biochem.* 60:757–794.

Sinnott, E. W., and R. Bloch. 1939. Changes in intercellular relationships during the growth and differentiation of living plant tissues. *Am. J. Bot.* 26:625–634.

Sluiman, H. J. 1983. The flagellar apparatus of the zoospore of the filamentous green alga *Coleochaete pulvinata*: absolute configuration and phylogenetic significance. *Protoplasma* 115:160–175.

———. 1985. A cladistic evaluation of the lower and higher green plants (Viridiplantae). *Plant Systematics & Evolution* 149:217–232.

Smith, F. A. 1985. Biological occurrence and importance of HCO_3^- utilizing systems: macroalgae (Charophytes), pp. 111–124. *In:* Lucas and Berry (eds.), *Inorganic Carbon Uptake by Aquatic Photosynthetic Organisms.* Am. Soc. Pl. Physiol., Rockville, Md.

Smith, G. M. 1950. *The Fresh-water Algae of the United States.* 2d ed. McGraw-Hill, New York.

Smith, G. M. 1955. *Cryptogamic Botany,* Vol. II. McGraw-Hill, New York.

Smith, R. C., B. B. Prezelin, K. S. Baker, R. R. Bidigare, N. P. Boucher, T. Coley, D. Karentz, C. MacIntyre, H. A. Matlick, D. Menzies, M. Ondrusek, Z. Wan, and K. J. Waters. 1992. Ozone depletion: ultraviolet radiation and phytoplankton biology in Antarctic waters. *Science* 255:952–959.

Smythe, C., and J. W. Newport. 1992. Coupling of mitosis to the completion of S phase in *Xenopus* occurs via modulation of the tyrosine kinase that phosphorylates p34^{cdc2}. *Cell* 68:787–797.

Southworth, D. 1974. Solubility of pollen exines. *Amer. J. Bot.* 61:36–44.

———. 1990. Exine biochemistry, pp. 193–212. *In:* Blackmore and Knox (eds.), *Microspores. Evolution and Ontogeny.* Academic Press, London.

Spence, D. H. N., and S. C. Maberly. 1985. Occurrence and ecological importance of HCO_3^- use among aquatic higher plants, pp. 125–144. *In:* Lucas and Berry (eds.), *Inorganic Carbon Uptake by Aquatic Photosynthetic Organisms.* Am. Soc. Pl. Physiol., Rockville, Md.

Stackebrandt, E. 1989. Phylogenetic considerations of *Prochloron,* pp. 65–70. *In:* Lewin and Cheng (eds.), *Prochloron. A Microbial Enigma.* Chapman and Hall, New York.

Staehelin, A. 1991. What is a plant cell? a response. *The Plant Cell* 3:553.

Stafford, H. A. 1991a. Flavonoid evolution: an enzymic approach. *Plant Physiol.* 96:680–685.

———. 1991b. What is a plant cell? *The Plant Cell* 3:331.

Staves, M. P., R. Wayne, and A. C. Leopold. 1992. Hydrostatic pressure mimics gravitational pressure in characean cells. *Protoplasma* 168:141–152.

Stebbins, G. L. 1992. Comparative aspects of plant morphogenesis: a cellular, molecular, and evolutionary approach. *Amer. J. Bot.* 79:589–598.

Stebbins, G. L., and G. J. C. Hill. 1980. Did multicellular plants invade the land? *Am. Nat.* 115:342–353.

Sterns, T., L. Evans, and M. Kirshner. 1991. γ-tubulin is a highly conserved component of the centrosome. *Cell* 65:825–836.

Sterk, P., H. Booij, G. A. Schellekens, A. Van Kammen, and S. C. De Vries. 1991. Cell-specific expression of the carrot EP2 lipid transfer protein gene. *The Plant Cell* 3:907–921.

Stewart, C.-B. 1993. The powers and pitfalls of parsimony. *Nature* 361:603–607.

Stewart, W. D. P., and G. A. Rodgers. 1977. The cyanophyte-hepatic symbiosis. II. Nitrogen fixation and interchanges of nitrogen and carbon. *New Phytol.* 78:459–471.

Stewart, W. N. and G. W. Rothwell. 1993. *Paleobotany and the Evolution of Plants.* Cambridge University Press, New York.

Stewart, K. D., and K. R. Mattox. 1980. Phylogeny of phytoflagellates, pp. 443–462. *In:* Cox (ed.), *Phytoflagellates,* Vol. 2. Elsevier/North-Holland, New York.

Stewart, K. D., K. R. Mattox, and G. L. Floyd. 1973. Mitosis, cytokinesis, the distribution of plasmodesmata, and other cytological characteristics in the Ulotrichales, Ulvales, and Chaetophorales: phylogenetic and taxonomic considerations. *J. Phycol.* 9:128–141.

Stewart, K. D., G. L. Floyd, K. R. Mattox, and M. E. Davis. 1972. Cytochemical demonstration of a single peroxisome in a filamentous green alga. *J. Cell Biol.* 54:431–434.

Strasburger, E. 1884. *Neue Untersuchungen über Befruchtungsvorgang bei den Phanerogamen, als Grundlagen für eine Theorie der Zeugung.* Gustav Fischer, Jena.

Stross, R. G. 1979. Density and boundary regulation of the *Nitella* meadow in Lake George, N.Y. *Aquatic Bot.* 6:285–300.

Strother, P. K. 1988. New species of *Nematothallus* from the Silurian Bloomsburg formation of Pennsylvania. *J. Paleont.* 62:967–982.

Sugiura, M. 1992. The chloroplast genome. *Pl. Mol. Biol.* 19:149–168.

Suzuki, K., K. Iwamoto, S. Yokoyama, and T. Ikawa. 1991. Glycolate-oxidizing enzymes in algae. *J. Phycol.* 27:492–498.

Sym, S. D., and R. N. Pienaar. 1991. Light and electron microscopy of a punctate species of *Pyramimonas, P. mucifera* sp. nov. (Prasinophyceae). *J. Phycol.* 27:277–290.

Syrett, P. J., and F. A. A. Al-Houty. 1984. The phylogenetic significance of the occurrence of urease/urea amidolyase and glycollate oxidase/glycollate dehydrogenase in green algae. *Br. Phycol. J.* 19:11–21.

Szymanska, H. 1989. Three new *Coleochaete* species (Chlorophyta) from Poland. *Nova Hedwigia* 49:435–446.

Takatori, S., and K. Imahori. 1971. Light reactions in the control of oospore germination of *Chara delicatula. Phycologia* 10:221–228.

Takeda, R., J. Hasegawa, and M. Shinozaki. 1990. The first isolation of lignans, megacerotonic acid and anthocerotonic acid, from nonvascular plants, Anthocerotae (hornworts). *Tetrahedron Letters* 31:4159–4162.

Tappan, H. 1980. *The Paleobiology of Plant Protists,* Chap. 11. Freeman, New York.
Taylor, A. O., and B. A. Bonner. 1967. Isolation of phytochrome from the alga *Mesotaenium* and the liverwort *Sphaerocarpos*. *Plant Physiol.* 42:726–766.
Taylor, T. N. 1982. The origin of land plants: a paleobotanical perspective. *Taxon* 31:155–177.
Taylor, T. N. and E. L. Taylor. 1993. *The biology and evolution of fossil plants.* Prentice Hall, Englewood Cliffs, NJ.
Taylor, T. N., W. Remy, and H. Hass. 1992. Parasitism in a 400-million-year-old green alga. *Nature* 357:493–494.
Tegelaar, E. W., H. Kerp, H. Visscher, P. A. Schenck, and J. W. de Leeuw. 1991. Bias of the paleobotanical record as a consequence of variations in the chemical composition of higher vascular plant cuticles. *Palaeobiology* 17:133–144.
Thomas, B. A., and R. A. Spicer. 1987. *The Evolution and Palaeobiology of Land Plants.* Dioscorides Press, Portland, Ore.
Thomas, R. J., D. S. Stanton, and M. A. Grusak. 1979. Radioactive tracer study of sporophyte nutrition in hepatics. *Amer. J. Bot.* 66:398–403.
Thomas, R. J., D. S. Stanton, D. H. Longendorfer, and M. E. Farr. 1978. Physiological evaluation of the nutritional autonomy of a hornwort sporophyte. *Bot. Gaz.* 139:306–311.
Thompson, R. H. 1969. Sexual reproduction in *Chaetosphaeridium globosum* (Nordst.) Klebahn (Chlorophyceae) and description of a species new to science. *J. Phycol.* 5:285–290.
Timell, T. E. 1980. *Compression Wood in Gymnosperms.* Springer-Verlag, New York.
Tolbert, N. E., H. D. Husic, D. W. Husic, J. V. Moroney, and B. J. Wilson. 1985. Relationship of glycolate excretion to the DIC pool in microalgae. *In:* Lucas and Berry (eds.), *Inorganic Carbon Uptake by Aquatic Photosynthetic Organisms.* Am. Soc. Pl. Physiol. Rockville, Md.
Trench, R. K. 1991. *Cyanophora paradoxa* Korschikoff and the origins of chloroplasts, pp. 143–150. *In:* Margulis and Fester (eds.), *Symbiosis as a Source of Evolutionary Innovation. Speciation and Morphogenesis.* MIT Press, Cambridge, Mass.
Tschermak-Woess, E. 1988. The algal partner, pp. 39–92. *In:* Galun (ed.), *CRC Handbook of Lichenology.* Vol. I. CRC Press, Boca Raton, Fla.
Tucker, M. E., and V. P. Wright. 1990. *Carbonate Sedimentology.* Blackwell, Oxford.
Turgeon, R., and D. U. Beebe. 1991. The evidence for symplastic phloem loading. *Plant Physiol.* 96:349–345.
Turner, F. R. 1968. An ultrastructural study of plant spermatogenesis. Spermatogenesis in *Nitella*. *J. Cell Biol.* 37:370–393.
———. 1970. The effects of colchicine on spermatogenesis in *Nitella*. *J. Cell Biol.* 46:220–233.
Turner, M. A., M. B. Jackson, D. L. Findlay, R. W. Graham, E. R. De Bruyn, and E. M Vandermeer. 1987. Early responses of periphyton to experimental lake acidification. *Can. J. Fish. Aquat. Sci.* 44:135–149.
Turner, S., T. Burger-Wiersma, S. J. Giovannoni, Luuc R. Mur, and N. R. Pace. 1989. The relationship of a prochlorophyte *Prochlorothris hollandica* to green chloroplasts. *Nature* 337:380–362.
Urbach, E., D. L. Robertson, and S. W. Chisholm. 1992. Multiple evolutionary origins of prochlorophytes within the cyanobacterial radiation. *Nature* 355:267–270.
Valero, M., S. Richerd, V. Perrot, and C. Destombe. 1992. Evolution of alternation of haploid and diploid phases in life cycles. *Trends in Ecology & Evolution* 7:25–29.

van den Ende, H. 1976. *Sexual Interactions in Plants. The Role of Specific Substances in Sexual Reproduction.* Academic Press, London.

van der Meer, I. M., M. E. Stam, A. J. van Tanen, J. N. M. Mol, and A. R. Stuitje. 1992. Antisense inhibition of flavonoid biosynthesis in petunia anthers results in male sterility. *The Plant Cell* 4:253–262.

Vaughn, K. C. 1992. Using immunogold protocols to investigate bryophyte development and subcellular organization. *Amer. J. Bot.* 79:14.

Vaughn, K. C., E. O. Campbell, J. Hasegawa, H. A. Owen, and K. S. Renzaglia. 1990. The pyrenoid is the site of ribulose 1,5-bisphosphate carboxylase/oxygenase accumulation in the hornwort (Bryophyta: Anthocerotae) chloroplast. *Protoplasma* 156:117–129.

Venverloo, C. J., and K. R. Libbenja. 1987. Regulation of the plane of cell division in vacuolated cells. I. The function of nuclear position and phragmosome formation. *J. Plant Physiol.* 131:267–284.

Vollbrecht, E., B. Veit, N. Sinha, and S. Hake. 1991. The developmental gene Knotted-1 is a member of a maize homeobox gene family. *Nature* 350:241–243.

Wada, M., and T. Murata. 1991. The cytoskeleton in fern protonemal growth in relation to photomorphogenesis, pp. 277–288. *In:* Lloyd (ed.), *The Cytoskeletal Basis of Plant Growth and Form.* Academic Press, London.

Waegemann, K., and J. Soll. 1991. Characterization of the protein import apparatus in isolated outer envelopes of chloroplasts. *The Plant J.* 1:149–158.

Wagner, G., and F. Grolig. 1992. Algal chloroplast movements, pp. 39–72. *In:* Melkonian (ed.), *Algal Motility.* Chapman and Hall, New York.

Wang, K., B. D. E. Chatterton, M. Attrep, Jr., and C. J. Orth. 1992. Iridium abundance maxima at the latest Ordovician mass extinction horizon, Yangtze Basin, China: terrestrial or extraterrestrial? *Geology* 20:39–42.

Wasteneys, G. O. 1992. The characean cytoskeleton: spatial control in the cortical cytoplasm, pp.273–295. *In:* Menzel (ed.), *The Cytoskeleton of the Algae.* CRC Press, Boca Raton, Fla.

Waters, D. A., M. A. Buchheim, R. A Dewey, and R. L. Chapman. 1992. Preliminary inferences of the phylogeny of bryophytes from nuclear-encoded ribosomal RNA sequences. *Amer. J. Bot.* 79:459–466.

Wayne, R. 1993. Excitability in plant cells. *Amer. Sci.* 81:140–151.

Webster, K. E., T. M. Frost, C. J. Watras, W. A. Swenson, M. Gonzalez, and P. J. Garrison. 1992. Complex biological responses to the experimental acidification of Little Rock Lake, Wisconsin, U.S.A. *Environmental Pollution* 78:73–78.

Wesley, O. C. 1928. Asexual reproduction in *Coleochaete. Bot. Gaz.* 86:1–31.

———. 1930. Spermatogenesis in *Coleochaete scutata. Bot. Gaz.* 89:180–191.

Wick, S. M. 1991. Spatial aspects of cytokinesis in plant cells. *Curr. Opinion in Cell Biol.* 3:253–260.

Wilcox, L. W., P. A. Fuerst, and G. L. Floyd. 1993. The phylogenetic relationships of four charophycean green algae inferred from complete nuclear encoded small subunit rRNA gene sequences. *Am. J. Bot.* (in press).

Wiley, E. O. 1981. *Phylogenetics. The Theory and Practice of Phylogenetic Systematics.* Wiley, New York.

Williams, E. A., R. B. Knox, V. Kaul, and J. L. Rouse. 1984. Post-pollination callose development in ovules of *Rhododendron* and *Ledum* (Ericaceae): zygote special wall. *J. Cell Sci.* 69:127–135.

Williams, G. C. 1975. *Sex and Evolution.* Monographs in Population Biology 8. Princeton Univ. Press, Princeton, N.J.

Williamson, R. E. 1991. Orientation of cortical microtubules in interphase plant cells. *Int. Rev. Cytology* 129:135–206.

———. 1992. Cytoplasmic streaming in characean algae: mechanism, regulation by Ca^{2+} and organization, pp. 73–98. *In:* Melkonian (ed.), *Algal Motility.* Chapman and Hall, New York.

Wright, V. P. 1985. The precursor environment for vascular plant colonization, pp. 143–145. *In:* Chaloner and Lawson (eds.), *Evolution and Environment of the Late Silurian and Early Devonian.* The Royal Society, London.

Xu, P., C. W. Lloyd, C. J. Staiger, and B. K. Drøbak. 1992. Association of phosphatidylinositol 4-kinase with the plant cytoskeleton. *The Plant Cell* 4:941–951.

Yamamoto, Y. T., C. G. Taylor, G. N. Acedo, C.-L. Cheng, and M. A. Conkling. 1991. Characterization of *cis*-acting sequences regulating root specific gene expression in tobacco. *The Plant Cell* 3:371–382.

Yang, H.-Y., and C. Zhou. 1992. Experimental plant reproductive biology and reproductive cell manipulation in higher plants: now and the future. *Am. J. Bot.* 79:354–363.

Yapp, C. Y. and H. Poths. 1992. Ancient atmospheric CO_2 pressures inferred from natural goethites. *Nature* 355:342–344.

Ye, Z.-H., and J. E. Varner. 1991. Tissue-specific expression of cell wall proteins in developing soybean tissues. *The Plant Cell* 3:23–37.

Yew, N., M. L. Mellini, and G. F. Vande Woude. 1992. Meiotic initiation by the *mos* protein in *Xenopus. Nature* 355:649–652.

Yoshinaga, K., Y. Kubota, T. Ishii, and K. Wada. 1992. Nucleotide sequence of *atpB, rbcL, trnR, dedB,* and *psaI* chloroplast genes from a fern *Angiopteris lygodifolia*: a possible emergence of Spermatophyta lineage before the separation of Bryophyta and Pteridophyta. *Plant Molecular Biology* 18:79–82.

Yu, L.-X., and C. M. Lazarus. 1991. Structure and sequence of an auxin-binding protein gene from maize (*Zea mays* L.). *Plant Molecular Biology* 16:925–930.

Zechman, F. W., E. C. Theriot, E. A. Zimmer, and R. L. Chapman. 1990. Phylogeny of the Ulvophyceae (Chlorophyta): cladistic analysis of nuclear-encoded rRNA sequence data. *J. Phycol.* 26:700–710.

Zelitch, I. 1992. Control of plant productivity by regulation of photorespiration. *BioScience* 42:510–516.

Zhang, W., H. Yamane, N. Takahashi, D. J. Chapman, and B. O. Phinney. 1989. Identification of a cytokinin in the green alga *Chara globularis. Phytochemistry* 28:337–338.

Zhang, W., D. J. Chapman, B. O. Phinney, C. R. Spray, H. Yamane, and N. Takahashi. 1991. Identification of cytokinins in *Sargassum muticum* (Phaeophyta) and *Porphyra perforata* (Rhodophyta). *J. Phycol.* 27:87–91.

Zheng, Y., M. K. Jung, and B. R. Oakley. 1991. γ-Tubulin is present in *Drosophila melanogaster* and *Homo sapiens* and is associated with the centrosome. *Cell* 65:817–824.

Zimmermann, W. 1959. *Phylogenie der Pflanzen,* p. 130. Aufl. Stuttgart.

Zuckerkandl, E., and L. Pauling. 1965. Evolutionary divergence and convergence in proteins, pp. 97–166. *In:* Bryson and Vogel (eds.), *Evolving Genes and Proteins.* Academic Press, New York.

INDEX

Abscisic acid, 212
Acetobacter xylinum, cellulose synthase in, 148
Acetolysis resistance, 216–218, 231
Achyla, sexual reproduction and, 179–180
Acid rain, 88
Actin, 44, 154, 157, 249
Actin-myosin systems, 80, 235
Actomysin, 214
Adaptation:
  defined, 85
  survival and, 89
Alternation of generations, 34–37, 57
Amino acid sequencing, 47
Angiopteris, DNA sequence of, 51
Angiosperms:
  Cu/Zn SOD in, 46
  early, 85
Antheridia, 124, 133, 136
Anthoceros:
  antheridia, 136
  placental cells, 130, 132
  pseudoelaters, 128
  wall ingrowths, 66
Antisense RNA inhibition, 227
Antithetic hypothesis, 29, 31–32, 35, 57, 190, 240
Aphanochaete elegans, plasmodesmata and, 165
Apical meristems, 170–172
Aquatic macrophytes, 1

Arabidopsis:
  embryo of, 2
  as model plant system, 234
  *tuf*A gene in, 53
Arabidopsis thaliana, transporter proteins, 114
Arabinoglactan proteins, 153
Arborescent lycopods, 23
Archegonia, 2, 124, 133, 135, 236
Aspergillus nidulans, 156
Astral microtubules, 119, 121
ATP, 80, 108
*atp*BE genes, 79
Autapomorphies, 165, 240
Auxins, 212–213

Bangia atropurpurea, sexual reproduction and, 20
Baragwanathia, 22
Beta-glucuronidase (GUS) reporter gene, 192
Bicarbonate ion, 105, 235, 238
Bicarbonate users:
  carbon and, 109, 112
  defined, 105
Bicentriolar centrosomes, 57
Biochemical gaps, 118
Blasia, 123, 142
Bloomsburg Formation, 17
Bold's Basal Medium agar, 199
Botrydium, 87

275

Bryophytes:
  apical meristem cells, 175
  bicarbonate and, 110
  character state changes of, 241
  cuticles on, 227
  cytoskeletal work, 158
  fossil evidence and, 13
  life cycle of, 32–33
  mature sperm of, 144
  microtubule systems, 154
  mitosis in, 44
  MLS in, 80
  morphogenesis in, 213
  nodal organization, 59
  p-hydroxyphenyl in, 224
  resistant biopolymers in, 47
  sporocytes of, 134, 197
  ultrastructural studies, 34

Calcareous stromatolites, 24
Calcification, 110, 240
Calcium carbonate, 104, 245
Calcium ion, 197, 210
Caledonian Ocean, 22
Callose:
  deposition of, 186
  function of, 133, 189, 236
  gametogenesis and, 66
  generation of, 149
  sporogenesis and, 66
  zygospores, 205
Calmodulin, 80, 197
Calmodulin mRNA levels, 210
Cambrian period, seawater during, 20
Carbon dioxide:
  diffusion of, 111–113, 244–245
  levels of, 16–17, 23, 105, 111, 234, 244
Carbonic anhydrase, 108–109
Carboniferous period:
  climate of, 23
  fossils during, 70
  oxygen levels, 22
Carbon limitation problems, 112, 202, 240
Carotenoids, 231
Catalase, 77, 106
Catskill Formation, 17
cdc2, gene, 212
Cell plates, development of, 235, 249–250
Cell theory:
  defined, 165
  vs. organismal theory, 166
Cellulose microfibrils, 147–148
Cellulose synthase, 147–149

Cell walls:
  development of, 146–153
  glycine-rich proteins (GRPs) in, 221
  nature of, 169
Centrin, 80
Centrioles:
  absence of, 240–241
  biogenesis, 247
  cytoskeletal features, 172
  development of, 139, 141–142
  separation and reorientation model, 143
Centrosomes, development of, 142
Cephaleuros parasiticus, cytokinesis in, 167
Ceratophyllum:
  fossils of, 85
  morphology of, 240
  size of, 59
Chaetosphaeridium:
  branched thalli in, 92
  cell division in, 61
  cell wall in, 64
  cladistic analyses including, 35
  classification of, 41, 61
  filaments, 104
  hair cells in, 63
  ultrastructural study of, 34
  zoospores of, 206
Chalcone isomerase, 225
Chalcone synthase, 225
Chaperonins, 81, 113
Chara:
  actin in, 154
  antheridium of, 138, 185
  branching and nodal organization of, 58–59
  calcified, 240
  cell walls, 147
  charasome production, 108–109, 111
  classification of, 41
  cytokinesis in, 162–163
  filaments, 238
  flavonoids in, 224
  freshwater habitat, 240
  gametangial development in, 137, 139
  graviperception studies and, 248
  meiosis in, 31
  phylogenetic tree and, 34
  plasmodesmata and, 165
  scales on, 75
  sporopollenin in, 46
  thalli of, 202
  *tuf*A gene in, 52
  ultrastructural study of, 34, 39
  zygote germination in, 69, 201
Chara globularis, 213

Charales:
    adaptation in, 102, 238
    bicarbonate use, 111–112
    calcification, 110
    characteristics of, 58–61
    chloroplasts in, 238
    cladistic analyses including, 35
    class of, 41, 58
    cortical microtubules, 154
    cpDNA in, 117
    gametangia, 185
    inorganic carbon and, 110
    life cycle, 57
    molecular analyses, 38
    morphology of, 100
    plastids, 103, 239
    pyrenoids, loss of, 238
    reproductive cells, 40
    spindles, 238
    ultrastructural studies, 34
Chara zeylanica:
    cell wall development, 151–152
    DNA levels in, 201
    maintenance of, 110
    morphology, 58
    peroxisome in, 77
    polyplastidic cells, 103
    spermatogenous filaments, 141
    sporelings, 114
    zygote, 201
Charasomes, 109, 111
Charophyceae:
    establishment of, 41–42, 45
    hypothetical ancestor of, features of, 73–74
Charophycean algae:
    classification, 56
    Cu/Zn SOD in, 46
    phenolics, 215–219
Charophytes, generally
    bicarbonate and, 240
    defined, 42
    physiological adaptations, 104–114
Chemical fixation techniques, 43
Chlamydomonas:
    carbon dioxide and, 113
    cell wall development, 151
    classification of, 42
    division cycle of, 247
    nuclear genome, 83
    resistant zygotes, 89
    stellate structure of, 80
    transformation of, 246
Chlamydomonas reinhardtii, chloroplasts of, 82
Chlorcoccum, 87

Chlorella:
    autospores, 12
    carbon loss, 105
    cell walls, 12
    classification of, 42
    fossil tetrads and, 11
    nuclear genomes of, 83
    sporopollenin in, 24–25
Chlorella kessleri, transporter proteins, 114
Chloroccum, sexual cycle, 179
Chlorococcales, ultrastructural study of, 34
Chlorogonium elongatum, 82
Chlorokybales:
    characteristics of, 72–73
    cladistic analyses including, 35
    classification of, 41
    zoospores in, 74
Chlorokybus:
    as bicarbonate nonuser, 107
    classification of, 41, 72
    cytokinesis in, 158, 166
    plasmodesmata and, 164
    reproductive cells, 40, 88–89, 180
    sexual reproduction and, 89, 179
    spindles and, 39
    zoospores:
        production, 86
        scales, 206
Chlorophyceae:
    establishment of, 42, 45
    land plant ancestry and, 43
Chlorophyll:
    shallow water habitats and, 237
    structure of, 79
Chlorophyte algae, 82
Chloroplast bundles, 157
Chloroplast genomes, 49, 51–54
Chloroplasts:
    genomes, 57
    protein import, 81
    zygote germination and, 193
Chloroplast tRNA, genes, 179
Chlorosarcinales, 86
Chromophyte algae, 82
Chroolepidaceae (= Trentepohliales), 87, 92, 166
Cinnamate 4-hydroxylase, 223
Cinnamoyl CoA NADPH oxidoreductase, 223
Cinnamyl alcohol dehydrogenase, 223
Cladistic analyses, 55
Cladonia, 26
Cladophora glomerata:
    sexual reproduction and, 20
    zoospore production in, 91

Class I aldolases, 46, 70
Cleavage furrow, 158, 160
Climate, influence of, 15
Closterium:
 plastids, 108
 zygospores, 205
Closterium ehrenbergii, chloroplast division, 157
Coleochaetales:
 branched thalli in, 92
 characteristics of, 61–67
 classification of, 41
 as ecophysiology model, 111
 life cycle, 57
 pH environment of, 107
Coleochaete:
 acetolysis-resistant material in, 217
 antheridia of, 185–186
 branched thalli in, 92, 96
 cell wall development, 152
 characteristics of, 66
 chloroplast in, 238
 classification of, 41
 cortical cells in, 65–66
 cpDNA in, 118
 cultures of, 107
 filaments, 96–97, 99, 101, 104, 238
 fossils, comparison to, 67
 hair cells in, 63, 149, 215
 introns in, 51
 life cycle of, 31–32
 meiospore development, 202, 205, 208
 meiotic divisions in, 66, 202
 meristems, 170, 172
 mitosis in, 44
 molecular analyses, 38
 morphological groups, 92
 occurrence of, 96
 Parka, distinguished from, 207–208
 phenolic compounds in, 217–218
 phylogenic tree and, 34
 placental junctions, 132
 placental regions, 208
 plasmodesmata and, 165
 plastids and, 124, 134, 198
 polygonal arrays, 249
 reproduction study of, 35
 reproductive cells, 40
 resistant biopolymers in, 47
 scale retention, 141
 sexual reproduction and, 180
 spermatozoid, 140
 spindles in, 119, 122
 sporopollenin in, 46
 surface material, 227
 thalloid species, 96, 99–101, 104, 166, 170
 *tuf*A gene in, 52–53
 ultrastructural study of, 34, 39
 wall ingrowths, 66
 zygotes:
  generally, 65
  germination, 202–203
Coleochaete circularis, meiospores, 205
Coleochaete nitellarum, 111
Coleochaete orbicularis:
 antheridia of, 190
 callose and, 189
 cell division in, 121
 cell wall structure, 64, 150
 cladistic analyses including, 35
 cortical cells in, 64–65
 mitosis in, 41
 morphology of, 62
 peroxisomes in, 45
 phragmoplasts in, 162
 placental transfer cells, 129, 131
 pyrenoid of, 109
 radial divisions, 163
 sexual reproduction in, 196
 spermatogenous cells, 186
 sperm production in, 189
 surface material, 229
 thallus, 95, 98, 201, 216–217
 wall ingrowths, 66
 zygotes of, 216
Coleochaete pulvinata:
 antheridial branches, 175, 186
 branched filaments, 92
 cladistic analyses including, 35
 fertilization in, 194
 filaments, 96–97
 hair cells in, 64
 meiospores, 133, 203, 206
 morphological characteristics of, 62
 scales, 75
 sexual reproduction and, 194
 sperm development, 142, 174, 187–188
 spore size, 207
 zoospore production in, 94
 zygotes:
  development in, 195
  germination in, 202
Coleochaete scutata:
 cellulose synthesis, 238
 cell wall development, 147–148
 morphology of, 62
 multicellular antheridia, 188

nuclear DNA levels, 203–204
phragmoplasts and, 161
scales, 74
seta cells in, 63
spermatogenesis in, 186, 188–189
surface material, 228
thallus circumference, 95, 98
zoospore production, 44, 92–93, 95
Coleochaete soluta;
 filaments, 96–97
 meristem, 171
 as morphological intermediate, 170
Colletotrichum magnum, 243
Confocal microscopy, 43
Coniferyl alcohols, 221, 223
Conjugation, 67–68, 88–89, 116, 179, 182, 198–199
Conocephalum:
 sporophytes, 127
 thalloid form, 60
Conochaete, 61
Cookson, Isabel, 6
Cooksonia:
 first appearance of, 5–7, 14, 234
 fossil discovery of, 22
Cooksonia pertoni:
 location of, 7
 spores of, 8
Copper/zinc superoxide dismutase (Cu/Zn SOD), 46, 118
Corallorhiza, 3
Cortical microtubules, 154–156
Cosmarium, 88, 179, 183
Coumaryl alcohols, 221, 223–224
cox II sequence information, 83
cpDNA, 52, 116–117
Cretaceous period:
 angiosperms and, 17
 climate of, 23
Cryofixation, 43
Cryptothallus, 3
C6-C3-C6, 225
Cutan:
 in fossils, 9
 precursors of, 248
 production of, 46, 118
 properties of, 231
Cuticles, 47, 227
Cutin:
 biosynthesis of, 215, 228–229, 231
 evolution of, 231
 precursors of, 248
 production of, 118

properties of, 227–229
structure of, 230
Cyanidium, 83
Cyanobacteria, 24–26, 245
Cyanophora, MLS in, 79
Cyanophora paradoxa, 79
Cyclic AMP, 197, 210
Cylindrocystis, 198
Cylindrocystis brebissonii:
 conjugation, 199
 cultured cells, 199
 zygote germination, 200
 zygospore wall, 199–200
Cymbomonas tetramitiformis, 78
Cytochrome oxidase subunit II, 83
Cytokinesis:
 of *Coleochaete*, 61
 evidence of, 169
 furrowing, 166, 168, 249
 phragmoplasts and, 157–164
 ultrastructural survey of, 39
 vesicle fusion during, 163
Cytokinins, 212–213
Cytoplasmic streaming, 44, 248
Cytoskeleton, 153–157, 235

DAPI, 201
Davis, Bradley Moore, 137
Dawsonia, 224
Degenerate MLS, 40
Delay in meiosis:
 advantage of, 204
 effect of, 29, 31, 187, 190–191
Dendroligotrichum, 224
Desiccation:
 effect of, 86
 resisting, 96
Desiccation-avoidance adaptations, 34
Desmids:
 filaments and, 67
 freshwater habitat, 87
 ligninlike compounds, 218
 wall development in, 163
Desmotubules, 43
Devonian period:
 carbon dioxide levels in, 17
 climate of, 23
 gametophytes of, 38
Dictyota:
 apical cell, division of, 173
 cytoskeleton, 172
 meristematic activity, 171
 phragmoplasts and, 172

Dictyota dichotoma, cell cycle, 171–172
Diploidy, 182, 236
Dissolved inorganic carbon (DIC) species, 104
Distance matrices, 55
DNA-DNA hybridization, 47
DNA replication, 185
DNA sequences:
　molecular studies of, 47–48
　phylogenetic tree construction and, 54
Double fertilization, 117
Drepanocladus, 110
Dunaliella, PAL in, 223

E. coli, 113
Ectocarpus, 137
EF-hand protein, 80
Elaters, 126–127
Electron microscopy, significance of, 27
Electrophysiology, 249
Electroporation, 177
Embryogenesis:
　callose in, 236
　early, 28
　generally, 248
Embryophytes:
　autapomorphies, 57, 116–117
　cladistic analyses of, 35
　life cycle of, 28
Embryos, generally, 2–3
Emergent gametangia, 133
Endoduplication, 204
Endoplasmic reticulum, 164–165, 222–223, 228
Endoreplication, 185
Endosymbiosis, 244
Enzyme distribution, patterns of, 46–47
Eocene impact event, 22
Eochara, 58
Eosynechococcus moorei, 26
EP2 gene, 228
Ephedra, 59, 117
Equisetum, 59
Ethylene, 212–213
Euglena, 247
Eukaryotic mitosis, 212
Eukaryotic signal pathways, 210–212
Exaptation, 117
Extensin, 153, 215

Fatty acids, 227–228, 231
Fenestrations, 39, 42
Ferrous iron, formations, 18
Finger Lakes, 100
Flagellar transition regions, 79–80

Flagellates, functions of, 80. *See also specific organisms*
Flavonoids:
　biosynthesis of, 47, 215, 225–227
　precursors to, 222
　production of, 118
　properties of, 224–225
　structure of, 226
Fluorescence microscopy, significance of, 27
Fontinalis, 110
4-coumarate:CoA ligase (4-CL), 223
4-methylmorpholene-N-oxide, 231
Fossils:
　atmospheric oxygen and, 18
　of Charales, 58
　cuticles and, 231
　microfossils, 10
　petrifaction, 13
　placental transfer cells and, 132
　significance of, 8
　types of, 8–9
Fourier transform infrared (FTIR) microscopy, 47, 65, 217
Four-way junctions, 175–176, 250
Freeze substitution, 43
Freshwater East Dyfed Pridoli, 7
Freshwater habitats, 235, 237
Fritschiella, 87, 92, 166, 168
Fritschiella tuberosa, 88
Funaria:
　archegonia of, 126
　cellulose synthase and, 148
　cytoskeletal involvement, 158
Funaria hygrometrica, placental region of, 129
Fungal symbiosis, 244
Furrowing, 158, 166, 168, 249

Gametangia:
　complexity of, 126
　development of, 95, 135
　female gametangia, 138
　origin of, 185–190
　selection pressure on, 137
　See also Jacketed gametangia
Gametes, 39
Gametogenesis, callose in, 66, 236
Gametophytes, photosynthate transport to, 192
Gamma-tubulin, 144, 156
GC/MS (gas chromatography/mass spectrometry, 213, 217
Genetic diversity, 125
Gene trees, 48–49, 238

Gibberellins, 212
Glaciation, 22–24, 184, 243
Glaucocystophyta, MLSs in, 79
Global redox balance, 221
Gloeothece coeruleus, 26
Glycine-rich proteins (GRPs), 221
Glycolate dehydrogenase, 46, 70, 83, 107
Glycolate oxidase, 46, 83, 106–107, 113, 118, 238
Glycoproteins, 153
Glyptographus persculptus, 22
GMP, 197
Gonatozygon, 49
Gondwana, 22
Gosslingia, 221
Gosslingia breconensis, 241
G-proteins, 211
Grand Canyon, 19
Gravitropism, 248
Gray, Jane, 9, 11, 38
Griffithsia, 113
Guanine nucleotide-binding proteins, 210
Gymnosperms, 23
Gyrogonites, 58

Halite (NaCl), seawater inclusions, 20
Halosphaera:
  features of, 75
  life cycle, 78
Haploidy, 182, 236
Haustorium, 245
Heat shock proteins, 81, 113
Herbivory, 96
Hexose transporter proteins, 114
Histogenesis, 120
Homeobox genes, 193, 197
Homeohydry:
  defined, 6
  transition to, 34, 241
Homologous hypothesis, 29–30
Homo sapiens, 1
Hornworts:
  antheridia, 136
  apical cells in, 175
  flavonoids in, 224
  lignan in, 47
  placental region of, 130
  Rubisco in, 124
  sporophytes of, 126, 128
  systematic position of, 128
HPLC (high-pressure liquid chromatagraphy), 213
Hydrogen peroxide/sulfuric acid, 231
Hydroxyacyl-CoA:cutin transacylase, 229

Hydroxyphenols, 218–219
Hydroxyproline-rich glycoproteins (HRGPs), 153
Hypochlorite/hydrochloric acid, 231

Iapetus Ocean, 22
Ice-albedo feedback, 23
Immunofluorescence, 43
Immunogold labeling, 156
Immunolocalization techniques, significance of, 27
In vitro fertilization, 177
Intercellular communication:
  phragmoplasts and, 157
  plasmodesmata and, 165
Interpolation hypothesis, 29, 190
Interspecific competition, 96
Intracellular second messengers, 197
Isoetes, 122, 132, 134, 169
Isogametes, 178
Isopentenyladenosine, 213
Isozyme analyses, 47

Jacketed gametangia:
  defined, 124, 133
  function of, 185
  homologous, 136
  stereotypic view of, 125
Jacoba, 78

Kidstonophyton, 36, 38
Klebsormidiales:
  as bicarbonate nonuser, 107
  characteristics of, 70–72
  cladistic analyses including, 35
  classification of, 41
  cytokinesis in, 158
  ultrastructural studies, 34
Klebsormidium (= *Hormidium*):
  cell walls, 147
  classification of, 41, 70
  cytokinesis in, 159, 166
  filaments, 71, 86
  mitosis in, 159
  occurrence of, 87
  phragmosomes and, 160
  plasmodesmata and, 164
  reproductive cells, 88–89, 180
  sexual reproduction and, 89, 179
  spindles and, 39, 119
  zoospores, 70, 141
$Km(CO_2)$, 114
Knotted-1 (Kn1) gene, 193
KOH, fused, 231

Lake Tahoe, 102
Laminaria, polygonal arrays, 249
Lamins, 157, 247–248
Land plants, generally:
　ambiguity of, 4
　defined, 3–4
　first appearance of, 4–14
　geologic time chart for, 5
　megafossil record of, 8
Langiophyton, 36–38
Laurasia, 21
Laurussia, formation of, 22
Lichens:
　appearance of, 24
　formation of, 26
Lignan:
　in hornworts, 47
　phenolic materials and, 218–219
Lignin:
　biosynthesis of, 215, 221–224, 241–242
　in fossils, 9
　production of, 46, 118, 232, 249
　properties of, 219–221, 241
　in *Staurastrum,* 69, 215, 217
　structure of, 220
Lilium, meiocytes of, 197–198
Lipids, 192
Lipid transfer protein, 228
Liverworts:
　elaters, 126–127
　fossils and, 13–14
　mitosis in, 122
　sporophytes of, 126–127
Lunularic acid, 213
Lycophytes, 117
Lycopodium:
　bicentriolar centrosomes, 249
　gametangial development, 136
　spermatozoid, 140
Lycopods, 249
Lyonophyton, 36, 38, 60

Marattia, 136
Marchantia:
　archegonia, 125
　chloroplast, 79
　DNA sequence of, 51
　introns in, 52
Marchantia polymorpha, mtDNA of, 54
MARCKS (myristolated alanine-rich C kinase substrate), 157
Maturation promoting factor (MPF), 197
Maule test, 218

Meiosis:
　in *Coleochaete,* 202
　control of onset/delay of, 197–198
　cost of, 182
　delay in, 29, 31, 187, 190–191, 204
　evolution of, 184–185
　nuclear envelope, 247
　zygotic vs. sporic, 31
Meiosis I, 204
Meiospores, 39
Meiosporogenesis, 203
Meristems, 165, 242
Mesostigma, 75
Mesotaenium:
　chloroplast reorientation of, 69
　growth of, 114
　phytochrome in, 214
　saccoderm desmids and, 70
　sexual reproduction and, 198
Micrasterias:
　cell wall development, 147–148, 151
　complex morphology of, 152
Microfilament bundles, 156–157
Microfossils, 13, 38, 234
Micromonadophyceae, establishment of, 45
Microtubule-organizing centers (MTOCs):
　antibodies to, 156
　function of, 121, 132, 154
　intermediate filaments, and, 248–249
Microtubule systems, 44, 154–155
Milankovitch cool-summer orbits, 23
Mitochondrial transit sequences, 83
Mitosis:
　closed, 40
　comparative, 44
　open, 41, 74, 87
Model systems:
　charophycean green algae, 247–248
　generally, 246–247
　noncharophycean protists, 247
　seedless plants, 249–250
Molecular gaps, 117–118
Molecular systematics, 47–54
Monoplastidy, 103, 122–124, 132, 169
Morphogenesis, 165, 248
Morphological characters, environmental variation and, 86–104
Morphological gaps:
　reproductive differences, 124–145
　vegetative structure, differences in, 119–124
mos proto-oncogene, 197
Mosses:
　development of, 119–120
　placental transfer cells, 129

sporophytes of, 126
See also specific mosses
Mougeotia:
  cell walls in, 146–147
  chloroplast reorientation of, 69
  classification of, 41
  cortical microtubules, 154, 156
  cytokinesis in, 158
  filaments, 88
  microtubules in, 119
  occurrence of, 87
  phenolic compounds in, 214–215
  phragmoplasts and, 160
  ultrastructure study of, 39
MPF, degradation of, 212
mRNA, 197
mtDNA, 83
Multicellular sporophyte generation:
  genetic diversity of, 125
  genetic recombination, 124
Multilayered structure (MLS):
  development of, 145
  microtubules in, 44
  modification of, 241
  production of, 87
  structure of, 139, 141, 144
  ultrastructural studies, 40–41
Mutation, 184
Mutualistic association, 243
Mycorrhizal association, 244
Myosin, 214

Nephroselmis (= *Heteromastix*), 75
Netrium, 198
Nicotiana, DNA sequence of, 51
Nitella:
  cellulose synthesis, 238
  cell wall development, 147–148
  centriole development in, 144
  classification of, 41, 58
  cortication, 103
  cpDNA in, 117
  filaments, 238
  flavonoids in, 224
  introns in, 51–52
  reproductive cells, 40
  sperm development in, 145
  thalli of, 202
  *tuf*A gene in, 52
Nitella tenuissima:
  antheridium, 137
  female gametangium, 138
Nitrogen, in soil, 24
Nodal cells, whorled branch production, 174

Notothylas:
  apical cell division in, 173, 175
  archegonia, 136, 190
  pyrenoids and, 108
  spermatozoid, 140
  sporophytes of, 128–129
Notothylas orbicularis, apical cell of, 175
Nucleic acid sequences, 4, 6
Nucleotide substitution:
  mitochondrial, 49, 51
  molecular studies of, 48
  significance of, 4

Oedogoniales, ultrastructural study of, 34
Oedogonium:
  conjugation and, 179
  resistant zygotes, 89
Old Red Sandstone, 22
Oligonucleotide primers, 227
Oligosaccharides, 213
Oligotrophic lakes, 20, 104
Olisthodiscus, 113
Ordovician period:
  carbon dioxide levels in, 17, 23, 111
  climate during, 22–23
  cuticles during, 231
  extinctions, 22
  geography of, 22
  high-latitude glaciation in, 24, 184, 243
  land plant appearance in, 5
  microfossils, 13, 234
  ocean chemistry, 20–21
  oligotrophic lakes, 104
  oxygen levels, 20
  soils, 24, 26
  spores, 11–12
  terrestrial biota, 24
  vegetation in, 14
Organismal theory:
  vs. cell theories, 166
  defined, 165
Origin of a Land Flora, The (Bower), 27
Origin of Species (Darwin), 28
Oryza, DNA sequence of, 51
Osmotrophic capabilities, 192–193
Oxygen, levels of, 18–19, 243
Ozone (O_3):
  depth of, 17
  holes, 16
  oxygen levels and, 18
  production of, 16
  screen, accumulation of, 243

Pachysphaera, life cycle, 78
Padina, 171

Paleochemistry, of lakes and oceans, 20–21
Paleoclosterium leptum, 69
Paleonitella, 58, 243
Paleosols, study of, 18, 26
Paleozic era:
  carbon dioxide levels, in, 17
  climate of, 23
Pallavicinia, conducting cells, 14
Pangaea, 21
Parenchyma:
  evolution of, 119
  morphology, 170
  organismal theory and, 165
Parenchymatous land plants, defined, 104
Parka decipiens:
  Coleochaete, distinguished from, 207–208
  defined, 67
Parsimony analysis, 55
p-coumaryl-CoA, 223, 225
Pectins, 149
Pedinomonas, 75
Peripheral furrowing, 162
Permian period, climate of, 23–24
Peroxisomes:
  in Chlorokybales, 72
  ultrastructural studies, 41, 45
Petrifaction, 13, 38
pH environment, 104, 107
Phaeoceros, centriole pairs, 142
Phagocytosis, 78
Phenylalanine ammonia lyase (PAL), 47, 222–223, 231
Phenylalanine, lignin formation and, 222
Phenylpropanoid pathway genes, 226
Phosphatidylinositol 4-kinase, 210
Phosphoenol pyruvate carboxylase kinase (PEPCK), 105
Phosphoinositides, 210
Phosphorus levels, 99
Photoautotrophs, 113
Photolysis, 17
Photorespiration:
  carbon loss, 105
  photosynthesis and, 106
Photosynthate, 192, 236, 238
Photosynthesis:
  bicarbonate ion and, 105, 110, 235, 238
  carbon limitations, 112, 202, 240
  inorganic carbon and, 112
  photorespiration and, 106
Photosynthetic carbon oxidation (PCO) cycle, 107
Phragmoplasts:
  absence of, 70
  defined, 39, 43
  formation of, 162, 166, 247
  function of, 157, 160
  microtubules, 155
  production of, 61
Phragmosome:
  cytoskeletal elements of, 160
  origin of, 161
*phy*A gene, 214
*phy*B gene, 214
Phycomata, 77–78
Phycopeltis, sporopollenin in, 24
Phycoplasts, 39
Phylogenetic hypothesis, 55
Phylogenetics, analysis methods, 55
Phylogenetic tree construction, 54–55
Physcomitrella patens, cytoskeletal involvement, 158
Phytochrome:
  defined, 74
  development of, 235
  photomorphogenesis, 214–215
Photomorphogenesis, 214–215
Phytohormones, 209, 212–213
Pickett-Heaps, Jeremy, 39
Pisum, 53–54
Placental junction, developmental interactions, 193–197, 243
Placental transfer cells, 129–130, 192–193
Plant embryogenesis, cell-to-cell interactions, 191–193
Plant life cycle, origin of, 190–191
Plant signaling systems, origin of, 209
Plant sugar transporters, 192–193
Plasmodesmata:
  absence of, 70
  concentration of, 176
  desmotubules and, 164–165
  intercellular communication and, 158
  production of, 61
Plastids:
  chlorophyll and, 74
  cytoskeletal system and, 247
  division of, 132
  evolution of, 247
  genes and, 52–53
  migration, 198, 249
  orientation of, 166, 169
  rotation of, 214
  significance of, 78, 124
  status of, 122
Pleurastrophyceae, establishment of, 45
Polyphenolics, 131

Polyplastidy, 103, 122–123, 169
Polysaccharides, 149
Polytrichum formosum:
  hexose sugars in, 192
  placental cells, 129
Potomogeton, 101
Porphyra, polygonal arrays, 249
Potassium dicromate/sulfurate acid, 231
Praesycidium siluricum, 58
Prasinophytes, 78, 153, 179, 247
Preadaptations, 117
Precambrian period:
  fossils of, 24
  soils in, 26
Preprophase band (PPB), 57, 121–122, 156–158, 169
Preprophase microtubule bands (PPBs), evolution of, 119–120
Pringsheim, 31
Prochlorococcus, 78
Prochloron, 78, 105, 113
Prochloron didemni, 79
Prochlorothrix, 78
Propagules, growth of, 184
Proteases, 81
Protein import systems, 81
Protein kinases, 211
Protein phosphatases, 211–212
Protists, evolutionary relationships of, 38
Pseudoelaters, 128
Pseudo-homologous alternation, 32
Psilophyton, fossil occurrence of, 22
Psilotum, 115, 221, 243
Pteridophytes:
  cytoskeletal work on, 158
  life cycle of, 33
  prothalli, 249
  sporophytes of, 126
Pterosperma:
  features of, 75
  life cycle, 78
Pterosperma cristatum, flagellate phase, 77
p34 kinase, 247–248
Pyraminomas, 75
Pyramimonas parkeae, 82
Pyrenoids:
  in Charales, 102
  in Chlorokybales, 72
  monoplastic charaphytes, 107
  in plastids, 124, 239
  Rubisco and, 108–109, 124
  thylakoids and, 108
Pyrite sulfur, 20

Quadripolar microtubular system (QMS), 132–134

Radioramus, 61
Raphidonema:
  classification of, 41, 70, 72
  filaments, 86
  occurrence of, 87
  sexual reproduction and, 179
  spindles and, 39
rbc L gene, 240
rbc S gene, 80
Receptor proteins, 211
Recombination, 184, 204
Red beds, 19
Red iron deposits, 18
Reflective luminosity, 16
Repetitive proline rich proteins (RPRP), 153
Resistant biopolymers, occurrence of, 46–47, 184, 248
Restriction fragment length polymorphism (RFLP) analysis, 47
Rhizobium, 244
Rhododendron, callose in, 205
Rhodophyte algae, 82
Rhynia, fossil occurrence, 22
Rhynie Chert deposits, 36, 38
Ribosomal RNA sequence, molecular studies of, 51
Ribulose biphosphate carboxylase (Rubisco):
  active site, 108
  carbon dioxide and, 113, 240
  chloroplast transit peptides and, 82
  nature of, 105
  nuclear coding of, 80
  pyrenoids and, 108–109, 235
  sequencing subunit genes and, 85
Ribulose 1,5-bisphosphate (RuBP), 105, 113
Ricciocarpus, branching structure, 13
Riella, 126
rpo C1 gene, 78–79
rRNA genes, 51, 57, 70, 118

Scale morphology, 206
Scanning electron micrographs, 68
Scenedesmus, sporopollenin in, 24–25
Scytonemin, 24–24
Seawater, chemical composition of, 20
Second messengers, 197, 210–211
Selaginella, 122, 136, 169
Sennicaulis hippocrepiformis, 241–242
Serology, 47

Sexual reproduction:
  loss of, 20, 89, 179
  origin, of, 178–185, 248
  See also specific plants
Shade adaptation, 102
Shallow water habitats, 99–101, 107, 111–112, 236–238
Silurian age:
  carbon dioxide levels in, 17, 111
  cuticles during, 231
  microfossils, 13
  oligotrophic lakes, 104
  spores in, 11
  vegetation in, 14
Sinapyl alcohols, 221, 223
Sirogonium:
  classification of, 67
  cpDNA in, 117
  gene trees and, 49
  *tuf*A gene in, 52
Sister chromatids, 184–185
16S rRNA, 78–79, 82
Soil:
  atmospheric chemistry and, 15
  in Ordovician period, 24, 26
  study of, 18
Solanaceous lectins, 153
South Pole, prehistoric location of, 23
Species trees, 48, 50
Spermatozoid development, 140–141, 144–145
Spermatozoid microanatomy, differences in, 139, 144, 235
Sphaerocarpales, 126
Sphaerocarpos, 126
Sphaeroplea, resistant zygotes, 89
Sphagnum:
  bogs, 87
  gametophyte, 62
  sporocytes, 134
Spirogyra:
  cell cycle, 154, 156
  classification of, 41, 67
  conjugation and, 179–181
  cpDNA in, 117
  cytokinesis in, 158
  filament of, 68, 88
  gene trees and, 49
  interphase nuclei, 160
  introns in, 51–52
  phragmoplasts and, 160
  sporopollenin in, 46, 89
  *tuf*A gene in, 52
  ultrastructure study of, 39
  zygotes in, 89

Spirotaenia, 198
Spore fossils, 10
Spore tetrads, 206
Sporic meiosis, 191
Sporogenesis:
  callose in, 66, 133, 236
  zygotes and, 31
Sporophytes, 126–127
Sporopollenin:
  biosynthesis of, 207, 231–232
  in cell walls, 11–12, 24, 69, 132
  composition of, 231
  deposition of, timing of, 206
  in fossils, 9
  identification of, 6
  phenolic compounds in, 215, 218
  production of, 46–47, 183, 236, 248
Starch production, 108, 192
Staurastrum, 68–69, 88, 215, 217
Stellate structures, 79–80
Sterome, 6
Stichococcus:
  absence of sexual reproduction in, 179
  classification of, 41, 70, 72
  filaments, 71, 86
  occurrence of, 87
  spindles and, 39
Streptophyta, 84
Suberin, biosynthesis of, 215
Submergent gametangia, 133, 136
Sucrose, 192
Sugar transporter protein (STP1) mRNA, 114
Sun:
  luminosity, 16
  ozone production and, 16–17
Symplesiomorphies, 54, 73
Synapomorphy, 55
Syngamy, 179

Tasmanites, fossil leiospheres in, 78
Telome theory, 32, 242
Temnogametum, occurrence of, 87
Terminal meristems, 170
Tertiary period, fossils from, 85
Thalassiophyta, 137
Thalassiophyta and the Subaerial Transmigration (Church), 34
Thallus:
  morphology, 96–97, 99–100, 104
  organization, 95–96
  See also specific plants
3-aminopropanol, 231
Three-dimensional plant bodies, 104
3-7 kD transit peptide, 80

Three-way junctions, 250
Timmiella, 122
Topoisomerase II activity, 184
Transfer cell morphology, 131, 243–244
Transfer RNA sequences, intron analysis of, 52
Transformation hypothesis, 29
Transmission electron microscopy, 43, 66
Transporter proteins, 114
Trentepohlia aurea:
    classification of, 87
    closed mitosis in, 40
    plasmodesmata in, 168
    plastids, 103
    stellate structures, 76
    terrestrial colonization, 88
Triassic period, Characeae in, 58
Trilete spores, 6, 11
tRNA Ala genes, 51
tRNA genes, 72, 116, 213
tRNA Ile genes, 51
Tryptophan, lignin formation and, 222
T-Tauri stars, 16
*tuf*A gene, 51–53, 236, 238, 241
2-aminoethanol, 231
2-keto sugar acids, 151
2,2,2-nitriloethanol, 231
28S rRNA gene, 82
Tyrosine, lignin formation and, 222

Ulothrix zonata, zoospore production, 91
Ulotrichales, ultrastructural study of, 34
Ultrastructure, light microscopy and, 39–46
Ultraviolet radiation:
    adaptive responses and, 235
    effect of, 16–18, 182, 235
    filters for, 224
    protection from, 24
Ulva:
    carbon dioxide and, 113
    cytokinesis and, 39
    life cycle of, 29–30
    meiosis in, 31
    polygonal arrays, 249
Ulvophyceae, establishment of, 42–43, 45
Urease, 46

Vaucheria:
    adaptation of, 87
    sexual reproduction and, 180–181
Vestigial connectives, 76

Viridiplantae, 84
Volvocales, ultrastructural study of, 34
Volvox, resistant zygotes, 89–90

Walled meiospores, origin of, 198–208

Xenopus:
    meiosis in, 197
    mitosis in, 212

Yangtze Basin, 22

Zamia, 83
Zea mays, 43, 83
Zonaria, 171
Zoospores:
    loss of, 240
    production, 87, 89, 91–95
    ultrastructural studies of, 39
zygDNA, 198
Zygnema:
    classification of, 67
    cytokinesis in, 158
    gene trees and, 49
Zygnemataceae, classification of, 41
Zygnematales:
    cell wall development, 148
    characteristics of, 67–70
    cladistic analyses including, 35
    conjugation, 179, 182
    cytokinesis in, 158
    life cycle, 57
    pH environment of, 107
    plasmodesmata absence from, 164
    sexual reproduction and, 179, 198
    *tuf*A gene in, 238
    ultrastructural studies, 34
    zoospore production absence from, 92
Zygospore wall, *Cylindrocystis brebissonii,* 199–200
Zygotes:
    in charalean algae, 60
    in *Coleochaete,* 61, 64–65
    cortication of, 180
    germination of, 193, 198, 200–203, 235
    influences of, 65
    meiosis, 236
    resistant, 182, 184
    See also specific plants
zygRNA, 197–198